ANALYZING FAILURES: THE PROBLEMS AND THE SOLUTIONS

*Proceedings of the Failure Analysis Program
and Related Papers presented at the
International Conference and Exposition on
Fatigue, Corrosion Cracking, Fracture Mechanics
and Failure Analysis*

2–6 December 1985
Salt Lake City, Utah, USA

Edited by
V. S. Goel

Organized by
 American Society for Metals
in cooperation with:

 American Society for Testing & Materials	Japanese Society for Strength and Fracture of Materials
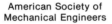 American Society of Mechanical Engineers	Japan Light Metal Association
 American Welding Society	**SVMT** Swiss Association for Material Testing
American Society of Naval Engineers	Institution of Engineers of Ireland
The Metallurgical Society of AIME	*AR* Danish Aluminum Council
National Association of Corrosion Engineers	**VDEh** Verein Deutscher Eisenheuttenleute
Institute of Corrosion Science and Technology	**SAE** The Engineering Resource For Advancing Mobility. Society of Automotive Engineers
SEM Society for Experimental Mechanics	Society of Naval Architects and Marine Engineers
National Society of Professional Engineers	American Society of Agricultural Engineers

International Association for
Bridge & Structural
Engineers

Published by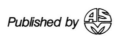

ORGANIZING COMMITTEE MEMBERS

V. S. Goel
Consultant
Littleton, Colorado, USA
Chairman, Organizing Committee

J. E. Cunningham
Oak Ridge National Laboratory
USA

R. W. Rohde
Sandia Laboratories Livermore
USA

E. Sommers
Fraunhofer-Institute fuer Werkstoffmechanik
West Germany

D. Hoeppner
University of Utah
USA

K. Iida
University of Tokyo
Japan

PROGRAM COMMITTEE

A. Kumnar
Army Materials Research Laboratory
Champaign, Illinois, USA
Chairman

V. Charyulu
Idaho State University
USA

Y. Saiga
Ishikawajima-Harima Heavy Industries Co., Ltd.
Japan

B. D. Liaw
Nuclear Regulatory Commission
USA

T. Yokobori
Japanese Society for Strength and
Fracture of Materials
Tohoku University, Japan

T. N. Singh
EG&G Idaho Falls
USA

M. Kebrin
Israel Naval Engineers
Israel

Wu MinDah
Shanghai Research Institute
China

H. Yoshikawa
University of Tokyo, Japan

SESSION CHAIRMEN

S. Aihara
Nippon Steel Corporation
Japan

R. Block
University of Oklahoma
USA

R. B. Tait
University of Witwatersrand-Johannesburg
South Africa

B. Leis
Battelle, Columbus Laboratory
USA

P. Riccardella
Structural Integrity Associates
USA

N. Pugh
National Bureau of Standards
USA

K. Mogani
National Research Institute of Police Science
Japan

T. Koyama
Babcock-Hitachi K.K.
Japan

J. Missimer
Tennessee Valley Authority
USA

H. Peterson
Colorado School of Mines
USA

H. P. Rossmanith
Technical University of Vienna
Austria

C. Jaske
Battelle Research Institute
USA

C. Czajkowski
Brookhaven National Laboratory
USA

F. Lin
Northwestern Polytechnical University
China

W. Bradley
Texas A&M University
USA

C. Smith
Rose-Hulman Institute of Technology
USA

A. Garlick
United Kingdom Atomic Energy Authority
United Kingdom

I. Smith
ICOM Steel Structures
Switzerland

A. Jones
Royal Aircraft Establishment
United Kingdom

M. Pevar
General Electric
USA

FOREWORD

This volume contains part of the total number of papers presented at the "International Conference on Fatigue, Corrosion Cracking, Fracture Mechanics and Failure Analysis," held in Salt Lake City, Utah, USA, from 2-6 December 1985. Response to this conference was so good that it resulted in a large number of papers. To satisfy the needs of different interest groups and to keep the proceedings of the conference in a manageable form, it was decided to publish it in four separate volumes:

Analyzing Failures: The Problems and The Solutions
Corrosion Cracking
The Mechanism of Fracture
Fatigue Life: Analysis and Prediction

The above paper collection volumes may be obtained from the American Society for Metals. This conference covered a wide range of topics, some of fundamental interest and some of application interest. To facilitate an early publication, the editing has been kept to a minimum. We hope the the technical merits of the papers outweigh any grammatical or minor stylistic deficiencies.

The advances in the concepts of design are pushing the operational limits of engineering materials and so maximum performance is expected out of the materials. Due to the general economic crunch, almost everyone wants the maximum life out of their equipment. The electric utilities want their plants to run more than the designed plant life (mostly 40 years), aircraft companies want their planes to fly longer, the transportation industry wants that its bridges last indefinitely, and the chemical industry wants their plants to keep on producing products. There is also an increased awareness on the part of the public for safety and reliability of components, because failure of components in large aircraft, nuclear plants or other large structures can lead to large-scale disasters like the Bhopal tragedy in India, the Three Mile Island accident in the USA and the string of airline disasters in 1985.

All of this shows that today materials are expected to show maximum performance, provide long life for maximum economy and at the same time ensure safety and reliability of components and systems. For all this, we need to understand the materials better and apply the principles of fracture mechanics, corrosion and fatigue to the solution of practical problems. This conference was planned to provide a forum for the exchange of ideas and allow a better understanding of the theory and applications of the materials science which can ensure safety in combination with the expected life and performance goals for materials.

The theme of this conference was "Technology Transfer" among the various groups who apply theory to the application of practical problems. There are many specialized meetings in this area which permit workers to come together and discuss problems in their specific application areas. However, there is no single meeting or conference which brings together workers in the various application areas such as Aerospace structures, Army-Navy Applications, Bridges and Architectural Structures, Transportation Industry and Nuclear Industry to learn what is being done in other areas which they may be able to utilize to their advantage. This conference was aimed at bringing together workers from different applications areas to give them a wider perspective. Hence, this conference was of interest to engineers, metallurgists and also to the engineering managers who remain concerned about product failure and liability.

The success of this conference was based on the contributions of the speakers, session chairmen and members of the Technical Review Committee and the Organizing Committee who generously supported this Conference. I would like to thank all the participants on behalf of the American Society for Metals and the co-sponsoring societies for their generous contribution of time and effort towards the success of this Conference.

Dr. V. S. Goel
Chairman, Organizing Committee

TABLE OF CONTENTS

Plenary Lecture

Failure Analysis

High Temperature Failure

Test Techniques

Engineering Applications

MATERIALS FAILURE PREVENTION AT THE NATIONAL BUREAU OF STANDARDS*

Lyle H. Schwartz, Daniel B. Butrymowicz
National Bureau of Standards
Gaithersburg, Maryland, USA

Abstract

As a Commerce Department agency, the National Bureau of Standards provides the measurement foundation that our industrial economy needs. Crucial to these needs are the safe, efficient, and economical use of materials. The NBS programs that support generic technologies in materials and the mechanisms by which fundamental information is transferred are analyzed. Specific examples are drawn from recent developments in fracture of materials.

THE OCCURENCE OF MECHANICAL FAILURES - has implications for efforts of all concerned with their prevention. Those involved range from basic scientists to government policy makers, as well as engineers, metallurgists, engineering managers and a host of workers in allied areas. The Federal Government has long been concerned with minimizing mechanical failures. This concern has come about for several reasons. The Government purchases and operates a great deal of equipment. Designing, building, and maintaining the most modern and sophisticated machinery that technology is capable of producing requires some assurance that it will perform efficiently and without failure. Furthermore, there are areas of broad public concern, such as safety and quality standardization which justify government concern. Here Congress has granted government agencies the authority to set mandatory standards that are designed to prevent mechanical failures in the private sector. In addition, there are several current national trends that call for enhanced concern with mechanical reliability. These trends relate to the conservation of national resources, and the continuing need for new methods of improving product quality and maintainability in the face of higher labor costs and international competition. There is also the growth of new electronic based technologies, such as computing and telecommunications, requiring new plateaus of operating reliability.

TECHNOLOGY TRANSFER

This conference has as its theme "technology transfer" among the various groups who apply theory to the application of practical problems. The conference organizers have brought together workers from a wide variety of applications areas with the aim of giving them a wider perspective. This contribution to that perspective deals with an overview of materials failure prevention at the National Bureau of Standards, and a brief description of how NBS transfers measurements, standards, data and understanding to U.S. industry, commerce, government, and science. Because these "transfers" are the business of NBS, the Bureau has a wide range of technology transfer activities which are an integral part of all NBS programs. These activities are assigned to and supported by the corresponding technical programs. In addition there are several NBS-wide program areas which are directly responsible for technology transfer.

MATERIALS RESEARCH AT NBS

The technical programs in materials fracture at NBS are centered in the major organizational unit directly charged with materials research, the Institute for Materials Science and Engineering. As the nation's foremost science and engineering measurement laboratory, the National Bureau of Standards has some of the premier research and testing facilities in the United States. Bureau scientists and engineers use these special facilities to pursue measurement-related work needed by U.S. science and industry. The Institute for Materials Science and Engineering is one of the four major technical units comprising NBS. The laboratories and staff of the Institute are providing the nation with a central basis for measurement, data, standards, and reference materials fundamental to the processing, structure, properties, and performance of materials. The program of the Institute supports generic technologies in materials in order to foster their safe, efficient and economical use. Research in the Institute also addresses the science base underlying new advanced materials technologies. Institute research is carried out by a permanent staff of 350 and approximately 150 guest workers and research associates. About 75% of the professional staff hold Ph.D. degrees in the physical sciences.

The Institute for Materials Science and Engineering is organized in five technical divisions focused in the areas of Metallurgy, Ceramics, Polymers, Fracture and Deformation, and Reactor Radiation. The NBS research reactor is a national center for the application of reactor radiation to a variety of problems of national concern in materials science and nondestructive evaluation. The Institute also manages a Bureau-wide, interdisciplinary program in nondestructive evaluation to study the basic interactions between various forms of penetrating energy and materials, to develop standard reference materials, calibration services and other means for achieving traceability to national standards for NDE measurements, and to demonstrate these results in selected generic applications. This program includes not only the well-established NDE methods such as radiography, ultrasonics, eddy currents, magnetic particles, liquid penetrants and leak testing, but also research in the emerging methods of thermal wave imaging, acoustic emission, neutron scattering and laser-based optical inspection. It is clear that new and improved methods of nondestructive evaluation can upgrade the quality of manufactured products and enhance the safety and durability of structures in service.

There are a number of NBS offices which directly support technology transfer, half of them materials related. The Office of Standard Reference Materials develops well-characterized, stable, homogeneous material samples having one or more of the physical and chemical properties certified by NBS. These standard samples are distributed to U.S. manufacturing, business, government, public safety, and research communities. The Institute collaborates with this office in the development of a number of reference materials. The Office of Standard Reference Data assembles, evaluates, and ensures dissemination of scientific and engineering data through collaboration with the user communities.

MATERIALS DATA

Materials data is one of the most important components of the Institute's program. Activities leading to the development of data bases can be found in all of the Institute's divisions. These activities provide critically evaluated numerical data on materials to the community of users, from programs relating to phase diagrams to corrosion. The key role of NBS in reference data is in the validation process. The entire set of data activities rests on the premise that scientists and engineers experienced in a particular measurement technique and familiar with relevant theories can examine a body of data and make value judgments about its accuracy or reliability. The cost of this validation effort is justified by the added value to the ultimate users. Leaving each user to make the value judgments individually or, alternatively, to make do with unreliable data, is clearly an inefficient approach. For the community to realize the maximum benefits of this validation process, the materials data must be accessible via the computer. Modern computer-based modeling enables extension of limited data bases into as-yet unmeasured regimes and more accurate interpolation in those regimes in which measurements have been made.

It should be noted that each of these data programs has industrial society cooperative arrangements (See Table 1), with the cooperating societies providing program guidance and dissemination of the information, and in the area of phase diagrams, providing fund-raising beyond the pilot stage. There are two programs of particular interest. The joint NBS/American Society for Metals program critically evaluates phase diagrams for binary and higher-order alloys, and presents this information in an interactive computer data base. Industry has supported this effort since 1978 with funding of more than four million dollars. Evaluation work is done at NBS and also at research laboratories in universities and industry. This program with ASM will continue on a self-sustaining basis after the initial evaluation program is concluded.

In the second program, an agreement with the American Ceramic Society is expanding an earlier cooperative effort to provide improved, evaluated phase diagrams to the ceramic community. As with the effort on metals systems, NBS is responsible for data evaluation, providing coordination with other phase diagram compilation centers and compiling evaluated phase equilibria data. The Ceramic Society develops funding support and is responsible for data dissemination.

CORROSION DATA - With colleagues from the National Association of Corrosion Engineers (NACE), NBS is about to embark on a major expansion of a corrosion data program, a data effort of particular interest to the conference. This effort will be modeled on NBS's successful earlier programs in phase diagrams, with the National Association of Corrosion Engineers and the National Bureau of Standards joined together to establish a collaborative program to collect, evaluate, and disseminate corrosion data. In this effort, NACE's responsibilities are for overall management of the program and dissemination of the products. Funds for

TABLE 1 - Industrial Society Cooperative Arrangements of the Institute for Materials Science and Engineering, National Bureau of Standards

SOCIETY COOPERATION

IMSE DATA CENTERS	GROUPS
Alloy Phase Diagrams	American Society for Metals
Phase Diagrams for Ceramists	American Ceramic Society
Corrosion Data	Association of Corrosion Engineers
Polymer Blends Phase Diagrams	Society of Plastics Engineers
Crystal Data	Committee for Powder Diffraction Standards-International Centre for Diffraction Data, Cambridge Crystallographic Centre for Diffraction Data
Powder Diffraction Data	Joint Committee for Powder Diffraction Standards-International Centre for Diffraction Data
Diffusion in Metals	American Society for Metals-International Copper Research Assocation
Welding Technology	Welding Research Council-American Welding Society-American Welding Institute
Wear and Friction Data	American Society of Mechanical Engineers-American Society of Lubrication Engineers

the pilot stage of the project have come from NACE and NBS, but the full realization of the proposed data program will require the participation of U.S. industry. A major fund raising program is being launched this winter. NACE will seek funds from those in industry and others who will benefit from the successful completion of this program. The program is designed to produce an easy-to-access, user-friendly, computer data base of evaluated corrosion data which can be retrieved in a number of graphical or tabular formats.

INDUSTRIAL RESEARCH ASSOCIATES

In addition to raising funds from industry to launch their portions of the program, a number of the cooperating groups (private companies, trade and professional associations) also support industrial research associate programs at NBS. Their research associates cooperate with NBS scientists in building the computer data bases. Whether in data centers, or in the laboratory, these industry people work side by side with NBS researchers and staff. These industrial research associates represent technology transfer in its most effective form--people-to-people interactions with both parties committed to a common goal.

One of the largest and most successful of these Industrial Research Associate Programs at NBS is one which has among its goals the elimination of defects that could result in materials failures. A steel sensor program was initiated at the National Bureau of Standards in mid-1983 with the American Iron and Steel Institute (AISI). The impetus for this program arose from a workshop sponsored by AISI and NBS in 1982 to review the detailed requirements of high-priority sensors identified by the steel industry (1). The objective of the present program is to develop and evaluate ultrasonic sensors to meet two of these high-priority needs of the steel industry: (a.) detecting pipe and gross porosity in hot steel blooms, or slabs and (b.) profiling the internal temperature of hot or solidifying bodies of steel. Research associates from AISI member steel companies are working closely with NBS personnel in this research with most of the work being carried out in NBS laboratories and collaboratively with universities.

FUNDAMENTAL RESEARCH

There are many specific examples of the fundamental research going on at NBS which will lead to new materials standards in the future. Of particular interest are those examples in the narrow area of fracture, the central theme of the conference. Environmentally-induced cracking appears in a wide range of materials. It occurs as stress corrosion cracking in engineering alloys, as hydrogen embrittlement in high strength steels, as water vapor-induced cracking in glass and ceramics, and as environmental stress cracking in ethylene-based plastics. Institute work in each of these areas is establishing the basic mechanisms and developing test methods to determine the resistance of materials to these forms of failure.

STRESS CORROSION CRACKING - To further illustrate the metals aspect of this phenomenon, research is being carried out in the Corrosion Group of the Metallurgy Division under the leadership of Neville Pugh. These are stress corrosion experiments which are yielding results that show how cleavage can be induced into underlying ductile material by local corrosion at the surface. These experiments are an example of the ability of ductile materials to cleave under dynamic conditions. Specifically, the group has been addressing the problem of transgranular stress corrosion cracking and how to go about improving the resistance of metals to failure. There are many common examples of materials that undergo transgranular stress corrosion cracking, but much of the discussion centers around homogeneous face-centered-cubic alloys. The failure of austenitic stainless steels in chloride media and the ammonia-cracking of admiralty metal represent major practical problems.

Both intergranular and transgranular stress corrosion cracking occur in service, and both are of practical importance although the mechanisms are fundamentally different. In the case of intergranular SCC, it is believed that cracking generally occurs by the film rupture or slip dissolution model where the crack propagates by preferential anodic dissolution at the crack tip. On the other hand, transgranular cracks are thought to propagate discontinuously by environmentally-induced cleavage. The propagation of transgranular stress corrosion cracks has been described in detail by Pugh (2,3). Typical transgranular stress corrosion fracture surfaces are displayed in the scanning electron micrographs shown in Figure 1.

(a)

(b)

FIGURE 1 - Transgranular stress corrosion fracture surfaces. Scanning electron micrographs of admiralty metal in aqueous ammonia: (a) taken perpendicular to the primary (011) cleavage facets; (b) an off-perpendicular view illustrating the formation of cleavage steps. Micrographs courtesy of Kaufman (4).

Environmental reactions take place during the periods of arrest. Fracture surfaces in all cases of transgranular SCC display the characteristic cleavage-like morphology shown schematically in Figure 2 and consist of parallel but displaced primary cleavage facets separated by steps which are generally also crystallographic in appearance. Although the mechanisms by which the environmental interaction causes brittle behavior in normally ductile alloys has not been determined, modeling studies of crack-tip chemistry by Bertocci (5,6) rule out mechanisms based on hydrogen embrittlement. Current ideas rely on a model where dealloying takes place in a shallow (5 to 20 x 10^{-6} mm) layer. A cleavage crack initiates in this dealloyed zone, and then moves into the unaffected substrate. This is shown schematically in Figure 3. The dealloyed layer which forms is coherent, epitaxial, and has a smaller lattice parameter which subjects it to a tensile stress vis-a-vis the unaffected substrate metal. The crack, forming in the dealloyed layer, then propagates through the ductile substrate. Theoretical work by Thomson and Lin (7) of NBS supports the view that cleavage can occur in unaffected FCC alloys such as austenitic stainless steel and alpha-brass under these dynamic conditions.

Step formation (shown schematically in Figure 4) occurs by plastic shearing, a process different than primary cleavage failure. However, step formation can control crack propagation rates. From studies of Cu - 30Zn in ammonia it was concluded that step formation lags behind the tip of the cleavage crack, so that the length of the unfractured ligaments

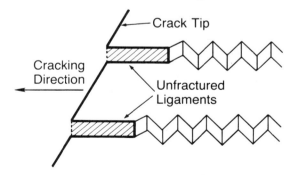

FIGURE 2 - Schematic illustration of a typical transgranular stress corrosion fracture surface showing serrated steps between primary cleavage facets and unfractured ligaments at the crack tip. After Pugh (2).

increase during each crack advance event and ultimately result in crack arrest. The latter follows because the ligaments become increasingly load bearing as they lengthen, until the stress at the crack tip falls below that required to sustain cleavage. Step formation is thought to continue during arrest, transferring the stress back to the crack tip. However, the tip is then blunted and further cleavage must be initiated by environmental reaction. Most important is that resistance to this form of cracking can be achieved by impeding the formation of these steps between parallel but displaced cleavage facets by ensuring that ready cross-slip can occur or by increasing the shear strength by precipitation hardening (8). This research program, although centered in the Corrosion Group at NBS, had important contributions from visiting faculty and students and was partially supported by grants from another government agency. The transfer of information via personal exchange was supplemented by the traditional external interactions of technical publications, technical talks, seminars, theses, conferences and workshops, visits to industrial and university staff and visitors received. Plans are to continue this research program in league with the steel industry, studying the cracking of austenitic stainless steels.

BRITTLE FRACTURE - This second example illustrates the well known fact that in many applications brittle fracture limits the use of ceramic materials. In order to understand mechanisms of fracture that are entirely brittle, Peter Swanson, a National Academy of Sciences-National Research Council Postdoctoral Research Associate in the Ceramics Division, has been performing crack-propagation experiments. His results (9) indicate that resistance to fracture increases with crack extension in certain ceramic materials. Swanson, a geophysicist, had been performing subcritical-fracture experiments on different rock types in air using fracture mechanics techniques originally perfected on glasses and fine-grained ceramics at NBS in the early 1970's. Swanson extended his fracture propagation studies in rock to aluminas and several

a. Plastic Shearing

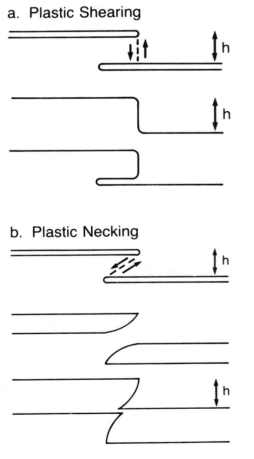

b. Plastic Necking

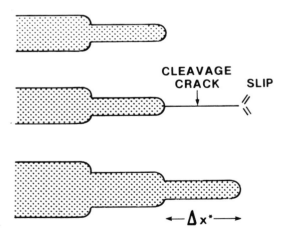

FIGURE 3 - Schematic illustration of successive events during the propagation of transgranular stress corrosion cracks. The views represent a section at the crack tip with Δx^* being the crack advance distance per event. After Pugh (3).

FIGURE 4 - Schematic illustration of the formation of cleavage steps by (a) plastic shearing and (b) plastic necking. After Pugh (3).

glass ceramics, where his in-situ (Figure 5) optical and scanning electron microscopy observations of propagating fractures show the presence of a zone of restraining forces acting <u>behind</u> the crack tip along the crack interface. The restraining forces result from both crack-plane friction, due to the interlocking nature of the fracture surfaces, and incomplete decohesion, which leaves "islands" of intact material behind as ligaments bridging the fracture surfaces. These restraining forces are illustrated schematically in Figure 6. These observed mechanisms of restraint are analogous to those observed in fiber

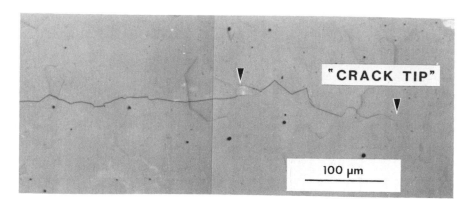

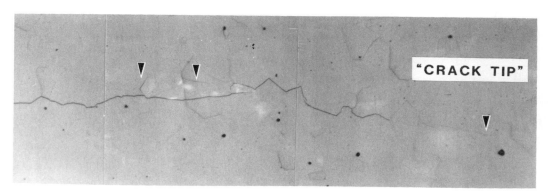

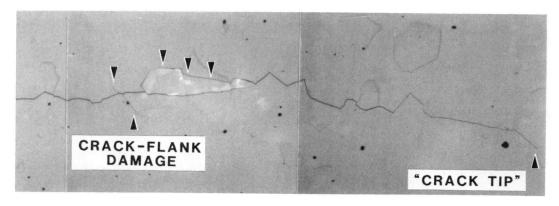

FIGURE 5 - Successive optical photomicrographs from the in-situ observation of a fracture propagating in an alumina specimen. The sequence of photos illustrates microfracture damage occurring behind a relatively well-defined crack tip, propagating from left to right in alumina. A zone of restraining forces is clearly visible along the crack interface, <u>behind</u> the crack tip. After Swanson (9).

reinforced composites (see Figure 7) where fiber "pull-out" and subsequent fiber fracture consume large amounts of fracture energy. In both materials the restraining forces serve to shield the crack tip from high levels of stress and must be overcome for fracture to proceed. The purpose of the in-situ fracture observation experiments is to identify the energy absorbing mechanisms acting along the crack flanks and unravel their relationships with the ceramic microstructure. Once this is established, Swanson hopes to prescribe the appropriate microstructural features to the ceramic processor in order to develop more fracture-resistant ceramics. The nature of these crack-plane processes and their role in providing toughness, rising fracture

resistance, and resistance to subcritical fracture in ceramics are currently under study.

DYNAMIC CRACK ARREST - A unique set of experiments is being carried out in the Fracture and Deformation Division's Time and Structure Dependent Properties Group under the direction of Richard Fields. It is well known that whether fracture occurs in milliseconds, as in the collapse of a hotel walkway, or over a period of years, as in

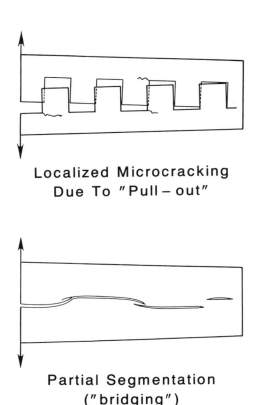

Localized Microcracking Due To "Pull–out"

Partial Segmentation ("bridging")

FIGURE 6 - Schematic illustration of crack-flank process-zone mechanisms. Fracture-surface geometry can lead to frictional contact and microcracking between locally-separated crack surfaces. Partial decohesion, or grain-localized microcracking, leaves intact "bridges" of material behind the "crack tip." After Swanson (9).

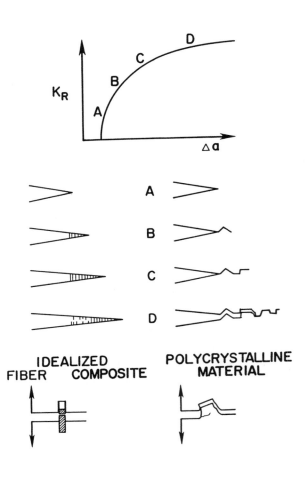

FIGURE 7 - Increasing resistance to fracture with crack extension. The history dependence of the fracture resistance can be characterized by an R curve, which relates the resistance, K_R, to the extent of crack growth, Δa. In fiber-reinforced composites, fiber "pull-out" and subsequent fiber fracture require large amounts of fracture energy. After Swanson (9).

the slow rupture of steam pipes, the time dependence or kinetics of fracture is always important. This group studies the kinetics, rather than the mechanics or energetics, of fracture. It has projects which include dynamic fracture of reactor pressure vessels, in which crack growth must be resolved during microsecond time increments, and creep crack growth in which failure may take years. The research is primarily concerned with measurement of fracture rates and the theoretical framework of fracture kinetics. One of their major accomplishments has been their success in making some rather exciting dynamic crack arrest measurements using ten-meter long specimens.

Crack arrest toughness is the ability of a material to stop a running crack; it occurs when dynamic stress/strain fields at the tip of a running crack are insufficient to cause the fracture to continue. If materials of sufficient crack arrest toughness can be placed properly in a structure, then fractures that initiate in brittle zones can be stopped before complete rupture of the structure occurs. Crack arrest considerations are important in assessing the margin of safety in a structure. To ensure the reliability of pressurized nuclear reactors, a study was undertaken of dynamic crack growth and arrest in wide plates of nuclear pressure vessel steel. The study will provide the basic data required by the nuclear industry to predict pressure vessel behavior during

the thermal shock that would accompany an emergency reactor shutdown. The experiments are carried out in collaboration with the Nuclear Regulatory Commission and the Oak Ridge National Laboratory's heavy-section steel technology program. Low temperatures and accumulated neutron exposure increase the tendency for flaws to propagate under abnormal loadings. In the case of pressurized-thermal-shock conditions, the severely irradiated material in a vessel is exposed to severe stresses at the relatively low-temperatures that result from the injection of low-temperature cooling water. What is the behavior of cracks that might exist in a reactor pressure vessel under these conditions? Prior studies of crack arrest have utilized small specimens and focused on reducing dynamic effects of the running crack. Small specimens, however, provide limited constraint of deformation in the crack-plane region and permit only the generation of data at temperatures below those where arrest is likely to occur in some scenarios. This information is summarized in Figure 8. With the unique facilities at NBS it is possible to investigate the crack run-arrest behavior in large plates with steep toughness gradients. These specimens' large dimensions (0.1 m x 1 m x 10 m) are dictated by the speed of sound in the steel, the speed of crack propagation, and the need to achieve plastic constraint (see Figure 9). A total of 10 to 12 tests are underway at loads ranging from 10 to 20 meganewtons (2 to 5 million pounds). These will be the highest-load tensile tests ever performed in the United States.

The wide-plate specimens possess a single-edge notch that initiates fracture at low temperature and arrests in a region of increased fracture toughness. The toughness gradient is achieved through a linear temperature gradient across the plate. One of the first objectives is to provide K_{Ia} data above the ASME K_{Ia} curve upper-limit criterion for prototypical pressure vessel steels. The steels include typical base metals and low-upper shelf materials representative of degraded weld materials. Other objectives include providing data from which dynamic fracture analyses can be performed. To obtain such data on wide plates fracturing at temperatures up to the upper shelf region, a total of twelve, single edge-cracked tensile specimens are to be tested in a thermal gradient that, in the most extreme

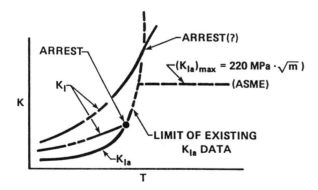

FIGURE 8 - Summary of available K_{Ia} data as a function of temperature for prototypical pressure vessel steels. The data do not extend to high enough temperatures for adequate failure assessment of pressurized-thermal-shock scenarios. After deWit and Fields (10).

case, might extend from -100 to 200 C across the 1m specimen width. Five tests have been completed (10) on the 26 MN Universal Testing Machine at NBS. The edge-notched plates in the middle of the specimen were supplied by ORNL. The pull plates and tabs were designed and constructed by NBS. These are welded together into a pull plate assembly.

An important part of this study is data acquisition from the numerous strain gauges, thermocouples, timing wires, crack mouth opening displacement gauges, and acoustic emission transducers that are mounted on the specimen (see Figure 10). Low reactance bridges, wide range dynamic amplifiers, high speed digital oscilloscopes, and a wide band FM magnetic tape recorder are used to record the

response to crack propagation and arrest. This equipment is controlled remotely by a microcomputer. In addition, a lead-wire junction box and data logger with digital voltmeter are used to monitor the temperature and load. The data are recorded on a second microcomputer. Details of the data acquisition system have been published (11) and some of this equipment has been loaned to outside laboratories, enabling those researchers to begin gathering dynamic strain data from their test systems.

Each test has been different with respect to conditions of testing, specimen configuration, and instrumentation used. The progressive changes in test procedure represent attempts to obtain the desired crack run and arrest behavior and to improve upon the quality of the data collected. In particular, efforts were made to initiate crack propagation at lower stress intensity factors. Also, strain gauge combinations and locations were optimized to better deduce the crack position as a function of time.

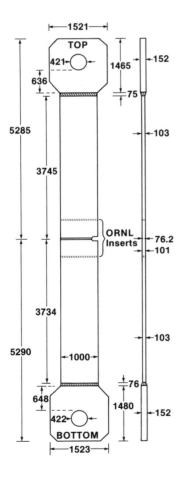

FIGURE 9 - Wide plate crack arrest test specimen. The schematic drawing of the specimen has all dimensions in millimeters. The photograph shows the specimen mounted in the testing machine (note truck in lower right corner for approximate scale). After Fields et al. (11,12).

Each test yields a wealth of information. Since using the tape recorder on the second test it has been possible to simultaneously record 28 signals. To analyze the results the data are played back from the recorder into the digital oscilloscope and from there into a microcomputer. Hence, a record of the complete loading history can be obtained (see Figure 11). The capability of the tape recorder is good enough that details of the run arrest events on the order of 10 microseconds can be resolved.

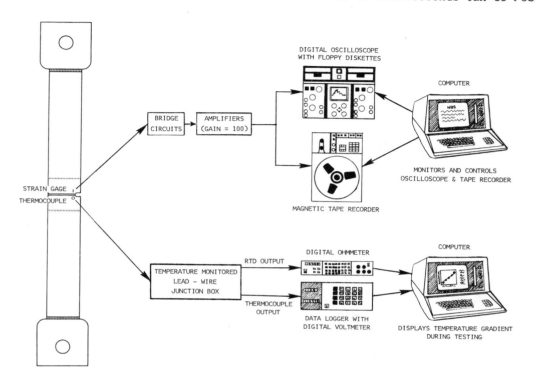

FIGURE 10 - Schematic illustration of the instrumentation for the data acquisition system employed in the wide plate crack arrest experiments. The two independent subsystems (one to monitor temperature, the other to monitor strain and other gauges) are monitored or controlled by a separate microcomputer. After Fields et al. (11,12).

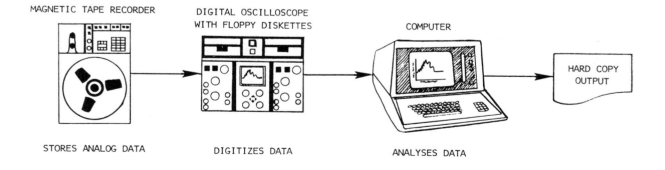

FIGURE 11 - Schematic illustration of the recording instrumentation used in the wide plate crack arrest tests. After Fields et al. (11,12).

The strain gauges near the crack plane show a peak as the cleavage crack runs past each gauge. This helps determine crack position as a function of time. Other clues are used to determine crack position when the crack growth becomes ductile. By differentiating these results the crack velocity is obtained as a function of time. Another result of great interest that can be deduced from these tests is the initiation-of-fracture toughness and the arrest toughness.

Examination of this wealth of data is no simple matter and efforts are underway for analysis methods to take into account the dynamic nature of crack propagation-arrest events. The Southwest Research Institute, the University of Maryland and ORNL are working on integrated efforts to develop elastodynamic fracture analysis finite-element programs and

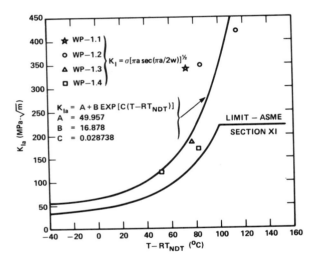

FIGURE 12 - Results of the first four wide plate crack arrest tests of a prototypical pressure vessel steel (A533B). Arrest toughness (K_{Ia}) versus temperature (where RT_{NDT} is the reference temperature relative to the nil-ductility temperature and is equal to -23C). The plain strain (K_I) equation was used to calculate an approximate arrest toughness. The arrest toughness (K_{Ia}) is determined from an empirical equation, fitted to low temperature data. This empirical relation (based on the low temperature data) is also used for the ASME curve. The tests are providing data above the ASME upper-limit criterion and compare data with theory to the results. After deWit and Fields (10).

viscoplastic analysis methods. Preliminary elastodynamic fracture analysis procedures have been applied to the analyses of the completed wide-plate crack-arrest experiments. The early results show that the dynamic behavior is being modeled as shown in Figure 12 (which compares theory to the results for K_{Ia}).

MECHANICAL FAILURES PREVENTION GROUP

The above-mentioned experimental crack arrest data were initially reported (12) last spring at the 40th Meeting of the Mechanical Failures Prevention Group (MFPG). This symposium had as its theme "the use of new technology to improve mechanical readiness, reliability, and maintainability" and met at the NBS Laboratories in Gaithersburg, Maryland. For nearly twenty years the Mechanical Failures Prevention Group has been meeting to stimulate voluntary cooperation among the segments of the scientific and engineering communities in efforts to reduce the incidence of mechanical failures and to develop methods to predict mechanical failures. Its objectives, then as now, have been to

1. Provide a focal point of awareness of mechanical failure problems in areas critical to the national needs.

2. Critically examine the field of mechanical failures and make recommendations regarding areas needing greater endeavor.

3. Stimulate interdisciplinary communication and provide a forum through meetings, conferences, and symposia to include academic, government and industrial research communities.

4. Encourage research and development directed toward both prior identification and reduction of mechanical failures.

5. Provide a common meeting ground for those who have mechanical failure problems and those who have the potential for solving mechanical failure problems.

No formal application is required for membership in the MFPG. Attendance and participation in MFPG activities is open to all by an expression of interest to:

T. R. Shives, The Executive Secretary,
MFPG, Materials Building, Room A-113,
National Bureau of Standards, Gaithersburg,
Maryland, 20899.

RESEARCH OPPORTUNITIES

Each year NBS provides literally thousands of opportunities for transferring the results of its research to those who need them. It is done in the interest of supporting the country's economy by expanding and strengthening the partnership among industry, universities and other branches of the government. For the first time, NBS is making some of its facilities available for private companies to use in conducting proprietary research. Increasing access to advanced measurement capabilities will help industry improve its products and processes and thus its international competitiveness. There will be a continuing expansion of cooperative research efforts. At this time industry is being asked to participate in a major new program in composite materials.

This newly initiated program is intended to turn the resources and expertise of NBS's metallurgists, ceramists, and polymer scientists toward the development of the underlying basic science of high performance composites. What is being developed is a multifaceted study of the complex relationships between processing, characterization, properties and performance of advanced, metal-, ceramic- and polymer-matrix composites.

The results from these upcoming investigations will aid in the development of the underlying basic science of these high-performance materials. The planned research program currently addresses:

1. The characterizaton of structure in fiber, matrix and composites of polymer, metal, and ceramic matrix materials.

2. Understanding the composite structure/property relationships. The fiber, or particulate/matrix interphase region has not been well characterized, although it is critical to composite performance.

3. Development of knowledge of failure mechanisms. Flaw identification, growth, and elimination will be examined. The response of composites to differing load conditions and environmental factors will be analyzed in fundamental terms.

Central to the theme of this program is the development of a fully effective nondestructive evaluation methodology. Contemporary nondestructive test methods are of undetermined effectiveness in detecting all materials defects because of the heterogenous anisotropic nature of composites. Process technology produces materials containing a large number of different, albeit characteristic defects. Hence, the development of detection and flaw criticality assessment methods must take place in parallel with the evolution of materials forms and manufacturing processes. As NBS bring its capabilities and necessary disciplines to bear on its composites research program, industry is being invited to share in the laboratories proven strengths, its broad-based expertise, and its unique facilities. Together both can identify those efforts needed to enhance productivity, reduce costs, and improve the performance levels of composites.

Acknowledgements

The authors thank E. Neville Pugh, Michael Kaufman, Richard Fields, Roland deWit, and Peter Swanson for making available their research results. Financial support for some of the research described here came from the the Nuclear Regulatory Commission and the Oak Ridge National Laboratory.

REFERENCES

1. R. Mehrabian, R.L. Whitely, E.C. van Reuth and H.N.G. Wadley, "Process Control Sensors for the Steel Industry," Report of a Workshop held at the National Bureau of Standards, July 27-28, 1982, Report No. NBSIR 82-2618, 40 pages, National Bureau of Standards, Gaithersburg, Maryland (1982).

2. E.N. Pugh, "On the Propagation of Transgranular Stress-Corrosion Cracks," Atomistics of Fracture, pp. 997-1010, Plenum Publishing Corp., New York, NY (1983).

3. E.N. Pugh, "Progress Toward Understanding the Stress Corrosion Problem," Corrosion 41, 517-26 (1985).

4. M.J. Kaufman, Unpublished Research, Metallurgy Division, National Bureau of Standards, Gaithersburg, Maryland.

5. U. Bertocci, "Modeling of Crack Chemistry in the Alpha brass-Ammonia System," Embrittlement by the Localized Crack Environment, pp. 49-58, American Institute of Metallurgical Engineers, New York, NY (1984).

6. U. Bertocci and E.N. Pugh, "Proceedings of a Conference on Corrosion Chemistry Within Pits, Crevices, and Cracks," October 1984, Teddington, England (in press).

7. R. Thomson and I. H. Lin, Unpublished Research, Fracture and Deformation Division, National Bureau of Standards, Gaithersburg, Maryland. Some of these results were summarized in the paper "Fracture Fundamentals: An Overview," Second International Conference on Fundamentals of Fracture, Gatlinburg, Tennessee, November 1985, Oak Ridge National Laboratory Report No. ORNL/TM-9783, Oak Ridge, Tennessee (1985).

8. M.J. Kaufman and E.N. Pugh, "A New Approach to Improving Resistance to Transgranular Stress Corrosion Cracking," Critical Issues in Reducing the Corrosion of Steels - Proceedings of a USA-Japan Seminar, March 11-13, 1985, Nikko, Japan.

9. P.L. Swanson, "A Fracture Mechanics and Nondestructive Evaluation Investigation of the Subcritical-Fracture Process in Rock," Proceedings of the Fourth International Symposium on the Fracture Mechanics of Ceramics, June 19-21, 1985, Blacksburg, Virginia (in press).

10. R. deWit and R.J. Fields, "Wide Plate Crack Arrest Testing," Transactions of the Thirteenth Water Reactor Safety Research Information Meeting, U.S. Nuclear Regulatory Commission Report No. NUREG/CP-0071, Superintendent of Documents, U.S. Government Printing Office, P.O. Box 37082, Washington, DC (1985).

11. R.J. Fields, G.A. Danko, S.R. Low III, and R. deWit, "Wide Plate Crack Arrest Tests: Instrumentation for Dynamic Strain Measurements," ASTM Standardization News, pp. 42-47, October 1985.

12. R.J. Fields, "Wide Plate Crack Arrest Testing: Instrumentation for Dynamic Strain Measurements," 40th Meeting of the Mechanical Failures Prevention Group, Symposium on the Use of New Technology to Improve Mechanical Readiness, Reliability, and Maintainability, April 16-18, 1985, National Bureau of Standards, Gaithersburg, Maryland (in press).

MICRO-MECHANICAL ANALYSIS OF THE INFLUENCE OF METALLURGICAL AND MECHANICAL INHOMOGENEITY ON CLEAVAGE FRACTURE INITIATION

Shuji Aihara, Toshiaki Haze

R&D Laboratories II
Nippon Steel Corporation
Fuchinobe, Sagamihara, Kanagawa, Japan

Abstract

The influence of strength and toughness inhomogeneity on critical CTOD was studied. The cleavage fracture stress criterion, which assumes that cleavage fracture is initiated when maximum tensile stress ahead of a notch tip attains the cleavage fracture stress of a material, was extended to inhomogeneous material. From finite-element analyses together with a simplified model of stress distribution for a material with strength inhomogeneity, the maximum tensile stress at the crack tip region was calculated for the variety of the strength inhomogeneity. When a region with low strength exists near the crack tip, the maximum tensile stress is decreased and the crack opening displacement is increased for the maximum tensile stress to attain the cleavage fracture stress. Experiment showed that critical CTOD was increased when the distance of the low strength region from the notch tip became less or the size of the region with high strength and low toughness is decreased. The results of experiment compared well with the estimation by the present criterion.

IN ORDER TO MAINTAIN SAFETY of welded steel structures such as offshore platforms against brittle fracture, the crack tip opening displacement (CTOD) concept is being increasingly adopted to measure toughness of weld heat-affected zone (HAZ). While critical CTOD is largely influenced by the toughness of the most embrittled zone, it is also influenced by the inhomogeneity of the material. A welded joint has mechanical and metallurgical inhomogeneity through the base metal, HAZ and weld metal. The microstructure of the base metal is completely destroyed in the HAZ by the weld thermal cycles, which cause distribution of strength and toughness from the base metal to the HAZ. Furthermore, the toughness and strength of the weld metal are different from those of the base metal or HAZ because of the difference in chemical compositions and thermal cycles during welding.

While the CTOD concept is a useful criterion to evaluate the safety of a structure, it having a theoretical background other than the Charpy impact energy approach, the applicability of the concept to materials which have mechanical and metallurgical inhomogeneity has not yet been fully discussed.

Recent studies[1],[2] on brittle fracture of welded joints show that critical CTOD is influenced by the strength of weld metal under the same toughness of HAZ when the notch tip is located close to the fusion line of the HAZ. Another experiment[3] shows that critical CTOD of the HAZ decreases as the size of the coarse grained zone which is intersetcted by the notch is increased, despite the toughness at the most embrittled zone in the coarse grained region remaining the same.

In order to solve these problems, Arimochi et al.,[1] proposed "Local CTOD" criterion to predict critical CTOD, δ_C, for a CTOD specimen which has mechanical inhomogeneity in the plane of geometry and a notch tip located along the fusion line. When the strength of the weld metal is lower than that of the HAZ, crack opening becomes asymmetrical, i.e., the crack opening displacement of the HAZ side, δ_u, becomes smaller than that of the weld metal side, δ_L. It was proved that the smaller displacement, δ_u, characterizes crack tip strain field and it was assumed that brittle fracture occurs when $2\delta_u$ reaches a critical value of HAZ. The conventional CTOD, $(\delta_u + \delta_L)$, is not a material constant but depends on the strength distribution.

On the other hand, Satoh et al.[4] analyzed such effect on critical CTOD from the statistical point of view, using a CTOD specimen having a toughness inhomogeneity in the through-thickness direction for a welded joint. By assuming probabilistic distribution of δ_C, depending on the position of the notch relative to the welded joint, and that δ_C of the full-thickness specimen is determined by the lowest value of δ_C along the crack front (weakest link model), they predicted δ_C of the most brittle region from the full-thickness specimen. The present authors[2] examined the influence of strength distribution on critical CTOD from the viewpoint of stress-controlled fracture criterion.

However, a general concept on the influence of the material inhomogeneity has not yet been established.

In the present paper, the effect of mechanical and metallurgical inhomogeneity on cleavage fracture initiation or critical CTOD is studied based on the "cleavage fracture stress" criterion.

BASIC CONCEPT

It has been shown by many authors[5],[6] that fracture toughness can be predicted from the cleavage fracture stress criterion. In the present paper, this criterion is extended to the specimen with mechanical and metallurgical inhomogeneity. In principle, cleavage fracture initiation or critical CTOD, δ_C, can be predicted from the distribution of cleavage fracture stress, σ_f, along the notch tip and the calculation of maximum tensile stress ahead of the notch tip, σ_m, for the specimen with strength and toughness inhomogeneity. Cleavage fracture is predicted to occur when and where σ_m reaches σ_f at a certain point along the notch tip.

Consider a welded joint which has mechanical and metallurgical inhomogeneity as illustrated in Figs.1-(a) and (b). Hardness or strength changes along the line through the base metal, HAZ and weld metal. When the notch of a CTOD specimen for a K-type welded joint is oriented in the through-thickness direction, the specimen has a strength distribution in the plane of geometry, Fig.1-(a). In the other case, the microstructure changes along the thickness direction when the fusion line is inclined relative to the notch line in the specimen for a X-type welded joint. Hence, the CTOD specimen has toughness and strength inhomogeneity along the notch tip, Fig.1-(b).

Two types of three-point bend specimens which model the material inhomogeneity of the HAZ are analyzed. Type-A and B specimens are simplified model of CTOD specimens for K- and X-type welded joint, respectively. Figure 2-(a) shows the one with strength inhomogeneity in the plane of geometry and with the notch located at the center of the high strength and low toughness material-I which is surrounded by low strength and high toughness material-II, Type-A. The other one is shown in Fig.2-(b). In this type, high strength and low toughness material-I is interposed between low strength and high toughness materials-II at the mid-thickness position, Type-B.

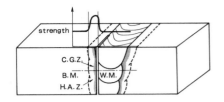

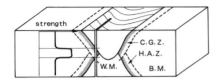

(a) Material inhomogeneity in the plane of geometry

(b) Material inhomogeneity in the through-thickness direction

Fig.1 Schematic illustration of CTOD specimens of a welded joint

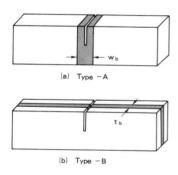

Fig.2 Simplified model of CTOD specimens of a welded joint

STRESS-STRAIN ANALYSIS

PROCEDURE OF CALCULATION - In order to predict δ_C for a specimen with material inhomogeneity from the cleavage fracture stress criterion, σ_m at the notch tip should be determined. In the case of homogeneous material, slip-line field theory was adopted to estimate for notched-bend specimen which deforms as a rigid-perfectly plastic material[7]. Since the finite-element analysis was adopted, the effect of work-hardening can be taken into account[5],[6]. In the present paper, the finite-element method is used to calculate σ_m for the two types of specimens.

Figure 3 indicates the coordinate system and stress components ahead of the notch tip.

σ_{yy} is the stress component which causes cleavage fracture to initiate. There are two definitions of σ_m. One is the stress at a characteristic distance from the notch tip. The distance depends on the microstructure of a material.[6] The other is simply the maximum stress at each stress distribution curve. Determination of the characteristic distance is rather ambiguous. So in the present study, σ_m is defined as the maximum value of σ_{yy} along y=0.

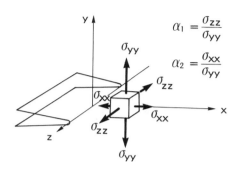

$$\alpha_1 = \frac{\sigma_{zz}}{\sigma_{yy}}$$

$$\alpha_2 = \frac{\sigma_{xx}}{\sigma_{yy}}$$

Fig.3 Definition of coordinate system and stress components at crack-tip region

Type-A - In this case, two-dimensional finite-element analysis can be applied directly. Although the thickness of the specimen is taken to be 10mm for comparison with the present experiments, plane strain condition is assumed because the plastic flow pattern is closer to that in plane strain than in plane stress condition. Yield stress of the surrounding material, σ_{YII}, and the width of the central material, w_b, are changed, while the yield stress of the central material, σ_{YI}, remains constant (Table 1).

Table 1 Analytical condition for Type-A and Type-B

σ_{YI}	σ_{YII}	w_b , t_b					
(N/mm²)		(mm)					
598	519 470	0.01 *	0.2	0.4 1.0	2.0	4.0 - **	

*:Type-B only, **:homogeneous

Type-B - Assume that the interposed material is more brittle than the surrounding material and cleavage fracture is initiated exclusively from the interposed material. Then, σ_m at the mid-thickness point of the interposed material (z=0) becomes critical for cleavage fracture initiation. In this case, three-dimensional stress-strain analysis is required to determine σ_m. However, it is costly and time-consuming to use the three-dimensional elastic-plastic finite-element analysis. So, it

is calculated based on the following assumptions together with the two-dimensional finite-element analysis. In this case, the thickness of the interposed material, t_b, and the yield stress of the surrounding material, σ_{YII}, are varied under the constant value of the yield stress of the interposed material, σ_{YI} (Table 1). It is also assumed that the surrounding material is thick enough that the surfaces of the specimen does not affect the constraint at the mid-thickness point.

At the strain-concentrated region ahead of the crack tip on y=0 and z=0 (mid-thickness of the interposed material), shear stress components are negligible, which leads to von Mises yield criterion expressed by

$$[\sigma_{YI}(\bar{\varepsilon}_p)]^2 = \tfrac{1}{2}[(\sigma_{xx}-\sigma_{yy})^2+(\sigma_{yy}-\sigma_{zz})^2+(\sigma_{zz}-\sigma_{xx})^2] \quad (1)$$

where, $\sigma_{YI}(\bar{\varepsilon}_p)$ is flow stress of the interposed material as a function of equivalent plastic strain. By defining

$$\alpha_1 = \frac{\sigma_{zz}}{\sigma_{yy}} \quad \text{and} \quad \alpha_2 = \frac{\sigma_{xx}}{\sigma_{yy}} \quad , \quad (2)$$

Eq. (1) can be rewritten as

$$[\sigma_{YI}(\bar{\varepsilon}_p)]^2 = [\alpha_1^2-(1+\alpha_2)\alpha_1+(\alpha_2^2-\alpha_2+1)]\sigma_{yy}^2 \quad . \quad (3)$$

From the slip-line field theory, α_1 is 0.81 for plane strain and 0 for plane stress. α_2 is 0.65 for plane strain and 0.55 for plane stress at the crack tip region[8]. From the present finite element analysis under plane strain condition, α_1 and α_2 have the value of 0.7-0.8 and 0.55-0.6, respectively. The value of α_2 is rather stable, approximately 0.6, irrespective of the condition of plane strain or plane stress. Hence, by fixing the value of α_2 at 0.6, Eq. (3) can be rewritten as

$$[\sigma_{YI}(\bar{\varepsilon}_p)]^2 = (\alpha_1^2-1.6\alpha_1+0.76)\sigma_{yy}^2 \quad . \quad (4)$$

Under plane strain condition, by substituting α_1=0.81, σ_{yy} =2.89 σ_{YI} is obtained. As the thickness of the interposed material becomes small, plastic constraint in the through-thickness direction is reduced and α_1 becomes small. Consequently, σ_{yy} is lower than in plane strain condition.

Weiss[8] showed the distribution of σ_{zz} along the thickness direction for a homogeneous material under small scale yielding, which is schematically shown in Fig.4-(a). While the plane strain condition is maintained at the inner part of the thickness, a transient zone from plane stress to plane strain exists between the plate surface and the distance Z from the surface, where Z is proportional to the plastic zone size, r_p, and expressed by Eq. (5).

$$Z = D \, r_P \quad (5)$$

where, D is approximately 4. r_p is related to crack opening displacement, δ, by $(mE/6\pi\sigma_{YI})\delta$, where m is a constant value of 1-2 and E is Young's modulus.

In the case of the Type-B specimen, distribution of σ_{zz} can be assumed from Fig. 4-(a) as shown in Fig.4-(b). Actual distribution of the stress is schematically shown by the broken line. Until the plastic deformation is not so large for the surrounding material as to affect the constraint at the mid-thickness point of the interposed material, plane strain condition is maintained at the mid-thickness, namely $\sigma_{zzm}=\sigma_{zzI}=0.81\sigma_{yyI}$, where σ_{zzm} stands for σ_{zz} at the mid-thickness point. Above a certain point of deformation, the constraint in the through-thickness direction is reduced. Because σ_{zz} approaches the plane strain value, $\sigma_{zz}=0.81\sigma_{yyII}$, with increasing z, it can be assumed for a simplicity that σ_{zz} takes the contant value, $\sigma_{zz}=\sigma_{zzII}=0.81\sigma_{yyII}$, in the surrounding material. σ_{zzm} changes between σ_{zzI} and σ_{zzII}, each of which is calculated for homogeneous material-I and II as a function of δ Because the constraint at the mid-thickness point is reduced by both sides of the surrounding material, triangular distribution of σ_{zz} is assumed shown by the solid line, where slope of σ_{zz} is assumed to be the same as the one for homogeneous material. From this assumption together with the finite-element calculation of σ_{yyI} and σ_{yyII}, σ_{zzm} can be estimated.

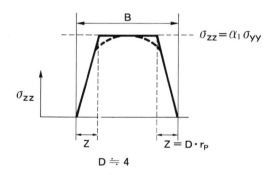

D ≒ 4

r_P = (plastic zone size)

(a) Homogeneous

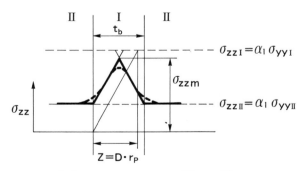

(b) Inhomogeneous (Type-B)

Fig.4   Schematic illustration of σ_{zz} distribution along the crack-tip

Satoh et al.[9] showed a uniform crack opening displacement for a specimen with mechanical inhomogeneity along the thickness direction by measuring the distribution of the stretched zone width. From the present finite-element calculations, it has been found that strain distribution ahead of the crack tip does not depend on yield stress when compared at the constant crack opening displacement. From these rusults, it can be assumed that crack opening displacement or strain distribution ahead of the crack tip is constant along the thickness direction even for the Type-B specimen. From this assumption, it is derived that equivalent plastic strains ahead of the crack tip for the specimen of homogeneous material-I only and the specimen of the interposed material (Type-B) are the same under the same crack opening displacement, which leads to the same equivalent stress or flow stress with these two specimens for a given crack opening displacement. Hence the following equation is derived.

$$(\alpha_1^2-1.6\alpha_1+0.76)\sigma_{yyI}^2=(\alpha_{1m}^2-1.6\alpha_{1m}+0.76)(\sigma_{zzm}/\alpha_{1m})^2 \quad (6)$$

The left hand side of the equation expresses the equivalent stress or flow stress at the crack tip for homogeneous material-I under plane strain condition and the right hand side for the interposed material at the mid-thickness. It is noted that $\sigma_{yym}=(\sigma_{zzm}/\alpha_{1m})$, where α_{1m} is α_1 at the mid-thickness. Eq. (6) has a quadratic form with respect to α_{1m} and can be solved if σ_{yyI} and α_1 are known from the finite-element analysis and σ_{zzm} from the above assumption.

Shear stress, τ_{xz} and τ_{zy}, are apparent near the interface of the two materials and the present method can not be applied to calculate σ_m at the point near the interface. However, the present method is expected to give a reasonable approximation of σ_m if the point of calculation is limited to the mid-thickness point of the interposed material.

RESULTS OF CALCULATION - Figure 5 shows an example of σ_{yy} distribution ahead of the notch tip of a three-point bend specimen obtained by the finite-element analysis. In the analysis, measured stress-strain curves are used. Initial notch root radius is 0.1mm. As the applied load increases, σ_{yy} increases and the point of the maximum stress departs from the notch root. Hereafter, σ_m, defined as the maximum value of σ_{yy} ahead of the notch tip, is compared in each case.

Type-A - Figure 6-(a) and (b) show the variation of σ_m with δ. In case that the central material-I is surrounded by material-II, which has lower yield stress, the rate of the increase of σ_m with δ is depressed. The point of the deviation from the curve for homogeneous material-I is coincident with the point where the plastic zone reaches the interface of the central and the surrounding material. The point of the deviation appears in the earlier stage of deformation as w_b decreases. In case that w_b is 0.2mm, σ_m coincides with that for homogeneous material-II after small amount of deformation. As σ_{YII} is lowered, increase of σ_m is more depressed.

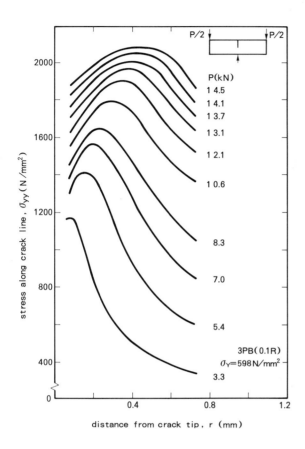

Fig.5 σ_{yy} distribution ahead of the notch-tip

From the cleavage fracture criterion, δ_C is determined from the point of intersection of the σ_m versus δ curve with the horizontal line which indicates σ_f of the central or interposed material. Thus, by assuming δ_C of the homogeneous material-I, the variation of δ_C with w_b and σ_{YII} can be estimated as shown in Fig.7. As is expected from Figs. 6-(a) and (b), estimated δ_C increases as w_b decreases and the amount of the increase is more considerable when σ_{YII} is lower. When δ_C is low, in other words when assumed σ_f is low, the effect of the surrounding material does not appear down to a small value of w_b, because cleavage fracture occurs before the plastic zone extends over the central material. However, once the plastic zone extends over the central material, δ_C increases drastically. On the contrary when δ_C of the homogeneous material is high, δ_C is affected in the wide range of w_b. However, in this case, the dependence on w_b becomes milder.

It can be easily expected that the plastic deformation is constrained by the surrounding material, the triaxiality is increased and σ_m is increased when the central material is surrounded by a material with higher yield stress, contrary to the present case. Hence δ_C is decreased.

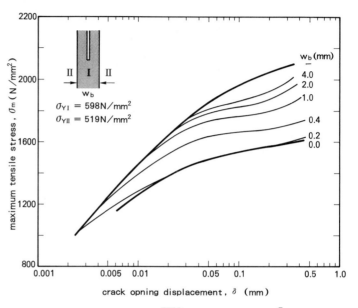
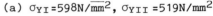

(a) $\sigma_{YI} = 598 N/mm^2$, $\sigma_{YII} = 519 N/mm^2$

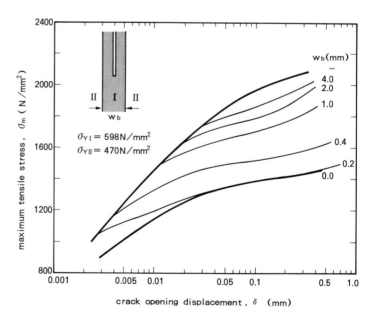

(b) $\sigma_{YI} = 598 N/mm^2$, $\sigma_{YII} = 470 N/mm^2$

Fig.6   Variation of σ_m as a function of δ
(Type-A)

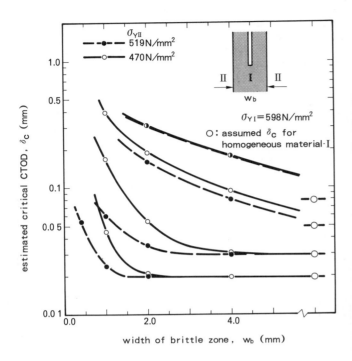

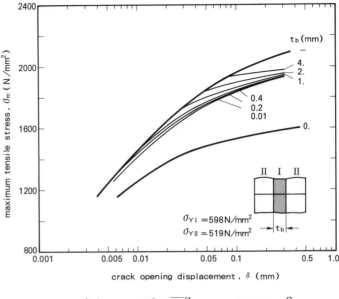

(a) $\sigma_{YI} = 598N/mm^2$, $\sigma_{YII} = 519N/mm^2$

Fig.7 Estimated change of δ_c based on the cleavage fracture stress criterion (Type-A)

Type-B - Figures 8-(a) and (b) show the variation of σ_m with δ. Until the decrease of the plastic constraint by the surrounding material appears, σ_m versus δ curve does not deviate from that for homogeneous material-I. After the surrounding material begins to lower the constraint at the mid-thickness point, increase of σ_m is depressed. Smaller t_b produces lower σ_m. σ_m versus δ curve is converged when t_b is less than about 0.2mm. Decrease of σ_m is more marked in case that σ_{YII} is lower.

When the toughness of the interposed material is very low, cleavage fracture is always initiated from the interposed material. In this case, δ_c can be estimated by the same procedure as in Type-A. By assuming δ_c for homogeneous material-I, δ_c for inhomogeneous material can be estimated as a function of t_b and σ_{YII}, as shown in Fig.9. As t_b is decreasd, δ_c increases. The amount of the increase is more considerable when σ_{YII} is lower. When the assumed δ_c for homogeneous material is higher, the effect of t_b on δ_c is more considerable in the wide range of t_b. δ_c is saturated to a constant value when t_b is less than about 0.2mm.

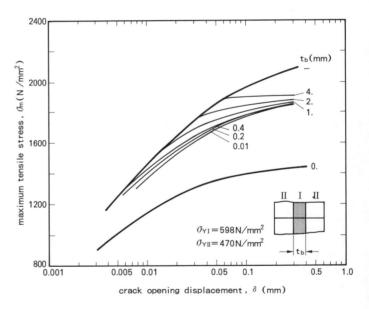

(b) $\sigma_{YI} = 598N/mm^2$, $\sigma_{YII} = 470N/mm^2$

Fig.8 Variation of σ_m as a function of δ (Type-B)

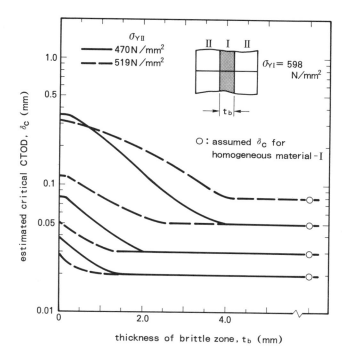

Fig.9 Estimated change of δ_C based on the cleavage fracture stress criterion (Type-B)

EXPERIMENT

PROCEDURE - Experiment was carried out to examine the effect of inhomogeneity on δ_C. Because it is rather difficult to control the difference of yield stress between the central and the surrounding material and the width or thickness of the central material in an actual welded joint, inhomogeneous material was made by hot-rolling a laminated slab, where the chemical compositions of the interposed and the surrounding material were varied as shown in Table 2. Nickel, molybdenum, niobium and a larger amount of manganese were added to the central material (SW01) to increase its strength and decrease its toughness. While the interposed material was kept constant, the surrounding material was varied to change σ_{YII}. t_b or w_b was aimed at 2 and 4mm by changing slab thickness of the interposed material. Combination of materials and w_b, t_b are listed in Table 3. Laminated slab 100mm in thickness was rolled to 20mm followed by normalizing. For Type-B, the specimen was extracted from the plate and

Table 2 Chemical compositions of inhomogeneous material

code	position	C	Si	Mn	P	S	Ni	Mo	Nb	Al
SW01	interposed	.11	.21	1.98	.003	.007	.99	.09	.022	.015
SW03	surrounding	.10	.21	1.54	.005	.006	-	-	-	.020
SW04	surrounding	.10	.19	1.06	.004	.005	-	-	-	.026

Table 3 Combination of materials and w_b, t_b

code	interposed		surrounding		w_b, t_b (mm)
	code	σ_{YI}(N/mm²)	code	σ_{YII}(N/mm²)	
LN1	SW01	598(-90°C)	SW03	519(-90°C)	0.0(SW03 or SW04 only) 2.0
LN2	SW01	598(-90°C)	SW04	470(-90°C)	4.0 - (SW01 only)

thermal cycles, the peak temperatures of which were 1400 and 720°C, were applied by a thermal cycle simulator to reproduce a grain-coarsened brittle microstructure at the HAZ close to the fusion line. CTOD specimen (B=10 mm, W=20mm) was machined. The radius of the notch tip was 0.1mm. For Type-A, a strip was cut from the plate and tab plates were welded in the through-thickness direction. The thermal cycles were applied after normalizing and the CTOD specimen was made. The notch was introduced parallel to the original plate surface. CTOD test was conducted at sub-zero temperatures. Figure 10 shows the microstructure near the interface of different materials for LN2. The microstructure of the interposed material consists of upper bainite with considerable amount of martensite-austenite constituent. On the other hand, the microstructure of the surrounding material consists of grain boundary ferrite and acicular upper bainite. It is also noted that defects at the interface of the two materials are very few and separation was not observed in Charpy and CTOD test.

In addition to the inhomogeneous materials mentioned so far, homogeneous materials, denoted by SW01, SW03 and SW04, were also hot rolled, normalized and thermal-cycled. Then CTOD test for the same configuration of the specimen as above was conducted to measure δ_C of each material.

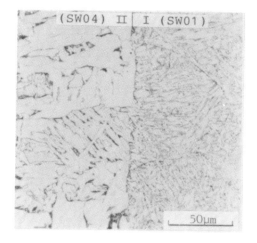

Fig.10 Microstructure of the inhomogeneous material (LN2)

Smooth and notched-round bar tensile tests were also conducted to measure the stress-strain curve and σ_f. The stress-strain curves so obtained were input to the finite-element analyses. σ_f was determined with the help of the finite-element analysis of the notched-round bar, notch root radius of which is 1mm, where the maximum tensile stress at the ligament section was calculated as a function of applied load.

Fractured surfaces were examined by scanning-electron microscopy and the fracture initiation points were examined by sectioning the specimens and etching the sectioned surface.

RESULTS - Figure 11 shows the yield stress and cleavage fracture stress of SW01. It can be seen that yield stress increases as the testing temperature is decreased. On the other hand, σ_f is almost kept at a constant 1300-1600 N/mm^2, in the wide range of testing temperature, from -60 down to -196°C. In addition, no indication of the difference is seen between σ_f values which were obtained in the longitudinal and the through-thickness directions of loading. The open circle shows σ_f obtained from CTOD specimen through finite-element analysis. It shows 300 N/mm^2 higher value than that obtained from the notched-round bar specimen. The same trend as to the influence of notch tip acuity on σ_f is obtained by Ohtsuka et al.[10] However, the difference is small and the cleavage fracture stress criterion seems to be reasonable for predicting cleavage fracture initiation.

Figure 12 shows δ_C as a function of temperature for each homogeneous material. Each plot represents on average of three δ_C values at each testing temperature. Clearly, toughness of SW01 is much lower than those of SW03 and SW04 and cleavage fracture can be expected to be initiated from the interposed material (SW01) for the specimen Type-B.

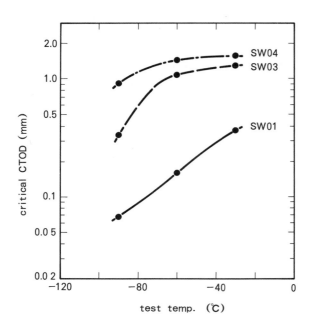

Fig.12 δ_C values of homogeneous materials (average)

Figures 13-(a) and (b) show the dependency of δ_C on w_b for the specimen Type-A. Test was conducted at -60 and -90°C. It is seen that δ_C increases with decreasing w_b both in LN1 and LN2 at -60 and -90°C. δ_C at -90°C increases to 2.8 times of δ_C for homogeneous material where w_b is decreased down to 2mm in LN1 and 1.6 times in LN2, if compared by the average values.

Figure 14 shows the sectioned surface, which is parallel to the specimen surface, close to the fracture initiation point for the specimen LN2 with 2mm of w_b. It is seen that the notch is located at the mid-point of the central material. Cleavage fracture is initiated from the corner of the notch root and propagates through the central material. Another example for the specimen LN2 with 4mm of w_b, Fig.15, shows that a subsidiary crack is seen to have initiated and propagated from the other corner symmetrically to the main crack.

Figures 16-(a) and (b) show the dependency of δ_C on t_b for the specimen Type-B. A general trend is seen as to the increase of δ_C with decreasing t_b. δ_C at -90°C increases to 4.0 times that for homogeneous material when t_b is decreased from 10mm to 2mm in LN1 and 5.8 times in LN2, if compared by the average values.

Figure 17 shows the fracture initation point of the specimen LN2 for t_b=2mm observed by scanning electron microscopy. From the comparison with the fracture surface of the surrounding material, which is shown in Fig.18, the surface at the initiation point consists of rather irregular cleavage surface. Furthermore, measurement of the distance from the surface of the specimen to the initiation point shows that the fracture is initiated from the interposed

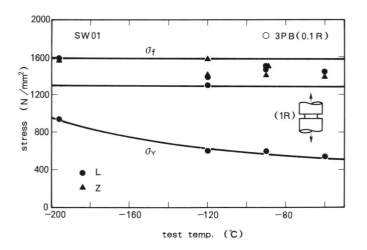

Fig.11 Yield and cleavage fracture stress (SW01)

material, 0.6mm inward from the interface of the two materials. This observation together with others shows that the fracture is initiated exclusively from the interposed material. It is also noted that the initiation point in Fig.17 is 550 μm inward from the notch root.

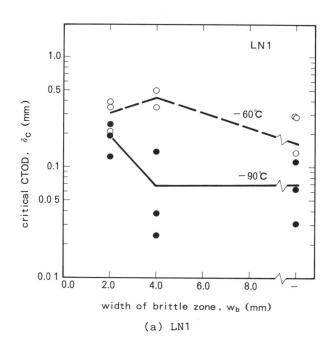

(a) LN1

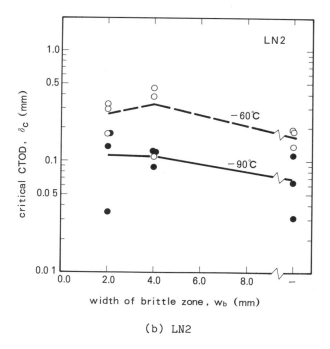

(b) LN2

Fig.13 Dependency of δ_C on width of brittle zone (Type-A)

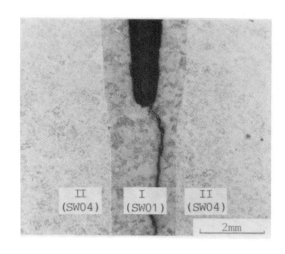

Fig.14 Sectioned surface close to the fracture initiation point (Type-A, w_b=2mm,LN2)

Fig.15 Sectioned surface which indicates a main and subsidiary crack at a notch root (Type-A, w_b=4mm,LN2)

Figure 18 shows the fracture surface at the interface of the interposed and the surrounding material. Fracture surface changes abruptly at the interface on the side of the interposed material-I, namely from the irregular cleavage, which is typical for bainitic microstructure, to plain cleavage on the surrounding material-II. Neither separation nor any indication of the fracture arrest is seen, despite the crack propagated into the tougher material.

23

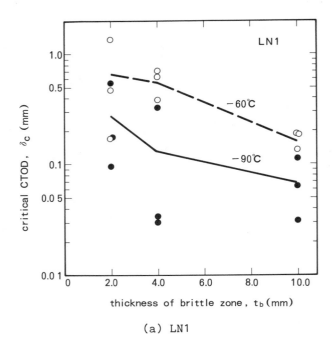

(a) LN1

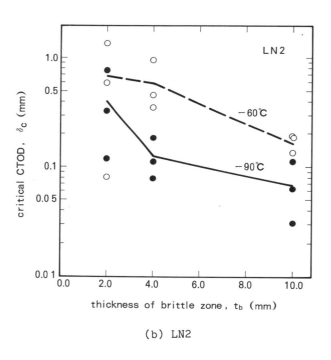

(b) LN2

Fig.16   Dependency of δ_c on thickness of brittle zone (Type-B)

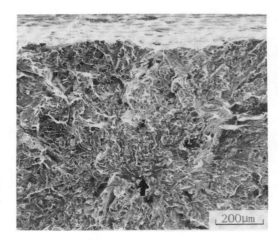

Fig.17   Fracture initiation point observed by SEM (Type-B, t_b=2mm,LN2)

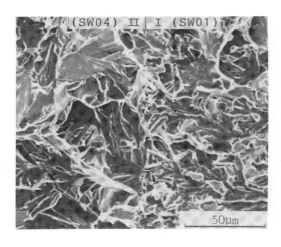

Fig.18   Fracture surface at the interface of the different materials (Type-B, t_b=2mm, LN2)

DISCUSSIONS

Because the conditions of the calculations of Figs.7 and 9 are the same as those of the experiment at -90°C, direct comparison with the experiment is possible. The dependency of δ_c on w_b or t_b by the experiment is almost the same with the calculation, despite some scatter in the experimental data. In Type-A, the estimation shows that δ_c for w_b=2mm increases to 4.0 times of δ_c for homogeneous material. Experiment shows that δ_c for the same w_b increases to 2.8 and 1.6 times of δ_c for homogeneous material in LN1 and LN2 ,respectively, which is roughly coincident with the estimation. In Type-B, the estimation shows that δ_c for t_b=2mm increases to

1.9 times of δ_C for homogeneous material in LN1 and no occurence of fracture in LN2. Experiment shows that δ_C for the same t_b increases to 4.0 and 5.8 times of δ_C for homogeneous material in LN1 and LN2, respectively. Also in Type-B, there is a rough coincidence between the experiment and the estimation, although the calculation shows more drastic influence of t_b on δ_C than the experiment.

The difference between the experiment and the estimation is presumed to be the ductile crack that occurs when δ_C becomes larger than about 0.3mm. Although no occurrence of cleavage fracture is estimated in LN2 at $t_b=2$mm, it is expected that high stress concentration occurs at the tip of the ductile crack and the cleavage fracture is initiated. From the above comparison, the present theory can be applied to predict the effect of material inhomogeneity especially in the range of low toughness, e.g. HAZ, where δ_C is not so large as to accompany ductile crack propagation.

If the cleavage fracture stress criterion is strictly applied, the fracture initiation point must coincide with the maximum stress point estimated from the numerical analysis. The present finite-element analysis shows that the distance of the maximum stress point from the notch root increases from approximately 200 to 500 μm in the range of CTOD level of our interest. It is noted that the stress distribution along y=0 is not steep but has a plateau. Although the experiment does not show systematic change of the distance with increasing δ_C, the distances lie between 250 and 650 μm. Fracture facet size near the initiation point is approximately 50 μm, about 1/3 of the prior austenite grain size. Only one or two prior austenite grain exists at the plateau. It seems that the grain which has preferred orientation for cleavage fracture does not always exist at the exact point of maximum stress but the fracture is initiated from the grain, which has preferred orientation, at the plateau. Hence, the systematic change of the distance of the initiation point is not observed. However, the range of the distance is almost the same as that of the maximum stress point. This fact supports the validity of the cleavage fracture stress criterion.

As shown in Figs.14 and 15, fracture is not always initiated at the plane of symmetry (y=0) but rather in the inclined plane. HRR stress singularity analysis for a crack with zero crack-tip radius[11] shows that the direction of maximum stress is in the plane y=0. In case that the notch has a finite notch-tip radius, the direction of the maximum stress may be inclined to the plane y=0 as speculated from the direction of cleavage fracture in Figs.14 and 15. In the present finite-element analysis, the element size is too coarse to account for the direction of the maximum stress (minimum element size is 0.05mm). For a more accurate analysis, finer mesh is necessary. In any case, it is concluded that the cleavage fracture stress criterion can be applied to inhomogeneous materials as well as to homogeneous materials.

The present method can be directly applied to some simple cases of actual welded joint such as shown in Fig.1-(a) and (b). For predicting δ_C of a welded joint which has toughness inhomogeneity in the through-thickness direction and fracture not always initiated from a particular microstructure, it is necessary to know the distribution of σ_f and σ_m along the notch tip and to compare these at each point of the notch tip. In this case, three-dimensional finite-element analysis may be necessary to obtain σ_m distribution. For the purpose of practical estimation, however, some simplifications are needed.

CONCLUDING REMARKS

The influence of mechanical and metallurgical inhomogeneity on critical CTOD was studied. By extending the cleavage fracture stress criterion to inhomogeneous material, the influence of the inhomogeneity was estimated irrespective of whether a material has the inhomogeneity in the plane of geometry or in the through-thickness direction. From the finite-element analysis with some simplifications, maximum tensile stress at the crack tip region was calculated as a function of strength inhomogeneity in a cracked specimen. By comparing the maximum tensile stress with the cleavage fracture stress, initiation of cleavage fracture, or critical CTOD could be estimated.

Experiment on three-point bend CTOD specimen, which simulated the inhomogeneity in a welded joint and had strength and toughness inhomogeneity, showed that δ_C was increased when the material close to the notch tip had low strength, despite the toughness of the notch tip material remaining the same and that δ_C was increased when the thickness of the material with high strength and low toughness was decreased along the notch tip. These results coincided with the estimation based on the extended cleavage fracture stress criterion.

The present theory can be applied to estimate critical CTOD for a welded joint which has strength and toughness inhomogeneity. However, extension of the theory is required for the application to more complicated case of welded joints.

REFERENCES

1) Arimochi, K., Nakanishi, M., Toyada, M. and Satoh, K., The 3rd German-Japanese Joint Seminar, August (1985), Stuttgart, West Germany.
2) Haze, T. and Aihara, S., Trans. Iron and Steel Institute of Japan, 25, 1, B-31 (1985)
3) Haze, T. and Aihara, S., to be published
4) Satoh, K., Toyoda, M. and Minami, F., IIW Doc. X-1064-84 (1984)

5) Ritchie, R.O., Fnott, J.F. and Rice, J.R., J. Mech. Phys. Solids, 21, p395 (1973)

6) Dormagen, D., Dunnewald, H. and Dahl, W., 6th Int. Conf. Fracture, New Delhi, India,2, p803 (1984)

7) Green, A.P. and Hundy, B.B., J. Mech. Phys. Solids, 4, p128 (1956)

8) Weiss, V. and Sengupta, M., ASTM STP590, p194 (1976)

9) Satoh, K. Toyada, M. Mutoh, Y. and Doi, S., J. Japan Welding Society, 54, 2, p766 (1980)

10)Otsuka, A. Miyata, T. and Arakawa, T. proc. National Meeting of Japan Welding Society, 31, p.236 (1982)

11)Hutchinson, J.W., J. Mech. Phys. Solids, 16, p13 (1968)

AN INVESTIGATIVE ANALYSIS OF THE PROPERTIES OF SEVERELY SEGREGATED A441 BRIDGE STEEL

Michael A. Urzendowski
Packer Engineering
Naperville, Illinois, USA

Frank J. Worzala
University of Wisconsin
Madison, Wisconsin, USA

ABSTRACT

Segregation is a common problem that occurs during the solidification process of cast steel plate. Typically, carbon and manganese migrate toward the center of the plate as the steel solidifies. Normally, the segregation is eliminated or reduced through hot rolling and post-heat treating, such as solutionizing or normalizing. If, however, proper processing procedure is not followed, a coarse grain structure can be formed and the segregation will not be eliminated.

This was the case with a tied arch bridge in western Wisconsin whose steel was the object of all our testing. Testing performed during this investigation indicates that the properties of such a steel are changed detrimentally. Toughness is greatly reduced, energy to initiate a crack decreases substantially and a fatigue crack will propagate much faster through the banded region.

The information obtained from this study is of value to bridge engineers, design engineers and the quality control personnel in the steel mill. These potential problems must be realized by all involved and suitable precautions or changes made.

IN 1979, DURING A ROUTINE BRIDGE inspection, a 4" fatigue crack was discovered in the top flange plate of one tie girder of a tied arch bridge crossing the Mississippi River. This bridge was relatively new, having been completed in 1974. The area surrounding the crack was removed and examined metallographically. This analysis indicated a banding or segregation problem in the middle of the plate, where the carbon content appeared to be twice what it should have been. Based on this and results of ultrasonic testing, which revealed that the banding occurred in 24-foot lengths, it was decided to close the bridge and replace the defective steel.

The scope of this paper examines how the mechanical and physical properties of the steel in the segregated region differ from the remainder of the steel plate. This was done by evaluating the microstructure, chemistry and hardness of the steel and measuring the dynamic toughness and fatigue crack growth rates in the banded region. By noting these differences, bridge engineers can re-evaluate existing designs and modify current bridge inspection procedures.

The steel used in the construction of this bridge was specified as ASTM A441, a steel commonly used in structural applications. Normally, the chemistry of this steel is as shown in Table I (appendix).

Experimental Procedure

The steel was removed from the bridge in the form of 4" diameter cores, 22 total, from various locations in different plates. The cores were sectioned, polished and examined metallographically under an optical microscope for the degree of segregation. Grain size measurements were made per ASTM E112.

Following the initial examination samples were categorized as to the severity of the segregation problem and representative samples were chosen for chemical and hardness evaluation. Charpy V-notch samples were then machined from the longitudinal direction of the central regions of all 22 cores and tested at various temperatures on an instrumented impact testing apparatus. Representative SEM photos were taken of the fracture modes of the fractured samples.

By instrumenting the hammer on this test device, information regarding load versus time can be obtained for the entire fracture process. This load versus time curve can give information regarding maximum load, yielding and fracture loads, and the times to crack initiation and failure. Using this information, the dynamic fracture toughness can be calculated per ASTM E399. All of the data from the impact tests were stored on disk, via a recording oscilloscope, and at a later date interfaced with an Apple microcomputer for all integrations and calculations.

An additional section of plate was removed for measurement of the fatigue crack growth rates through the segregated region of the plate. Specimens were prepared per ASTM E647-81, machined such that the crack would propagate transverse to the rolling direction. In this test the fatigue crack length, a, is determined as a function of fatigue cycles, N. This data is plotted in the form of da/dN versus K, where da/dN is the fatigue crack growth rate in in./cycle and ΔK is the crack-tip stress intensity range in ksi in. The crack length was determined using the compliance method.

Results

MICROSTRUCTURE - Figure 1 represents a typical banded microstructure. The area here was found to contain 100% pearlite and was coarse grained with a grain size of #1. A fine grained microstructure is what ASTM specifies for this particular grade of structural steel. Although the term fine-grained is somewhat undefined, it is commonly accepted that a grain size smaller than ASTM No. 5 is considered "fine".

Figure 2 represents the microstructure of the non-banded cores removed from the bridge. This structure contains approximately 60-70% ferrite and has a grain size of 6.

CHEMISTRY - The results of the chemical analysis on the central areas of the banded cores show a marked increase in carbon and manganese levels in the banded areas (Table 2 - Appendix). The carbon level approached .4 and the manganese level was at 1.3, both levels exceeding ASTM specification A441. The nonsegregated cores meet all chemical requirements.

HARDNESS - A typical hardness profile through a segregated sample is shown in Figure 3. The non-segregated cores had a relatively uniform hardness through the thickness.

DYNAMIC TOUGHNESS - The results of the impact tests and the resultant dynamic fracture toughness calculations are shown in Figures 4-5. Figure 4 shows the amount of energy needed to fracture versus the testing temperature. Figure 5 is a plot of dynamic fracture toughness versus temperature for both sets of samples. The non-banded samples show an increase in toughness over the banded cores.

The fractured CVN samples were examined with a scanning electron microscope for the mode of fracture. The segregated samples, as illustrated in Figure 6, fractured entirely by cleavage. No shear was evident even at the notch area, which typically experiences the highest shear stress.

The non-banded samples at the same temperature, 75° F, failed almost entirely by shear (Figure 7).

Fatigue crack growth rates are plotted in Figure 8. Data for tests at room temperature (75°F), -40°F, and the known value for growth rates through ferrite-pearlite steel are all plotted together for comparison. The growth rates calculated from bridge core samples are much greater than the known value. The growth rate for the banded cores at -40° F was $.278 \text{ E} -9 \Delta K^{3.94}$ in/cycle and $.907 \text{ E} -9 \Delta K^{3.54}$ in/cycle at 75° F. The accepted value at room temperature is $3.6 \text{ E} -10 \Delta K^{3.0}$ in/cycle.

Discussion

The results of the micro-structural, chemical and hardness analysis all correlate. The higher than normal carbon level corresponds with the amount of pearlite observed and also with the hardness evaluation.

The most important factor to note in this initial analysis is the large grain size, present in the center region of the plate. It has been pointed out (1-3) that grain size is the main factor controlling the strength and toughness of a ferrite-pearlite steel. In order to control the grain size during processing, one must both control the temperature during the final stages of rolling and perform post roll heat treating to cause recrystallization of the grain structure throughout the plate. An increase in grain size means an increase in transition temperature and nil ductility temperature of approximately 2.3 C per $\Delta d^{-1/2}$. In addition, the tensile ductility and toughness will decrease.

Segregation has always been a problem in cast steel plate. In general, the structure and properties of hot-worked metals are not uniform over the cross section as in metals which have been, for instance, cold worked and annealed. Since the defor-mation is always greater in the surface layers, the plate will have a finer recrystallized grain structure in this region. Because the interior will be at higher temperatures for longer times during cooling, grain growth can occur in the interior of large pieces. This is especially true if improper working and finishing temperature are used.

In order to refine the grain structure and to homogenize the microstructure. a normalizing heat treating process is used on A441 rolled plate It is apparent that neither of these two desired responses occurred in the center of our banded plates. Because of this, and the segregation problem, it seems as if the temperature used for normalizing was too low or times were too short. Therefore, the center of the plate did not get up to temperature.

The results of our impact testing revealed poor impact properties for the banded cores. The transition temperature, as defined by the change in slope of the energy curve was approximated at 145°F, while the non-banded cores had a transition temperature almost 90°F lower. This corresponds to what was stated earlier regarding the effect of the grain size on the transition temperature.

AASHTO toughness requirements for bridge steels (4) indicate that A441 steel should have a minimum fracture energy of 15 ft-lbs. at 40°F. The segregated core were markedly inferior. At 40° F, the fracture energy was approximately 4 ft-lbs. The 15 ft-lb. value was not reached until a temperature of 110° F. The non-banded samples tested meet this specified criterion.

By instrumenting these tests, it was possible to observe how the specimen reacted to the loading, prior to, during and following impact. At temperatures below 105° F the banded samples did not yield prior to fracture. Above 105° F, the banded samples did experience yielding. Once the crack initiated in these samples, crack propagation was rapid. The energy to initiate the crack was much larger than to propagate the crack.

The non-banded samples yielded at temperatures above 32° F and the energy required to initiate a crack was substantially greater than in the banded samples. In both cases once the crack initiated, propagation occurred instantaneously.

The dynamic fracture toughness value calculated represents the stress intensity needed to produce fracture under dynamic loading or high strain rate conditions. The dynamic toughness values are considered to be close to the minimum values of fracture toughness that can be obtained and thus are useful design values for structures where either dynamic loading or propagating cracks are present. The results reveal a difference in toughness between the segregated and the non-segregated areas of the bridge. In a non-segregated area, a much higher impact load would have to be applied to propagate an existing crack.

The fracture appearance of the CVN specimens corresponds to the amount of energy absorbed to fracture the samples. Cleavage is typically a low energy fracture mechanism, while shear requires a high energy level to occur.

When discussing the toughness of steel, it is necessary to consider grain size and carbon level as significant variables. The majority of the segregated samples tested had an undesirable grain size and carbon level, which resulted in poor charpy and dynamic toughness properties.

The carbon also is directly related to the yield strength of the related steel. Generally, as the yield strength increases, the amount of ductility associated with fracture decreases and the amount of cleavage, or brittle fracture, increases. As

pointed out earlier, the center of 16 of the 22 cores had higher than permissible carbon levels. These samples all had poor impact properties.

Because of this effect on the yield strength, the transition temperature was raised dramatically in the center of the segregated plates. These tests, and actual experience, demonstrated that a beam made of steel from a segregated plate in service, at room temperature, would fail in a brittle manner under the appropriate stress conditions; whereas, the non-segregated plates would fail in a ductile manner. It took a much greater, dynamic level to cause failure in the non-segregated plate because of the lower notch sensitivity and yielding that would occur.

High strain rate testing on the bridge samples is of interest to us because many structural components are subjected to high loading rates while in service, or must survive high loading rates during accident conditions. From the instrumented charpy results and the dynamic toughness data, it can be observed how the bridge samples reacted during dynamic loading conditions. During dynamic fracture of the segregated samples, the majority of the energy used to cause failure was consumed initiating the crack. Propagation followed with little energy required. Thus, these components should be designed against crack initiation under high loading rates, or designed to arrest a rapidly running crack. Furthermore, the fracture resistance of a material that is loaded rapidly, is generally lower than when the load is applied slowly; consequently, the dynamic fracture toughness is a more conservative value than the static value for design calculations.

The results of the fatigue crack growth rate testing reveal two phenomenon: The first is the increase in crack growth rates as the temperature lowers. The second is the fact that our room temperature results indicate a much faster growth rate than the literature indicates is acceptable for this material.

The temperature dependence of the growth rate was expected and has been documented and explained (3, 5). The deviation in the growth rate from the accepted literature value, however, needs further discussion.

There are several possible explanations for this phenomenon. Chemical analysis has shown that the carbon content and consequently the pearlite content is much greater than expected in our samples. Because of this increase in carbon level, the yield strength of this material is raised. Assuming that we have plane-strain conditions, the size of the plastic zone in front of the advancing crack tip is proportional to: $(1/8 \ [\ K/\sigma y]^2 \ )$. (ref. 3). For our samples, therefore, the plastic zone size is less than would be expected for A441 steel due to the inverse relationship between yield strength and size of the plastic zone.

The stress distribution is such that, with a smaller plastic zone, there is a higher actual stress intensity at the crack tip due to the fact that the plastic zone cannot "relieve" as much stress as its larger counterpart in a standard ferrite-pearlite steel. This increases the fatigue crack growth rate.

In addition, there was less crack tip blunting occurring with a smaller plastic zone. This includes the opening, advance and blunting of a crack tip in the loading portion of the cycle and the resharpening of the crack in the unloaded portion of the cycle. With less crack tip blunting, the stress intensity at the crack tip was much greater during the loading portion of the cycle. This also contributed to a faster growth rate.

Static values of fracture toughness, K_{1C} were estimated from the results of our crack growth rate tests. At $-40^{\circ}F$, the K_{1C} value is approximately 35 $ksi\sqrt{in}$, and at $75^{\circ}F$, it is 52 $ksi \sqrt{in}$. This is quite a difference from the 130-150 $ksi \sqrt{in}$ fracture toughness value specified by A441. This toughness value is inversely proportional to the yield strength which in our case, was raised due to the increased carbon level present.

Summary

Failure to exercise proper rolling and/or heat treating can lead to problems such as excessive grain size and chemical segregation in the center region of structural plate.

Our testing has shown an increase in hardness and weight percent carbon and manganese in the banded region. Further testing revealed that the area containing the segregation and coarse grain structure had a lower than expected toughness and a transition temperature $90^{\circ}F$ higher than specified by the ASTM standards. The fatigue crack growth rate through this area was much faster than expected. All of these property changes can be explained due to increased carbon levels, higher yield strength and larger than normal grain size.

This information is of use to bridge engineers, in that, the amount of physical and mechanical property changes can be realized and corrected for, by altering existing designs or creating new ones.

Conclusions

1. Microstructural examination indicated a banding problem in 16 out of 22 cores taken from the bridge. The microstructural evaluation corresponding to the results of the instrumented impact tests. The banded plates have poor impact properties and poor toughness values, while the non-banded plates exhibited acceptable properties.

2. Hardness testing revealed a sharp increase in hardness in the center region of the banded plates, while in the non-banded plates, there was little or no rise in hardness. Chemical segregation was also indicated by the results of the chemical analysis.

Both carbon and manganese levels increased in the center region where pearlite content was increased. Pearlite is much harder than ferrite. The chemistry in the center did not meet the chemistry specifications of the ASTM A441 in the segregated plate.

3. Either the temperatures involved in the rolling procedure were not correct, or the heat treating process was either not timed properly, or the temperatures used were too low. This is revealed by both the lack of homogeneity through the plate and the excessive grain size in the center of the plate.

4. This increased grain size contributed to the increase in the transition temperature which lowered the toughness of the material. It takes more energy to fracture small grained plates than large grained plates.

5. These findings are useful because of the large number of bridges that incorporate A441 or a similar type steel in the structure. If segregation is a common occurrence through this type of plate, bridge engineers will have to re-evaluate current design criterion and modify inspection intervals.

In addition, the steel industry must bring about tighter quality control measures to insure that all material leaving meets specifications.

References

1. Tetelman, A.S. and McEuily, A.J. *Fracture of Structural Materials*, John Wiley & Sons, 1967.

2. Hertzburg, R.W., *Deformation and Fracture Mechanics of Engineering Materials*, John Wiley & Sons, 1976.

3. Barsom, J.M., and Rolfe, S.T., *Fracture and Fatigue Control in Structures*, Prentice Hall, 1977.

4. ASM Metals Handbook, Ninth Edition, Vol. 8, ASM, 1985.

5. Tobler, R.L., and Cheng, Y.W., "Midrange Fatigue Crack Growth Data Calculations for Structural Alloys at Room and Cryogenic Temperatures," NBS, Boulder Co., Unpublished.

Appendix

Table 1 - Chemistry of ASTM A441 Steel.

C	Mn	Si	Cu	S	Ph	V
.22	.85-1.25	.4 max	.2 min	.05	.04 max	.02 min

Table 2
Chemical analysis of the central region on one of the banded cores.

C	Mn	Si	Cu	S	Ph	V
.36	1.30	.04	.34	.042	.015	.05

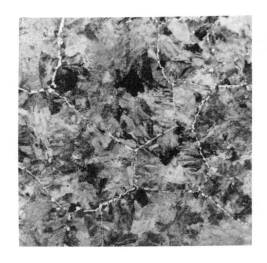

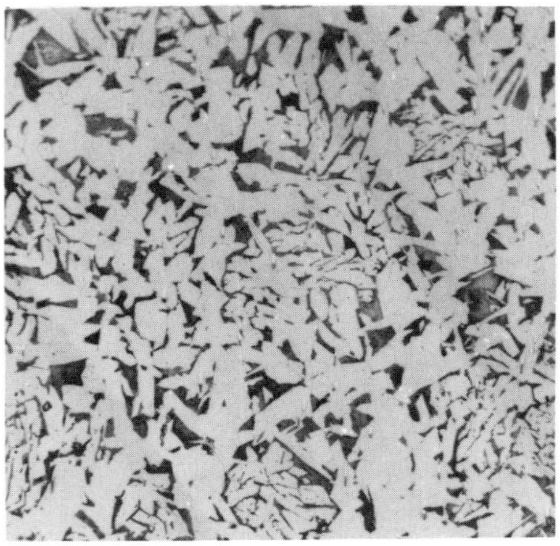

Figure 1 - Typical microstructure of a banded core. (Picral) 10 micron –

Figure 2 - Non-banded core microstructure. (Nital) 10 micron –

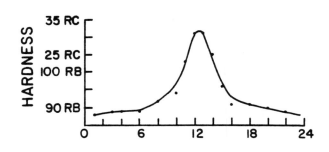

HARDNESS
35 RC
25 RC
100 RB
90 RB
0
6
12
18
24
DISTANCE (1/16 in. INCREMENTS)

Figure 3 - Typical hardness traverse through banded plate.

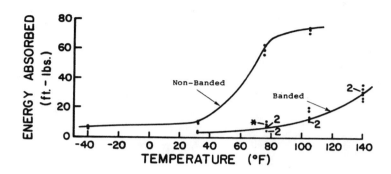

Figure 4 - Charpy impact energy plotted versus temperature
for banded and non-banded cores.

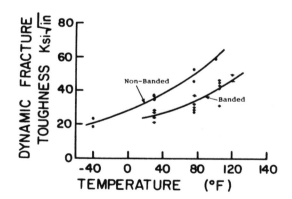

Figure 5 - Dynamic fracture toughness date plotted versus
temperature for banded and non-banded cores.

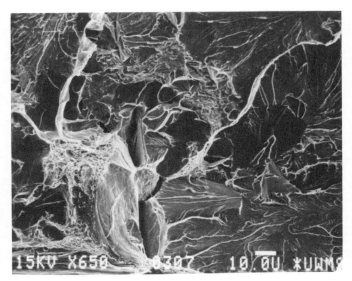

Figure 6 - SEM photomicrograph of the
fracture surface of a segregated CVN sample - sample
failed entirely by cleavage
75 F.

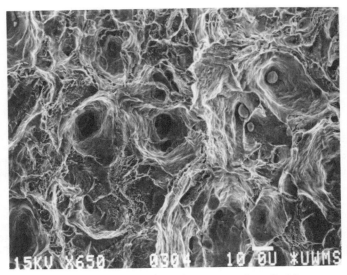

Figure 7 - SEM photomicrograph of the
fracture surface of a non-
segregated CVN sample-the
failure mode was entirely
by shear at 75 F.

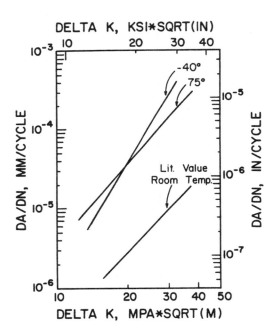

Figure 8 - Fatigue crack growth rates through the segregated
region compared to known growth rate values.

FAILURE ANALYSIS
OF A LARGE CENTRIFUGAL BLOWER

R.B. Tait, G.G. Garrett, D.P. Spencer
Department of Metallurgy
University of the Witwatersrand
Johannesburg , South Africa

ABSTRACT

A fracture mechanics based failure analysis and life prediction of a large centrifugal fan was undertaken following shortcomings in attempts to explain its fatigue life from start stop cycles alone. Measurements of the fracture toughness and flaw size at failure, coupled with quantitative SEM fractography using striation spacing methods, revealed that the cyclic stress amplitudes just prior to failure were much larger than expected, in this particular case. Subsequent improvements in fan design and fabrication have effectively alleviated the problem of slow, high cycle fatigue crack growth, at normal operating stresses in similar fans.

CONSTRAINTS OF relatively high altitudes and poor quality coal in South Africa necessitate the use of particularly large draft fans to provide air for the combustion process, in this country's coal fired electrical power stations. Such fans can measure over three meters in diameter add weigh more than thirty tons.

In 1981 one of these fans (of a double inlet centrifugal type) failed catastrophically under seemingly normal operating conditions causing extensive damage (Fig. 1). This paper describes an aspect of the failure analysis that was undertaken in an attempt to understand the present fracture and to try to prevent failures in similar fans in future.

The fan had been in service for only approximately 10 days following an outage for maintenance. Prior to that it had been in service for some 2 1/2 years. Preliminary investigation of the fractured pieces (1) revealed evidence of extensive fatigue cracking in the shroud plate as well as, but to a lesser extent, in the blade roots. Since any cyclic tensile stress in the shroud plate was nominally not more than 10 to 15 MPa under normal operating conditions (and thus below the fatigue limit), it was initially thought that the cyclic fatigue stress source arose from start stop cycling. The steady state stresses developed from start stop cycling were estimated to be approximately 110 to 120 MPa. An approximate fracture mechanics based fatigue life estimate of the service life of a fan under these cyclic stress conditions indicated that nearly half a million cycles would be required to grow the crack to failure proportions. In reality, however, the fan had experienced at most approximately 100 start stop cycles. Thus start stop cyclic stresses were ruled out as the major cause of fatigue cracking and a more rigorous analysis, including quantitative fractography, was undertaken.

MATERIAL TESTING

Compact tension samples of full thickness shroud plate material were machined and used for both fatigue crack propagation rate tests as well as fracture toughness tests, in accordance with ASTM E647-78T and BS.5447 (and BS.5762) respectively. A plot of the rate of crack growth per cycle, da/dN against cyclic stress intensity amplitude, ΔK, is shown in Fig. 2. The best fit straight line (the so called "Paris equation") to this data is:

$$\frac{da}{dN} = 4.8 \times 10^{-12}(\Delta K)^3 \quad \ldots \quad (1)$$

(where da/dN is in m/cycle and ΔK in MPa$\sqrt{m}$)

For the fracture toughness it was presumed that COD fracture toughness tests (in accordance with BS.5762, 1979 (2)) should be undertaken since the low carbon, medium strength steel may be expected to behaved plastically at ambient temperatures.

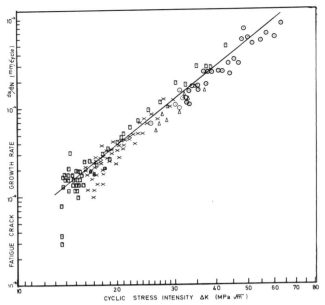

Fig. 2   Plot of rate crack growth versus cyclic stress intensity amplitude.

Unfortunately the steel exhibited elastic-brittle behavior in the operational temperature range of 0 to 40°C with a (lower shelf equivalent fracture toughness of 65 ± 9 MPa$\sqrt{m}$. At a temperature of 95°C the equivalent toughness was 135 MPa$\sqrt{m}$ in accordance with the reported (1) ductile to brittle transition temperature (DBTT) of this material of approximately +53°C. This illustrates that the steel was on the (brittle) lower shelf during normal operating conditions.

FRACTOGRAPHIC STUDIES

The fractographic studies were undertaken to try to determine the source and mechanism of failure as well as some measure of the cyclic stress amplitude.

FAILURE MODE - Chevron markings on the fracture surfaces, which point back to the origin of the crack, located the crack source in the shroud plate in a region near a blade trailing edge. There was also evidence that the crack may have originated from a welded on balance pad (or perhaps an

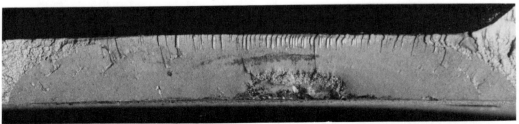

Fig. 3(a) Shroud plate fracture surface illustrating crack initiation from the defect and subsequent fatigue fracture.

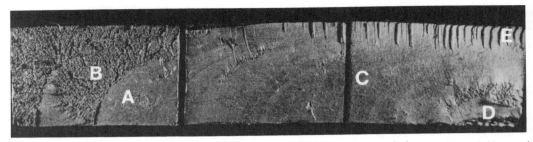

Fig. 3(b) Left hand side of the fracture surface shown in Fig. 3(a) sectioned through the defect origin and subsequently cut into three sections to facilitate SEM study. Note clamshell markings indicating progressive crack fronts.

arc strike). Distinction between these two possibilities could not be defined categorically, however, because the other (mating) half of the initiating crack fracture surface was never found.

The critical fracture surface is shown in Figs. 3(a) and (b) and was clearly indicative of bending fatigue in view of the bowed clamshell markings, and otherwise flat, featureless surface. Fracture appearances of this sort are normally characteristic of high cycle fatigue. The initiating flaw, approximately 15 mm long and 1.8 mm deep is apparent in Fig. 3(b).

A higher magnification of the microscopically flat surface of regions A and B in Fig. 3(b) is shown in Figs. 4(a) and (b) and clearly illustrates the presence of characteristic fatigue striations associated with conventional high cycle fatigue.

QUANTITATIVE SEM AND CYCLIC STRESS EVALUATION - Figure 5 shows a low magnification fractograph of the region at the fatigue origin (marked D in Fig. 3(b)) where the fatigue fracture has started from the 1.8 mm deep defect. Higher magnification SEM indicated fatigue striations similar to those of Figs. 4(a) and (b). A transverse was then undertaken from region D through to E (see Fig. 3(b)) - and subsequently for regions A and C - and multiple striation

spacing measurements made, together with the actual crack depth at these points. Using the fatigue crack propagation data determined earlier in the laboratory (Fig. 2), it was possible to read off values of ΔK corresponding to particular fatigue crack growth rates (da/dN). This follows since, in the linear region of the fatigue crack propagation

Fig. 4(b) Clearly defined fatigue striations from region B (of Fig. 3(b)). Magnification 1800X.

Fig. 5 Low magnification fractograph of the origin of failure (area D of Fig. 3(b)). Fatigue fracture initiated from the right hand lower corner of the main fracture surface. Magnification 18X.

Fig. 4(a) Fatigue striations typical of the flat fatigue surface (region A of Fig. 3(b)). Magnification 2000X.

curve at least, each striation corresponds to one fatigue cycle.

If environmental effects are small, it is normally reasonable to assume that a particular growth rate is associated with a particular value of ΔK in a given steel. For a given crack length and with appropriate corrections for crack shape and compliance factor (i.e. finite width corrections (3,4)), the prevailing alternating stress can be estimated. A more rigorous discussion of the applicability of this technique, is given elsewhere (5).

STRESS ESTIMATES - From the above quantitative SEM study it was apparent that the cyclic stress was not trivial, being in excess of 100 MPa, peak to peak, as the crack grew from a surface "thumbnail" crack to a "through thickness crack" condition.

The stress at failure was estimated in two different ways and found to be comparable. On the one hand conventional fracture mechanics methods were employed, utilizing the actual defect size and the material toughness, to give a peak stress at failure of approximately 156 MPa. On the other hand the failure stress was evaluated as the sum of the mean (steady state) stress predicted by finite element methods (of 110 to 120 MPa) plus half the cyclic stress amplitude derived from quantitative fractographic studies, to give values in excess of 160 to 170 MPa.

DISCUSSION AND ADDITIONAL TESTS

The evidence of the previous section indicates that, in the case of this particular fan failure, the cyclic stress amplitudes (>100 MPa) were much higher than expected (<20 MPa). At such high cyclic stress amplitudes the fan would develop a critical sized fatigue crack in approximately 300,000 cycles, or about 6 hrs. running time. This implies that such high cyclic stresses could not have been present for the whole 10 days since maintenance outage, but must only have developed either intermittently, or more probably, only in the last 12 to 6 hours of the fan's life. Such high stresses could not normally exist in regular fans as there would be rapid and widespread failures, which is not the case. The origin of the particular high cyclic stresses could not be determined, however, and to investigate possible failure scenarios other tests were conducted, which are outlined below.

Fans similar to the one that failed were strain gauged and run at operating speeds of 750 rpm (12.5 Hz) and strains measured and resonances evaluated using a telemetry system and modal analysis techniques (6). Under ordinary operating conditions such studies confirmed that normal dynamic cyclic bending stresses in the shroud plate were low, typically less than 20 MPa, but could potentially run to 30 or 40 MPa if resonances were approached. The closest approach to resonance conditions arising from, say aerodynamic excitation, was not at rotational frequency (12.5 Hz) but at blade passing frequency (150 Hz), but even this could not be sustained, under normal operating conditions.

The implication is that there may well have been some unusual event, other than resonance, which precipitated the high cyclic stress development. A possibility is a change in compliance or balance of the fan if cracking had occurred elsewhere--for example in blade roots or in the balance pad, and that the fatigue crack in the shroud (as a result of bending stresses) was a subsequent event. There was, however, little conclusive evidence in support of any such high cyclic stress scenarios.

It is of interest, however, to note that although the failure occurred in the early hours of the morning, an increase in bearing temperature had allegedly been recorded some hours before its failure, but no action was taken.

In addition, in view of the concern that was expressed about fatigue cracks possibly emanating from the balance pad weights in this and other fans, their behavior was investigated more fully. The fan construction is such that component pieces are carefully weighed before fabrication to minimize subsequent balance pad requirements. Final balance can only be achieved on commissioning and is effected by means of a balance pad welded onto the appropriate place on the shroud. Improvements to the balance pad design and attachment, from rectangular stitch welded to rounded pads, continuously welded, with the welds ground and peened, have resulted in an increase in the cyclic stress amplitude to initiate and propagate a fatigue crack in bending, from approximately 50 MPa to over 110 MPa (7).

Other improvements that have been made include increasing the fracture toughness of the plate material and setting more stringent acceptance levels for defect sizes. The fan design has been modified so that its natural resonant modes are far from blade passing frequencies, and the stresses have been reduced.

CONCLUSIONS

(1) This paper has described a fracture mechanics based failure analysis of a large industrial fan. The failure occurred from high cycle bending fatigue and appeared to have originated from a weld defect.
(2) The fracture toughness of the plate was relatively low, equivalent to $K_Q = 65 \pm 9$ MPa$\sqrt{m}$ over the operating temperature range of 0 to 40°C.
(3) By means of quantitative fractog-

raphy using striation spacing methods, the cyclic stress amplitude in the vicinity of the crack prior to failure was estimated and was much higher than expected.

(4) A simple fracture mechanics analysis relating measured plate toughness and final (critical) flaw size yielded limiting failure stresses consistent with finite element analysis models of limiting steady state stress, coupled with half the estimated cyclic stress amplitude.

(5) The precise origin of the high cyclic stresses could not be determined. Resonance and strain gauge tests revealed that, under normal operating conditions, cyclic stresses are typically less than 20 MPa.

(6) Improvements to fan design, material toughness, fabrication and weld detailing has resulted in apparently safer operation of this type of fan in service.

ACKNOWLEDGEMENTS

Support for this study from Messrs. Airtec Davidson and ESCOM (South Africa) is gratefully acknowledged.

REFERENCES

(1) C.A. Parr ESCOM Report OM-133, Nov. (1981).

(2) British Standards BS.5762 (1979).

(3) R. C. Shah and A. S. Kobayashi. ASTM-STP 513, pp. 3-21 (1972).

(4) G. R. Marrs and G. W. Smith. ASTM-STP 513, pp. 201 (1972).

(5) R. B. Tait, D. P. Spencer and G. G. Garrett. Fracture Research Report - FRP 82/9, Metallurgy Department, University of the Witwatersrand (1982).

(6) S. Franco. ESCOM Report, TRR/H83/005 (1983).

(7) C. A. Boothroyd, Ph.D. Thesis, Department of Metallurgy, University of the Witwatersrand, 1985.

A FRACTURE MECHANICS BASED FAILURE ANALYSIS OF A COLD SERVICE PRESSURE VESSEL

R.B. Tait, D.P. Spencer, P.R. Fry, G.G. Garrett

Department of Metallurgy
University of the Witwatersrand
Johannesburg, South Africa

ABSTRACT

Following a fracture mechanics "fitness for purpose" analysis of petroleum industry cold service pressure vessels, using the British Standard PD 6493, it was realised that an analogous approach could be used for the failure analysis of a similar pressure vessel dome which had failed in service some years previously. Examination of the fracture surfaces suggested, from fatigue striations manifected by SEM, that the vessel was subject to significant fatigue cracking (which was probably corrosion assisted). From COD measurements at the operating temperature of $-130°C$, and a finite element stress analysis, a fracture mechanics evaluation using BS PD6493 yielded realistic critical flaw sizes (in the range 51 to 150mm). These sizes were consistent with the limited fracture surface observations and such flaws could well have been present in the vessel dome prior to catastrophic failure. For similar pressure vessels an inspection programme based on a leak before break philosophy was consequently regarded as acceptable.

AN EXTENDED 'FITNESS FOR PURPOSE' evaluation of cold service pressure vessels (CSPV) used in the oil from coal industry was undertaken in 1982 (1). This study made extensive use of the British Standard document PD 6493 (2) and yielded critical flaw sizes for vessels in a range of stresses and temperatures from operational ($-60°C$) to ambient ($+25°C$). The critical opening displacement (COD) fracture toughness (for the particular 10 mm thick steel plate) was measured as a function of temperature from $-196°C$ on the lower shelf to $95°C$ on the upper shelf. The steel was a low strength, low carbon steel, (trade named IZETT) developed by KRUPP in Germany during the second world war, ostensibly for cryogenic application for liquid oxygen propellants in the V1 and V2 rocket programme. The steel had a fine grain structure, high aging resistance and low impurity levels resulting in significant low temperature toughness, further enhanced by a double heat treatment procedure. Further details of the steel are given elsewhere (3). The fitness for purpose study also considered the fatigue lives of these CSPVs to quantify safe inspection periods, which under the given operational stresses, turned out to be several years.

In view of the success of this study and various overlapping features it was considered worthwhile to apply the PD 6493 fracture mechanics analysis to an actual failure which represented another class of CSPVs. One of these vessels, used as carbon dioxide adsorber and operating at $-130°C$, had failed catastrophically through a crack in its domed head some two years previously. A failure analysis report (4) attributed the failure largely to stress corrosion cracking (SCC), but no fracture mechanics type analysis was undertaken at the time. Particularly in view of the safety of similar carbon dioxide adsorber pressure vessels and that they are subject to some (nominally) minor but periodic pressure loading, it was believed that this failure should be more carefully examined. The following questions were of particular concern: (i) whether there was any evidence of fatigue or whether the crack growth was purely of an SCC nature? (ii) if there was evidence of fatigue what was the cyclic stress amplitude? (iii) an estimate of the critical flaw size (from a fracture mechanics standpoint)? and (iv) could safe inspection intervals be recommended to obviate catastrophic failure of such vessels?

Fig. 1 Photograph of the failed pressure vessel dome illustrating the initiating crack region.

Unfortunately because the failure had occurred two years previously and had already been analysed the only material available for re-examination was in the form of a small (10 mm x 20 mm) microscope specimen piece of fracture surface, previously mounted and polished for metallographic examination! All other material had been scrapped.

Nevertheless it was possible to perform an experimental failure analysis. This paper describes that analysis and illustrates particularly the technology transfer theme of this conference through the use of PD 6493, to answer questions of present pressure vessel safety.

EXPERIMENTAL CONSIDERATIONS AND BS PD 6493 ANALYSIS

The failed pressure vessel, with a diameter of 2.5 m and several meters tall, had been made of 12 mm thick IZETT steel plate of the same type and heat treatment as used in the earlier fitness for purpose study. Its COD fracture toughness had thus already been measured, at its operating temperature of $-130°C$ to be $\delta_i = 0.05 \pm 0.02$ mm (For conservatism reasons the COD at crack initiation δ_i, was employed rather than the less conservative, but now widely accepted (5), COD at maximum load, δ_m . δ_m is almost always larger than δ_i). The origin of the fracture in the dome of the CO adsorber (Fig. 1) was identified by means of chevron markings in subsequent fracture surfaces pointing back towards the crack origin. The original critical flaw size at failure was not specifically measured at the time of the failure and initial report (3), but was believed to be approximately 40 to 130 mm long and located transverse to the hoop stress direction. A small sample of the fracture had been cut from the dome at the intersection of two major cracks as shown in Figure 1 and this had been duly mounted in resin for metallographic examination. The resin mounted metallographic sample was first examined microscopically and only limited evidence of SCC cracking observed.

This resin mounting was then carefully removed and the specimen thoroughly cleaned to expose the original fracture surface. Small samples of the steel were removed for hardness and compositional tests (which confirmed that the steel was indeed IZETT type of the appropriate heat treatment) and the original fracture surfaces mounted for scanning electron microscope (SEM) observation.

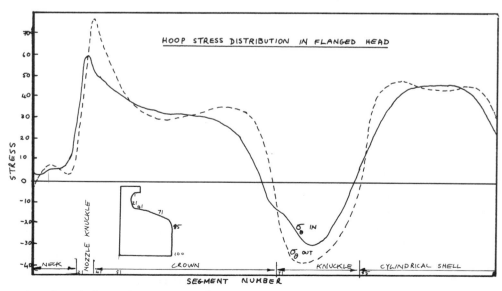

Fig. 2 Finite element stress analysis model of the adsorber dome.

The PD 6493 code provides, among other capabilities, a formalised means of determining flaw sizes in a structure given the material's (COD) fracture toughness and the effective stresses to which the structure is subject. As a first estimation of the critical flaw size the steady state stresses, as determined by a finite element model, (FEM) were considered (Fig. 2). The critical flaw occurred, not surprisingly, in the vicinity of the highest hoop stress as determined by the FEM model. This stress was approximately 80 MPa on the outside and 59 MPa on the inside surface. Using the appropriate ratio of applied stress to yield strength, Figure 14 of PD 6493 (presented here as Fig. 3) is entered, the corresponding value for the parameter C read off and the tolerable defect parameter, a, determined. This value is corrected for bulging, in the case of longitudinal defects in curved pressure vessels and resulted ultimately in a measure of the critical defect size. Various fracture toughness cases were considered (in view of the scatter in measured toughness), and resulted in critical flaw sizes in the range 60 to 176 mm which are approximately consistent with the size of the "actual" observed fracture.

A second approach to estimate critical flaw size made use of the limited fracture surface itself. High magnification in the SEM of the fracture surface revealed the presence of fatigue striations characteristic of fatigue, (although this may have been corrosion assisted), Figures 4 (a) and (b). Measurements of striation spacing together with measurements of the actual crack length were made. Using fatigue crack propagation data determined earlier in the laboratory, it is possible to read off the value of ΔK (and hence $\Delta\sigma$) corresponding to particular fatigue crack growth rates (da/dN). This follows since, in the linear portion of the fatigue crack propagation curve at least, it is normally reasonable to regard one striation as corresponding to one fatigue cycle. From this study the cyclic stress amplitude appeared to be in the range 60 to

Fig. 3 Values of constant C for different loading conditions (Fig. 14 of BS PD 6493 (2))

90 MPa (depending on the value of the assumed initial crack length). In the worst case the stress thus ranged from zero on unloading to 90 MPa under load.

The PD 6493 analysis relating this new stress and the COD fracture toughness was then repeated to yield critical flaw sizes in the range 51 mm to 150 mm, again approximately consistent with observations of the flaw at failure.

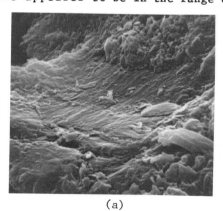

(a)

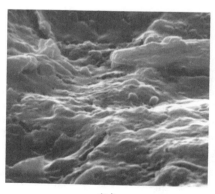

(b)

Figs. 4 (a) and (b). Fatigue striations of the fracture surface characteristic of cyclic fatigue and used in the cyclic stress evaluation.

DISCUSSION

Evaluation of the critical flaw size at failure from three viewpoints yielded approximately self consistent results. The estimate of flaw size from its fracture surface (but which was no longer available) was in the range 40 to 130 mm, while estimates based on a finite element model and on stress derived from fatigue striation measurements gave critical flaw sizes in the range 60 to 176 mm and 51 to 150 mm respectively. Finally a fatigue life estimate based on the cyclic stress amplitude and using appropriate values for initial flaw size from NDE criteria, was undertaken, using an integrated form of the Paris equation. This fatigue service life estimate depends significantly on the cyclic stress input and on the initial (and final) flaw sizes, but in any event was of the order of several years, much longer than present normal inspection periods.

Since the critical defect sizes are relatively large in comparison with pressure vessel wall thickness of 12mm, the vessel, if regularly inspected, may be regarded as fairly safe. For example, a through thickness defect approaching critical dimensions would result in an easily detectable "leak before break" situation, even taking into consideration the lagging of these vessels. As long as such leaks are detected and the appropriate preventative action taken, catastrophic failure can be avoided.

CONCLUSIONS

(1). This paper illustrates an aspect of the "technology transfer" theme of the conference by the use of the fracture mechanics based code, BS PD 6493, in the failure analysis of a cold service pressure vessel.
(2). The pressure vessel had experienced significant fatigue crack growth (which was probably corrosion assisted) as evidenced by fatigue striations on the fracture surface.
(3). The critical flaw sizes determined from (i) a finite element stress analysis and (ii) fatigue striation spacing were self consistent and comparable to limited direct fracture surface observations.
(4). Since the critical flaw sizes were large in relation to vessel wall thickness, and "worst case" situations have been assumed throughout the investigation, the present thin walled carbon dioxide adsorber vessels still in service may be considered to be safe, since leak before break would occur and be detected, as long as regular conventional inspections for such leaks are undertaken.

ACKNOWLEDGEMENT

The authors wish to acknowledge the suport of SASOL One and particularly the assistance of Mr. M.L. Holland and Dr. M.F. Ball of SASOL One on this project from which this paper is taken.

REFERENCES

1. R. B. Tait, D. P. Spencer, P. R. Fry and G. G. Garrett, FRP report FRP 82/14 Department of Metallurgy, University of Witswatersand (1982).

2. British Standard PD 6493:1980 "Guidance on some methods for the derivation of acceptance levels for defects in fusion welded joints".

3. M. L. Holland, M. F. Ball, R. B. Tait and G. G. Garrett" A fracture mechanics assessment of an unusual carbon steel used in low temperature petrochemical process plant, "Proc.," Fracture and Fracture Mechanics Case Studies", Eds. R. B. Tait and G. G. Garrett, Pergamon Press, September 1985.

4. M. L. Holland, SASOL, Private Communication 1983.

5. J. D. Harrison, "The State-of-the-Art in Crack Tip Opening Displacement (CTOD) Testing and Analysis". Metal Construction, September, October, November, 1980.

A DISCUSSION CONCERNING FIELD FRACTURES
IN HEAVY EQUIPMENT

Mitchell P. Kaplan
Willis, Kaplan & Associates, Inc.
Arlington Heights, Illinois, USA

ABSTRACT

This paper discusses the fracture behavior
and their ramifications in several pieces of
heavy equipment. This equipment includes a
crane, and a hi-rail device, i.e., a motorized
vehicle that can travel on both rail and on
highway. The failure that occurred in these
pieces of equipment was catastrophic in the
sense that it took its toll in human life and
left the equipment unusable. As lawsuits
arose from these accidents, both the plaintiff and the defense, through their attorney
hired engineering experts to determine the
mechanism of failure. In the case of the
crane, theories of failure centered upon
whether the welding technique utilized by
the manufacturer was susceptible to lamellar
tearing or was the crane severly abused in
both maintenance and usage. In the case of
the hi-rail, a hydraulic cylinder that raised
and lowered the rail wheels fractured. The
mechanical design, especially of a critical
component was the issue in this instance.
In these two cases, the various failure
theories forwarded by the experts as well as
their underlying foundations are discussed.

THE PURPOSE of this paper is twofold. First,
to relate to us as engineers what may arguably
be perceived as critical problems involved in
detail design, and secondly to discuss the
ramifications of accidents involved in the
failures of selected components. The loss in
manpower, and increasing cost due to litigation, is an extremely important issue that
engineers must face in todays society.

These issues are becoming more prevalent
in today's society. However, with the advent
of micro- and minicomputers that today are
found on many desks, the present day engineer
has tools to perform design and analysis that
the engineer ten years ago could not accom-
plish. But with these tools come added responsibilities. Society now dictates that
engineers design and manufacture products
that are without error.

This may seem like a bold statement;
nevertheless, it is true. In the court system
today, a product may be found "unreasonably
dangerous" if the product fails and is the
proximate cause of an accident. To find the
product "unreasonably dangerous," then from
the engineering perspective it may be enough
to show that the failure was due to a design
defect or a manufacturing defect. If the
court finds this to be true, then the manufacturer of the product could be responsible
for the product's failure.

Two examples will be used to illustrate
this. The examples selected are a hi-rail
device and a truck mounted crane. Each of
these pieces of equipment had undergone a
material/structural failure of a critical
component. As a result of the failure of
this component, the operator was severely
or fatally injured, and a lawsuit followed.

The design intent of the structures,
the facts relating to their particular accident, the components that failed including
the mechanism of their failures and alternative design solutions that may have alleviated these accidents will be shown. In
addition, the role that the engineers retained by the various parties involved in the
lawsuit play, will be briefly discussed.

HI-RAIL DEVICE

A. DESIGN

A hi-rail device, shown in Figure 1, is a
vehicle that is designed to travel both on

roads and on rails. This vehicle may be a truck outfitted with rail wheels or may be specifically designed to perform particular functions. In this instance, a truck was modified to accept the wheels for rail loco-motion. The rear wheel/axle set was attached, as shown in Figure 2, to the frame of the truck. Both the front and rear wheel/axle sets were raised by means of a hydraulic cylinder driven off the PTO of the truck. The controls to raise or lower the wheel/axle set were located on the side of the truck. These locations may be seen in Figure 1.

The hydraulic cylinder rod was attached to a clevis. By extending this rod, the clevis rotated and lifted the rail wheel/axle set. Please refer to Figure 3. When the rod was contained within the cylinder, the wheel/axle set was in a down position, and the vehicle was able to move on the rails, Figure 3A. When on rail wheels the inside truck tire on the rear axle would be sitting on the rail and would be the driving force for the vehicle, Figure 2.

B. FAILURE MECHANISM

In this design, the wheel/axle set was rigidly fixed into an up or down position by the use of locking pins. It was assumed by the manufacturer that there would be no load on the cylinder once the wheel/axle set was in its locked position. However, as the cylinder pivoted about its mounting trunnion and extended during its motion, it interfered with a frame member. This was due to alignment and fabrication problems. This caused both a bending load and a rotational movement. These effects, unacceptable for a two force system, caused a combination of fretting, galling, and fatigue to the internal thread structure of the clevis. As a result of these deleterious effects, the failure of the thread structure of the clevis occurred.

The cylinder rod/screw thread interface, because of its small fillet radius, has a high stress concentration factor. As the threads wear away, the clamping motion of the clevis is lessened and the compliance of the system increases. This adds addi-tional stresses to the already highly stressed area. Furthermore, while the cylinder rod is extended, the cylinder itself bears onto a frame member, Figure 3B. This static load plus an additional vibratory ben-ding load will occur when the truck is moving down the highway. It was the combination of bending loads that initiated and propagated the crack at the cylinder rod/clevis inter-face.

The cylinder used on this unit is 14 3/8 inches long and 5 1/2 inches in diameter. Although the cylinder rod is 1 3/4 in diameter, the threaded area was no greater than one inch in diameter.

The failure occurred where the cylinder rod screws into the clevis. This is shown in Figure 3. The rod was manufactured from 1045 steel. The tensile and yield strength of the steel was 110 and 58 ksi respectively. This type of steel has a toughness of approx-imately 10 ft-lbs at 0°F. The microstructure was typical for rolled 1045 steel, i.e., ferrite and pearlite.

As a result of these forces, a fatigue crack, Figure 4, initiated at the cylinder rod/clevis intersection area and progressed until failure occurred. Figure 4A, shows a low magnification photograph of the failed area, while Figure 4B, shows the fatigue area and Figure 4C, is the overload area, in this case cleavage fracture.

C. ACCIDENT DESCRIPTION

The failure was noted by the employee when he attempted to remove the wheel/axle set from the rails and store them under the truck frame. Noting that operating the valve to raise the wheels did no good, he crawled under the vehicle and saw the failed cylinder rod. Placing the cylinder rod against the clevis, he had his co-worker operate the system to raise the wheels. In doing so, the rail/wheel set moved abruptly. As the cylinder rod acts not only as the driving linkage but also as the control mechanism, its absence allowed the wheel/axle set to swing freely. It was during this motion that the operator was injured.

D. ALTERNATIVE DESIGNS

As stated above, a number of detail design and manufacturing problems allowed the failure to initiate and propagate. The design problems include both material selec-tion and detail design. The alloy, although structurally strong, had poor notch toughness and a high ductile to brittle transition temperature. If an alloy had been chosen with equally high tensile properties and a structure of tempered martensite, both these properties could have been improved. In addition, the threaded area which was signif-icantly less in diameter than the cylinder rod had a sharp radius connecting the two cross sections. This led to a high stress concentration factor, and a high stress on the smaller threaded area. Finally, there was the manufacturing fit-up which allowed the frame member of the truck to act as a driving force for vibratory loads during road travel. Greater attention to tolerances may have prevented this undesirable dynamic force.

TRUCK MOUNTED HYDRAULIC CRANE

A. DESIGN

The truck mounted hydraulic crane, shown in Figure 5, is a multipurpose vehicle used in the construction industry. The vehicle, for this discussion, consists of a carrier which is non-rotating and a crane which is able to rotate freely. Between the rotating and non-rotating structures is a horizontal thrust bearing that has one race attached to each of the two structures. The outside race of the bearing is driven by a pinion gear, and it is through this mechanism that the crane body rotates about a vertical axis.

There are generally two mechanisms available to firmly attach the race to its particular structure. These are welding the races or attaching them by the use of mechanical fasteners such as bolts. In this instance, the outer race was bolted to the carrier, and the inner race was welded to the turntable. The inner race of this bearing is shown in Figure 6. The thickness of the bearing is 3/4 inch. The manufacturer chose to weld this race to the carrier in a single pass.

B. ACCIDENT DESCRIPTION

After several years of service, the crane was being used to move material, and the attachment weld between the bearing inner race and the turntable failed. As a result of this failure the rotating portion of the crane was no longer attached to the carrier. The rotating structure containing both the boom and the operator's compartment fell seriously injuring the operator.

C. FAILURE MECHANISM

Examination of the fracture surface indicated that the weld failure occurred in the area adjacent to the heat affected zone. This is shown in Figure 7. Engineers who examined the failure disagreed amongst themselves regarding the rationale underlying the failure. The three mechanisms forwarded by the various investigators were; poor welding which allowed the initiation of fatigue cracks after prolonged usage, improper welding techniques which led to lamellar tearing, or failure due to overload resulting from misuse of the crane. Figure 8, shows the fracture face which one investigator described as indicative of fatigue failure.

Figure 7, shows cracking in an area that is just at or below the heat affected zone which one investigator used as evidence of lamellar tearing. Finally, Figures 9-12, show examples of deformation on the crane itself which may be taken as indications of misuse. Figure 9, shows the area where the lift cylinders for the boom are located. Deformation may be noted in this area. Figure 10, shows the plate upon which the rotating structure rests, and deformation is again noted. Figure 11, shows examples of cracks in gusset locations on the crane, while Figure 12, shows buckled or bent booms.

These mechanisms are different enough from one another to cause one to question their relationship to this failure. For this reason, additional facts need to be discussed. When a bearing of this type is loaded (note that the inner race which was attached to the superstructure was the failed member), one area on the ring is always loaded in tension and another area is always loaded in compression. Furthermore, it should be noted that if a fatigue failure initiates and propagates, unloading the crane will cause the failed surfaces to bear together. This effectively eliminates any observation of striations. One may, however, find coarser representations of fatigue such as beach marks or arrest lines.

After the failure has occurred, and the crane components removed for examination, it was noted that the fracture zone where there was the greatest tension was heavily oxidized. As one approached the neutral axis, the oxidation level decreased. In the zone where the bearing would have been in compression, there was a clean surface indicating recent fracture. Finally, there were areas where the weld did not meet AWS specifications for convexity or concavity. These areas may be sufficiently weak to allow fatigue cracks to initiate.

The investigator who opined that the misuse and abuse of the crane was the mechanism underlying failure noted the areas of distress on the crane. These included the buckled boom sections, the heavy brinelling, deformation of the base plate, and cracking. Analysis indicated that loads required to cause these deformations were sufficient to cause failures in the weld section.

D. ALTERNATIVE DESIGNS

The crane, similar to the hi-lift, can be redesigned in a manner that may reduce the likelihood of this accident. These changes would include material selection and detail design. These include; the use of bolts in lieu of welding, a welding schedule that reduces the propensity of lamellar tearing, and finally the use of an alloy that precludes lamellar tearing. However, if abuse of the crane were the primary factor in failure causation, none of these techniques would have been useful in preventing the deterioration of the machine to an extent that would have rendered the failure improbable.

LITIGATION

As a result of these accidents, the injured party or his estate hired an attorney, who in turn, filed a lawsuit against the manufacturers of the equipment. The lawsuit alleged, in both cases that the equipment was "unreasonably dangerous" because of defects (design and manufacturing). Furthermore, these suits alleged that the failure of this equipment, thus the injury that the plaintiff suffered, was due to this defect.

With the lawsuit served, the lawyers (plaintiff and defense) go through a discovery process. The defendant, the company who manufactured the product, may be required to turn over drawings, engineering change orders, engineering change notices, calculations, analyses, test results, memoranda, etc. This information is studied by engineers for plaintiff, as is usage history of the equipment, the failure itself, and the written records of all people remotely involved in the accident. The attorneys hire engineers to examine these documents and components, conduct analyses and tests, and render their opinions. This process is both lengthy and expensive.

With which side will the court agree? If the parties themselves don't agree on some sort of settlement, then the court will make a decision based upon the evidence presented. The engineers hired by the plaintiff and defense will become a part of this process and will be required to present their opinions. But, the court will make the decision.

In the cases presented above, the court never made a decision because the parties settled the suit prior to the court date. In these instances, no responsibility was assigned; however, the cost of the lawsuit in money due to attorney fees, engineer fees, and settlement fees were high.

CONCLUSION

As stated in the introduction, a purpose of this paper was to discuss the ramifications of accidents. This has been through discussion of the accidents, the design problems alleged, and cost. What may a company do to minimize the number of accidents and lessen the number of lawsuites? A company can do the following:

1. Pay careful attention to detail design,

2. Conduct a thorough design analysis, stress analysis, fatigue and fracture analysis, vibration analysis, etc.,

3. Conduct appropriate tests,

4. Conduct a safety analysis, including reasonably forseeable use and misuse of equipment,

5. Be aware of and take full account of applicable codes and standards,

6. Prepare users and maintenance manuals carefully, and in detail, and

7. Warn against hazards as appropriate.

Figure 1 Side View Of Hi-Rail

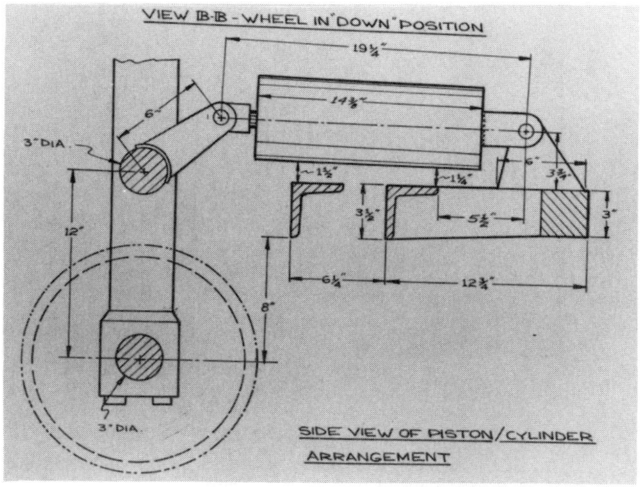

Figure 2 Rear View Of Hi-Rail

VIEW B-B - WHEEL IN "DOWN" POSITION

SIDE VIEW OF PISTON/CYLINDER ARRANGEMENT

3A

VIEW B-B - WHEEL IN "UP" POSITION

SCHEMATIC SIDE VIEW OF PISTON/CYLINDER ARRANGEMENT

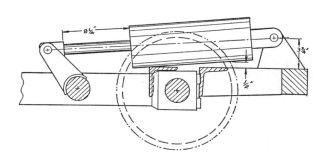

3B

Figure 3 Piston Cylinder Arrangement

A. Macro

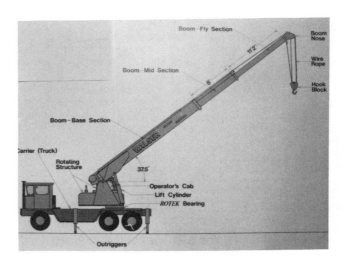

Figure 5   Truck Mounted Crane

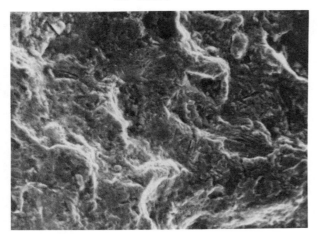

B. Fatigue Area

Figure 6   Inner Race Fracture Area

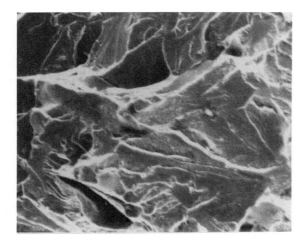

C. Fast Fracture

Figure 4   Fracture Surface

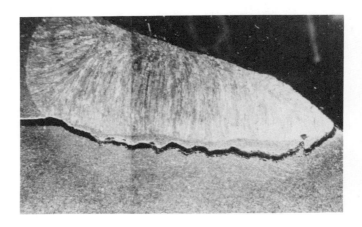

Figure 7   Failure at Weld

Figure 8 Fatigue Failure

A

Figure 9 Deformation in Lift Cylinder Area

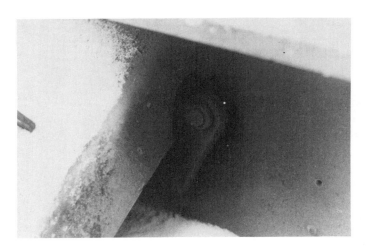

B

Figure 11 Typical Cracks in Carrier

Figure 10 Deformation of Carrier

Figure 12 Boom Deformation

FAILURE ANALYSIS OF PRESSURIZED ALUMINUM CYLINDERS AND ITS APPLICATIONS TO A SAFER DESIGN

I. Roman, D. Rittel
Graduate School of Applied Science and Technology
Materials Science Division
The Hebrew University of Jerusalem
Jerusalem, Israel

ABSTRACT

Recent catastrophic ruptures of several aluminum-magnesium pressurized air containers, used for diving purposes promoted an investigation as to the cause(s) of the failures. The results of the analysis revealed that catastrophic failure occured when a subcritical crack that grew by a stress corrosion mechanism in the presence of sea water, reached a critical size. In addition, the analysis indicates that critical crack size for unstable propagation was reached prior to wall penetration (which could have led to subsequent loss of pressure) resulting in explosion of the cylinder.
Consequently, it is proposed that the design of these cylinders should consider the employment of more stress corrosion resistant alloys for sea diving applications. Futhermore, cylinders should have a reduced wall thickness that can be determined employing the "leak before break" design philosophy, developed using fracture mechanics, in order to eliminate the possibility of catastrophic ruptures, when subcritical stable cracking occurs due to unexpected reasons.
The incorporation of both alloy and design modifications should increase safety significantly, when utilizing highly pressurized aluminum cylinders in marine applications.

THE EMPLOYMENT OF LIGHT CYLINDERS made of aluminum alloys for storage of compressed breathing gasses, increases constantly . The design of most of these cylinders is based on classical "strength of materials" considerations i.e. employing thin wall cylinder equations and requiring that the proof hoop stress (1.5 x service hoop stress) in the thinnest section be less then the yield strength divided by some safety factor. No consideration is given in currently available design specifications (e.g. ASME, RID etc.) to the fact that stable cracking may occur in service.

The analysis of some recent catastrophic failures of high pressure cylinders which were due to subcritical crack growth prompted a study of subcritical cracking in aluminum alloys used in cylinder fabrication and demonstrated the need for a critical review of the criterion utilized when designing high pressure aluminum cylinders.

FAILURE ANALYSIS

Several failed high pressure aluminum cylinders were examined. The failure analysis of only one group , made of a 5% Mg - non heat treatable aluminum alloy , is detailed next. These cylinders contain compressed air (200 BARS @ 15^0c) and are used by divers .

Characterization of Cracking

All failed cylinders ruptured longitudinally, as shown in Figure 1. One main crack , along the wall section , developed in each failed cylinder. When examining these fracture surfaces , the presence of a region ,A in Figure 2 , having a relatively rough surface topography that is less reflective than the rest of the fracture surface, B in Figure 2, is evident. Chevron patterns evident on all B type fracture surfaces that are indicative of the direction of crack propagation point to region A as the origin of cracking.

Typical SEM fractographs of regions A and B are shown in Figure 3, and indicate that the semi - elliptical region A, which is of an intergranular nature, Figure 3a, was formed as a result of the operation of a stable mechanism, stress-corrosion cracking. The dimpled topography in region B, Figure 3b, is characteristic of unstable crack propagation, by rupture.

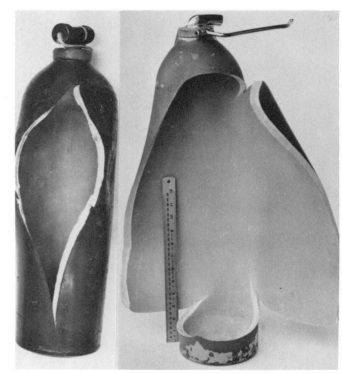

Fig. 1- Typical appearance of the failed cylinders.

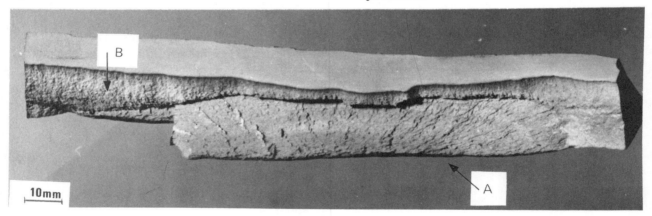

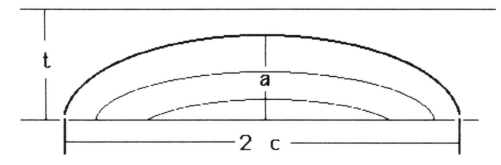

Fig. 2- Macrograph and schematic representation of origin area.

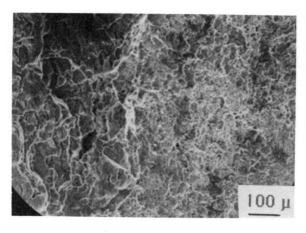

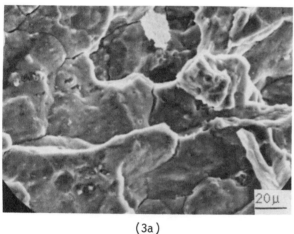

(3a)

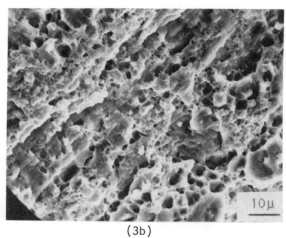

(3b)

Fig. 3- Typical SEM fractographs of the two fracture surface topographies observed on failed cylinders. (3a) Intergranular failure. (3b) Dimpled fracture.

Physical and Mechanical Testing

Mechanical properties, chemical analysis and microstructure of cylinders material were characterized and are given in Table 1 and Figure 4 respectively. No significant deviations from the design requirements (as specified by the manufacturer) were noted. Surface damage to coating was observed around origin and in other areas in some of the cylinders, Figure 5.

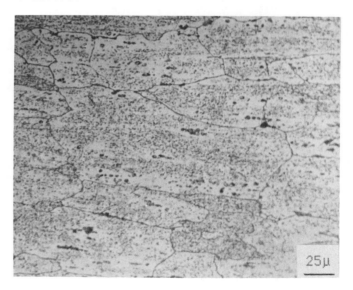

Fig. 4- Typical micrograph of cylinders material microstructure (Kellers etch).

Table 1. Average tensile properties and composition of the failed cylinders material (average of 5 failed cylinders, longitudinal and transverse specimens).

	0.2% YIELD STRENGTH [MPa]	ULTIMATE TENSILE STRENGTH [MPa]	ELONGATION [%]	REDUCTION OF AREA [%]
MEASURED	260 ± 12	350 ± 5	14 ± 2	28 ± 4
REQUIRED	269	340	≥ 11	NONE

	w/o Fe	w/o Mn	w/o Mg	
MEASURED	0.2 – 0.22	0.75 – 0.84	4.86 – 4.95	
REQUIRED	≤ 0.3	0.5 – 1	4.5 – 5.1	

Note: Composition determined by atomic absorption.

 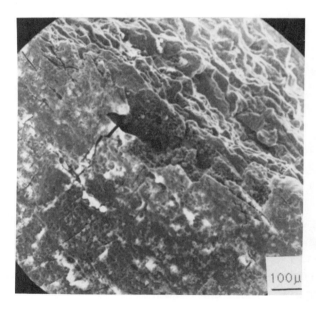

Fig. 5- Typical micrographs showing surface damage found on outer surface of some cylinders.

DISCUSSION AND SIGNIFICANCE
OF THE FAILURE ANALYSIS

The results of the study described above indicate that the cylinders were manufactured according to design specifications, and the failures investigated are not due to deviation from these requirements, i.e. strength deficiencies, neither to an incidental overpressurization. All catastrophic failures of the cylinders were found to be the result of in-service, stable growth of longitudinal flaws, due to stress-corrosion cracking upon exposure of cylinders material to sea-water. Such stable propagation of cracks, that grow longitudinally, due to the hoop stresses, and from the outer face of the cylinder wall inwards, can terminate by either one of the following failure modes:

a) in a sufficiently thick wall, the stably propagating crack will reach critical size, while the cylinder is still under pressure, resulting in a catastrophic explosion.
b) when the wall thickness is smaller then the critical crack size, the crack will propagate in a stable fashion until it penetrates the wall, and this constitutes a "leak before break" (LBB) situation.

Proper employment of LEFM (linear elastic fracture mechanics) based failure criterion can provide guidlines for design and material selection which would ensure desirable course of events, i.e., "leak before break". This was originally done in 1963 by Irwin [1] who have proposed the LBB criterion as means of estimating the necessary fracture toughness, K_C, of material so that a surface flaw can grow through the wall (of thickness t) and the vessel "leaks" before fracturing. Simplified LBB criterion can be derived as follows:
assume the critical crack size, a_C, to be twice the wall thickness, t , Figure 2. The critical crack size, in its simplest form, is given as [2]:

$$a_C = C_1 (K_C/\sigma)^2 \qquad (1)$$

thus

$$2t \leq C_1 (K_C/\sigma)^2 \qquad (2)$$

or, the wall thickness that will ensure LBB is given by:

$$t \leq C_2 (K_C/\sigma)^2 \qquad (3)$$

where C_i (i=1 or 2) is a geometrical correction factor, and is the hoop stress.

Consequently, the concept expressed by equation (3), i.e. limiting the wall thickness, should be blended with traditional design procedures which are based, solely, on strength of materials conciderations and favor increasing wall thickness. This will yield an optimal wall thickness, which in turn will ensure a safer pressure vessel.

CONCLUSIONS

1. Failure of the cylinders analyzed is due to improper material selection and deficient design procedure.
2. A safer product will result if LBB criterion is employed when designing pressurized aluminum cylinders.

REFERENCES

1. G.R. Irwin, "Materials for Missiles and Spacecraft", p. 204, Mc Graw Hill, New York (1963).
2. R.W. Herzberg, "Deformation and Fracture Mechanics of Engineering Materials", New York (1976).

BOLT FAILURE STUDIES
AT ABERDEEN PROVING GROUND

Will C. Simmons
U.S. Army Combat Systems Test Activity
Aberdeen Proving Ground, Maryland, USA

ABSTRACT

It has been noted that approximately 70% of all mechanical failures occur due to fastener failures. Fastener systems are the essential elements which hold a component together. Despite the common knowledge that fasteners are probably the single most important aspect of a mechanical design, bolt failures are a common everyday problem. Bolt fracture occurs by several mechanisms... tensile overload, shear overload, hydrogen embrittlement and fatigue. Failures result from improper design, manufacturing practices, improper installation and by a combination of these factors. Although the bolt is a relatively simple device in practice, it can provide headaches to anyone who ignores its importance.

Aberdeen Proving Ground is a major testing facility for U.S. military equipment. Defense contractors and failure analysts can attest to the vigorous shakedown testing at Aberdeen which provides ample evidence of the problems involved in bolt failures. This paper will cover several bolt failures, the methods used for their evaluation, and some simple procedures for preventing failures before they happen.

FATIGUE FAILURE OF USMC LOGISTICS VEHICULAR SYSTEM (LVS) FRAME BOLTS

DESCRIPTION - The LVS is a large 8-wheeled vehicle with an articular steering system for greater mobility. This steering system requires that the frame of the vehicle pivot about a point approximately in the mid-section of the vehicle. The engine compartment was joined to the cargo section by a bolted frame. In an attempt to speed up testing, the LVS was loaded with additional weight. After several thousand miles of testing the sixteen 3/4-inch bolts which held the bolted frame together failed. Crack arrest marks and racket marks were found on three of the recovered bolts. The crack arrest lines ran across better than half the cross-section of two bolts. The hardness of the bolts corresponded to grade 8. The maximum recommended preload based on SAE grade 8 coarse threaded bolts was 401 ft-lbs while the torque recommended by the manufacture was 225-275 ft-lbs. The use of the higher preload increases the clamping load from 20,000 lb to 32,000 lb per bolt.

COMMENTS - The manufacture decided to remove the additional weight which would have sped up the testing process. The recommended increase in torque was therefore never instituted. The indications are that the vehicle would have completed testing if the maximum design loads of the bolts had been utilized.

HYDROGEN EMBRITTLEMENT OF 105-MM GUN MOUNT BOLTS

DESCRIPTION - The 105-mm is the main battle gun for U.S. Army tanks. Four 1-in. diameter gun mount bolts from two 105-mm gun mounts fractured after substaining a 750 ft-lb torque load for three days. The bolts appeared to have failed ductilely about the head and body joint. Hardness measurements of 40-41 HRC met military standard MS 24677. Scanning electron microscopy revealed intergranular fracture which is characteristic of embrittled steel.

A cadmium plate of 0.00053-0.00062 in. was measured on the bolts using the microscopic examination. Federal specification QQ-P-416 calls for a class 3 cadmium plate thickness of 0.00020 inches. The cause of failure was found to be hydrogen embrittlement. The deleterious effects of cadmium plating on high strength bolts due to hydrogen embrittlement is thoroughly documented.

COMMENTS - It was recommended that the gun mount bolts undergo static load embrittlement tests until it was acertained that hydrogen embrittlement was not a problem. The test consisted of a 200 hour static load at the recommended torque value (750 ft-lbs). The test fixture consisted of two steel plates which allowed easy removal of the bolt body in the event of bolt fracture. All tested bolts were examined for cracks and flaws after testing.

FATIGUE FAILURE OF THE M1-E1 TANK HUB AND SPROCKET BOLTS

DESCRIPTION - During testing of the Army's new M1-E1 tank, five of eleven bolts which fastened the right track guide to the hub and sprocket failed. The tank logged approximately 700 km before the failure occurred. Replacement bolts of specified grade 8 quality failed after approximately 20 km of testing. A second tank experienced similar failure of its factory installed bolts after logging approximately 1,000 km. A total of 6 failed bolts, 6 unfailed bolts and 7 replacement bolts were submitted for evaluation.

The fracture surfaces of the factory installed bolts from the first vehicle showed signs of cyclic fatigue. The final fracture region was small indicating a low nominal stress. Several ratchet marks were apparent which indicated multiple crack initiation sites. The fracture surface of the replacement bolts indicated fracture by single cycle shear overload. Non-destructive magnetic particle testing indicated hairline cracks at the root of the threads of the factory installed unfailed bolts. The root radius of the threads was found to be 0.007 inch which corresponded to industrial standards for 1/2 inch threaded steel. The average hardness fell within specifications for grade 8 bolts. Microstructural examination revealed no unusual features.

Preload stress calculations showed the specified 350 ft-lb applied preload torque was below the maximum preload torque of 401 ft-lbs. There were no signs of necking or of tensile failure to indicate yield strength of the bolts had been exceeded.

The addition of the guide plate to the wheel assembly was a design change to stop the track from coming off the sprocket. The design change accomplished it's intended purpose; however, the track knocked against the inside of the guide as evidenced by the wear marks that were found on the side of the guide. The banging of the track against the guide induced fatigue failure of the original factory-installed bolts. Failure of the replacement bolts was attributed to improper preloading of the bolts.

COMMENTS - The failure of these bolts was attributed to a combination of factors. Final recommendations included applying a 400 ft-lb preload torque to the assembly and redesigning the guide plate to reduce the severity of loads experienced by the bolts.

THREAD DEFORMATION FAILURE OF A GROUND EMPLACED MINE SCATTERING SYSTEM (GEMSS) TRAILER

DESCRIPTION - The GEMSS is a mine dispensing system which sends out a disperse pattern of mines while sitting on a moving trailer. During testing the trailer endured considerable slow cross-country mileage. A surge-brake mechanism was built into the trailer to prevent sudden movements of the trailer by activating the brakes. The surge brake anchored a lunette which attached the trailer to the prime-mover. The failure occurred when the retainer nut which holds the lunette into the surge brake pulled out of the threaded sleeve of the surge brake. The threads of the sleeve were slightly damaged along the entire threaded cross-section. Bright, clean metal indicated some abrasive wear on the sides and peaks of many of the threads. The last five threads were damaged to the point that only half of the thread height remained. Damage to all the threads of the retainer nut was revealed using an optical comparator.

The thread damage to the interior retainer nut and the exterior sleeve indicated the threads suffered deformation as the nut worked its way out of the sleeve. The damaged thread pattern on the sleeve and the retainer nut indicated the major damage occurred in the vertical plane of the assembly. The last four or five threads on the sleeve were almost totally destroyed as the retainer nut finally pulled out of the sleeve. The dimensional measurements of the inside diameter of the threaded portion of the sleeve was slightly larger than specified. The diameter of the threaded portion of the nut was smaller than specified. The net affect was the engaged length per thread was reduced from the specified 0.0392 inches to 0.0050 inches after failure. The measurements were somewhat misleading since they were taken on the best threads in four planes after the threads were damaged. The thread engagement was limited to a class 2A fit to allow for ease of assembly.

The lack of preload on the nut and sleeve fastener allowed the lunette to vibrate inside the sleeve which resulted in thread wear. By seating the retainer nut against the sleeve, all movement between the nut and sleeve would be eliminated. The nut and sleeve fastener was not used to its optimum capabilities. A preload on the assembly creates a force which must be exceeded to allow any movement of the retainer nut.

A roll pin that was installed to prevent the retainer nut from backing out of the sleeve popped out during testing of the vehicle. The roll pin was slotted to allow removal of the pin by squeezing and due to design restrictions it was not deeply seated into the retainer nut. These two factors facilitated the loss of the pin. A combination of vibration, lack of preload and relatively loose fit between the sleeve and retainer nut caused the pin to pop out.

After the pin was lost, there was nothing to prevent the nut from backing out except the preload applied during installation and a rubber protective sleeve. The preload applied to the retainer nut was negligible, because the improper positioning of the sleeve and nut roll pin holes prevented installation of the pin after preloading. The rubber protective sleeve was intact and securely fastened after the failure. This showed the retainer nut did not back out of the sleeve but instead it was pulled out of the sleeve.

COMMENTS - It was concluded that the failure of the surge brake assembly of a M979 trailer resulted from a lack of preload on the retainer nut and the loss of the roll pin during testing. It was recommended that:

1. A preload of several hundred foot-pounds be applied to the retainer nut during assembly of the surge brake.

2. A locknut be added to the assembly to safeguard against any backing out of the retainer nut.

3. If a roll pin was used, that the holes in the retainer nut and sleeve be drilled such that they are lined up at the recommended preload.

OVERLOAD FAILURE OF M1-E1 GUN MOUNT BOLTS

DESCRIPTION - An M1 tank was traveling at approximately 25 mph or greater with its gun barrel pointed at a right angle to the direction of travel. The end of the tank's gun barrel struck a tree and eight bolts which secured the gun mount remained intact. Replacement bolts for the gun mount were checked using a magnetic particle technique (MT) and found to contain faint MT indications. Other replacement bolts that did not show MT indications could not be torqued to the recommended preload of 750 ft-lbs. Eight failed bolts and five replacement bolts were submitted for material evaluation.

The microstructure and the hardness of the failed bolts met grade 8 specifications. The ductile mode of failure indicated failure by overload. The weight and speed of the tank combined with a long gun barrel running into a semi-immovable object (the tree) provides sufficient energy to cause overload of the bolts.

MT indications on some of the replacement bolts were brought about by laps in the material formed during rolling. Although the laps did not appear to be extensive enough to cause deterioration of bolt properties, they did represent a concern for the quality of machining.

The bolts that could not be torqued up to the recommended load of 750 ft-lbs were found to be free from cracks or necking which indicated the bolts did not fail during torquing. The maximum torque achieved on these bolts was 650 ft-lbs. The surface finish or possibly a fine film of oil on these may have caused the bolt behavior to change during loading.

COMMENTS - It was concluded that the failed bolts met grade 8 specifications and that the failures occurred by overloading. The bolts that could not be torqued to 750 ft-lbs showed no signs of failure. It was recommended that these bolts be loaded to 750 ft-lbs in an assembly which allowed easy removal if they failed.

FATIGUE AND PRELOAD

DISCUSSION - The number one source of failures of bolted assemblies at Aberdeen Proving Ground was by fatigue. A cursory survey of failures over the past six years revealed nine out of 21 documented bolt failures occurred by fatigue. Fatigue and bolt preload are innately linked together by the function of a bolt. The primary culprit in fatigue failure is a lack of preload on the bolt. In a properly designed bolt assembly, the design does not allow the calculated service loads to exceed the bolt preload. The bolt will not experience any appreciable variation of the applied stress on the bolt during service. Without stress variation, the bolt will not fail in fatigue. Thus a general principle has been established, that the preload of a bolt should be the maximum permissable. The maximum preload accepted by many designers is 80% of the proof load to account for variations in lubrication and applied torque. The proof load can be acertained from SAE J429K for various sizes and grades of bolts. The applied preload torque can be found by the following equation:

$$T = KDF \qquad (1)$$

where T = torque, K = friction factor, D = diameter of bolt, and F = tensile force on the bolt. The friction factor (K) equals 0.2 for average friction conditions. The K factor however varies with material and lubricant. The following average values for normal steel bolt threads have been found:

Graphite lubricated*	K = 0.08
Well lubricated, smooth surfaces**	K = 0.10
Unlubricated, smooth surfaces**	K = 0.12
Well lubricated, surfaces not smooth**	K = 0.13-0.15
Unlubricated, surfaces not smooth**	K = 0.18-0.20

Therefore, higher friction factors call for decreasing applied preloads. The tensile force applied is usually 60 - 64% of the ultimate tensile strength. The greater the applied preload to the bolt, the greater the clamping force for the same size bolt. This brings up an important point. Often the solution to fatigue failure of a bolt is simply to increase the torque. A bigger and better bolt will do little to solve the problem unless an equivalent increase in preload is used. It should be noted that the above discussion does not apply if the bolts are clamping non-rigid bodies (ie. soft materials).

*see ref 1   **see ref 2

HYDROGEN EMBRITTLEMENT AND ENVIRONMENTAL PROTECTIVE COATINGS

One of the more significant recurrent problems at Aberdeen Proving Ground is hydrogen embrittlement of bolts. Hydrogen embrittlement occurs almost exclusively is bolts above 34 HRC in hardness or above 100,000 psi in tensile strength. Therefore the failure of these bolts is limited to grade 5 and grade 8 bolts which are often used in critical applications. This characteristic is sometimes referred to as the static fatigue limit; a stress value below which failure will not occur by hydrogen embrittlement. Applied stress as low as 40% of the yield strength can cause a failure in a short period of time.

The hydrogen embrittlement problems encountered at Aberdeen Proving Ground have been associated with cadmium plating of bolts for corrosion protection. Zinc electroplated coating are also known to cause hydrogen embrittlement. Before electroplating, the ferrous bolts are cleaned using an acid pickling bath. The cadmium plating operation is done in a cyanide bath of dissolved cadmium oxide. Both operations supply hydrogen which can be absorbed by the bolt. Both zinc and cadmium prevent the evolution of hydrogen from the surface of the part. Strong acid pickling solutions and long exposure times during pickling increase the probability of hydrogen absorption. In the case of high and medium strength bolts, bake the bolts at 350°F - 400°F for three hours after pickling and for an additional 24 hours after plating (ASTM B242). The elevated temperatures speed the hydrogen diffusion process. A visual inspection of the cadmium plating may reveal a bright, golden finish which indicates a thick, continuous coating of cadmium. A dull finish indicates a thinner coating which is less susceptible to hydrogen embrittlement. Hydrogen tends to diffuse to regions of highest stress (ie. the threads of bolts). The time and temperature dependence of diffusion therefore can delay failure of the bolts several hours or even days after the initial stress is applied.

MANUFACTURING DEFECTS

SEAMS IN THE THREADS - "Seams are crevices in the surface of metal that have been closed, but not welded, by the working of the metal." Seams make a bad stress concentration situation worse by intensifying the stress in a smaller, more localized area. It has been noted that significant improvement in fatigue life have been obtained by cold working the threads rather than cutting them. This process creates a smooth unbroken flow of surface metal. An extremely useful technique for non-destructive testing of surface imperfections in bolts is magnetic particle inspection. The technique is simple and quick and has been used at Aberdeen Proving Ground to detect a 0.02-mm wide seam in the root of bolt threads. This was confirmed by metallographic sectioning.

HEAD TO SHANK RADIUS OF CURVATURE - The head-to-shank fillet is a common site for fracture initiation due to the stress concentration factors involved in sharp corners. The effect of stress concentration due to changes in the fillet of a typical 1/2-in. nominal bolt is shown in Figure 1.

STRESS CONCENTRATION FACTOR
FILLET OF A TYPICAL ½ INCH BOLT

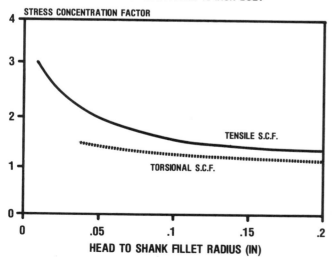

Figure 1 - Stress concentration factor of fillet radius in a 1/2-in. nominal bolt.

Once again hot or cold forging of a bolt provides for a smoother, unbroken flow of metal at the head-to-shank fillet and thus improves fatigue life.

CONCLUSION

Although the nuts and bolts of any assembly are often simple and overlooked, they are what holds the whole assembly together. Fatigue and overload can be prevented by proper joint design and preload specifications. Hydrogen embrittlement originates with inadequate precautions during electroplating operations. Bolts can be easily tested for hydrogen embrittlement if necessary. Finally, small things like laps and seams in the threads of the bolts, and exceedingly sharp head-to-shank fillets can cause premature failure.

REFERENCES

(1) Haviland, Girard S., Mechanical Engineering, Oct 1983, pp. 17-31

(2) Rothbart, Harold A. (ed), "Mechanical Design and Systems Handbook", pp. 21-6, McGraw-Hill, Inc., New York, 1964

(3) Boyer, Howard E. (ed), "Metals Handbook", vol. 10, 8th ed., American Society for Metals, Metals Park, Ohio, 1975.

(4) Cob, Bernie J., Product Engineering, August 19, 1963, pp. 62-66

(5) Federal Specification QQ-P-416

(6) Military Standard 1312

(7) Military Standard 24677

(8) SAE Specification J429K

FAILURE OF SHIP HULL PLATE ATTRIBUTED TO LAMELLAR TEARING

R. A. Myllymaki
Defense Research Establishment Pacific
Victoria, B.C., Canada

ABSTRACT

During a refit of a twenty-year-old Naval destroyer, two cracks were found on the inside of the hull plate at the forward end of the boiler room. The cracks coincided with the location of the top and bottom plates of the bilge keel. An investigation was launched to determine the root cause of the cracking and to see if there were any design or ship class implications.

Metallurgical examination of sections cut from the cracked area identified lamellar tearing as the principle cause of the cracking, rather surprising in 6-mm thick hull plate. Corrosion fatigue and general corrosion had also contributed to the perforation of the hull plate.

Although it is probable that more lamellar tears exist near the bilge keel in other ships and may be a nuisance in the future, the hull integrity of the ships is not threatened and major repairs are not needed.

DURING A REFIT of a Naval destroyer, two cracks were found on the inside of the hull plate near the bulkhead at the forward end of the boiler room. The cracks coincided with the location of the top and bottom plates of the bilge keel. Similar cracks in the same location had been found previously on three other ships from two different classes of ships. Thus the incidence of cracking seemed to be rather widespread and a detailed investigation was in order to investigate and determine the causes of the cracks to see if there were any design or ship class implications.

GENERAL OBSERVATIONS

The cracks in the hull plate were first observed when steam blown into the bilge keel was seen emitting from the edge of the bilge keel on the outside and from near a bulkhead on the inside of the hull. Figure 1 shows the outside surface of a section (460 mm wide by 800 mm long) removed from the ship. The remains of the top and bottom plates of the bilge keel where they are welded to the hull plate can be seen in Figure 1.

Fig. 1 - Outside hull surface showing remnants of horizontal bilge keel.

There was considerable corrosion on the boiler room hull plate but very little on the fuel oil side of the bulkhead. The hull plate had thinned as much as 3 mm of the nominal 8 mm thickness. Very little corrosion was evident on the outside surface of the hull attesting to the integrity of the coating systems used on the hull for the past 20 years.

MATERIAL

A chemical analysis of the hull plate is: 0.16%C, 0.68%Mn, 0.2%Si, 0.02%S, 0.007%P. The carbon equivalent is 0.28% using the widely accepted carbon equivalent formula for carbon-manganese steels:

Carbon Equivalent =

$$\%C + \frac{\%Mn}{6} + \frac{\%Cr + \%Mo + \%V}{5} + \frac{\%Ni + \%Cu}{15} \qquad (1)$$

Figure 2 shows that the microstructure of the hull plate is a ferrite/pearlite structure. Some banding can be seen in some of the micro-sections as well as numerous elongated inclusions.

Fig. 2 - Microstructure of hull plate.

RESIDUAL STRESS DETERMINATION

Four strain gauges; two at the bottom of the bilge keel and two at the top were positioned as near as possible to the toe of the bilge keel weld and about 50 mm to each side of the bulkhead. The rectangular rosette strain gauges (M-MEA-06062RF-350) were attached with Eastman 910 adhesive. A Vishay Model P350A strain indicator and Model SB-1 switch and balance unit were used to measure the micro-strain. The plate was sectioned into progressively smaller pieces until a 10 x 20 mm piece of plate containing each strain gauge was left. One cut, which removed part of the bilge keel, caused noticeable change in the distortion as the reaction stresses changed.

In addition to the destructive strain gauge method, a Rigaku Denki "Strainflex" X-ray diffraction apparatus designed for residual stress measurements was used to determine the residual stress near the bulkhead/bilge keel welds. Both the strain gauges and the X-ray diffraction produced results that varied from approximately zero residual stress to a peak

value of about 250 MPa tensile residual stress perpendicular to the bilge keel weld.

METALLOGRAPHY

Figure 3 shows a double fillet weld between the hull plate and bulkhead 36 while Figure 4 shows a crack running through the heat-affected zone of such a weld. Near the toe of the weld considerable corrosion has taken place.

Fig. 3 - Double fillet weld between bulkhead and hull plate.

Fig. 4 - Weld between bulkhead and hull plate showing crack in heat-affected zone.

A crack running in the heat-affected zone at a skewed T-butt partial penetration weld between the hull plate and the bilge keel is shown in Figure 5. Also seen on the bottom edge is part of the bulkhead-hull plate weld. Figure 6, a cross-section of a hull plate/bilge keel weld shows a crack running outside of the weld heat-

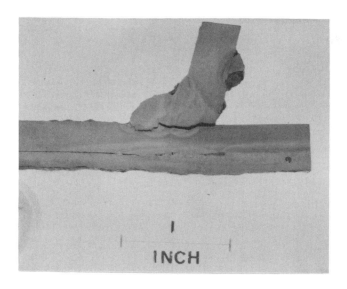

Fig. 5 – Skewed T-joint showing crack in heat-affected zone.

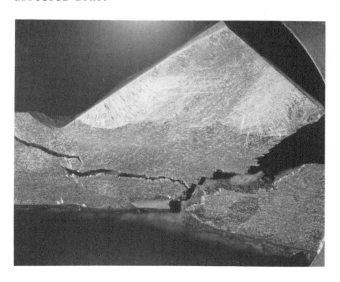

Fig. 6 – Skewed T-joint showing crack in heat-affected zone and corrosion.

Fig. 7 – Cracking in heat-affected zone.

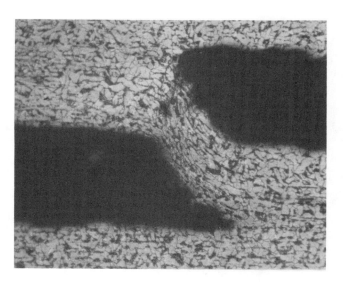

Fig. 8 – Vertical shear wall between two terraces.

affected zone.

Closer examination of the cracks shows that much of the cracking is made up of horizontal terraces and vertical shear planes as shown in Figures 7 and 8. A vertical shear wall which has nearly broken and joined two terraces is seen in Figure 8. Yet another crack running through and beyond the heat-affected zone under a bilge keel weld is seen in Figure 9 along with several other cracks. Transgranular cracks at the end of the main crack in Figure 9 are shown at higher magnification in Figure 10. These cracks cross both ferrite and pearlite grains and have blunt ends. A few, very small transgranular cracks were also seen in some of the other specimens examined. The crack sur-

faces were badly corroded and examination with the scanning electron microscope did not reveal any useful information.

Microhardness measurements were made at the toes of the welds shown in Figures 4, 5, and 9. The maximum hardness recorded was 270HV.

LAMELLAR TEARING TEST

To assess the susceptibility of steel plate to lamellar tearing the slice bend test[1,2], a simple and inexpensive destructive test was used.

Figure 11 illustrates the procedure for preparing the slice bend test. This form of the test ensures that reasonably uniform strain is applied across the whole sample width i.e. the test-plate thickness and both the mid-thickness

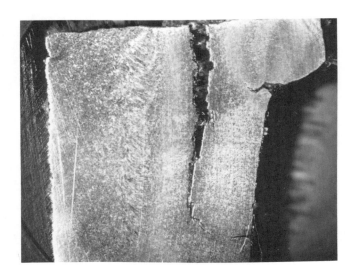

Fig. 9 - Crack under bilge keel, with two secondary cracks.

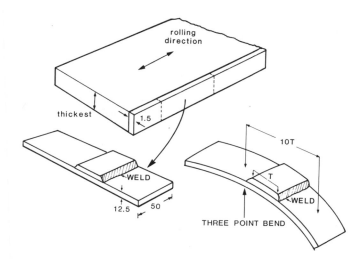

Fig. 11 - Sketch of slice bend test.

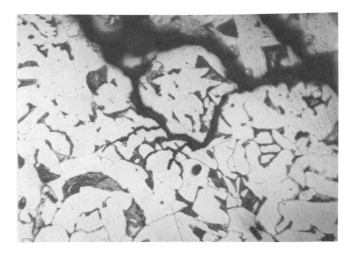

Fig. 10 - Transgranular cracks from the end of the main crack shown in Figure 9.

and near surface regions are tested. Cracking identical in appearance to lamellar tearing can occur in any plane from the surface to the mid-thickness.

Figure 12 shows tearing around the inclusions near the centre of the section. At a surface strain of about 5% (a rough estimation from the radius of curvature of the test specimen) a number of shear walls formed between the terraces. The first evidence of separation at the inclusion/metal interface was seen at less than 5% strain.

Figure 13 shows the fracture surface on the terraces. The pattern is fern-like in appearance with many fragmented non-metallic inclusions in the base of the elongated depressions.

Fig. 12 - Slice bend test showing lamellar tears.

The inclusions are composed mainly of Mn and S and from their flat elongated form can be identified as Type II MnS inclusions. A shear wall between two terraces is seen in Figure 14. At higher magnification the fracture on the shear wall is seen to be primarily micro-void coalescence which is evidence of plastic deformation.

Figure 15 shows the decohesion and separation along the inclusion/metal interface and the initiation of shear between the two terraces. Many MnS fragments can be seen in the hole on the left side. Plastic deformation, which is similar to that seen in Figure 8, is evident at the ends as well as between the two holes.

Fig. 13 - Fracture surface showing fern-like pattern of inclusions.

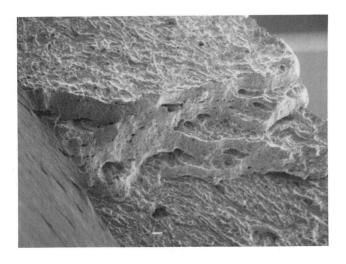

Fig. 14 - Vertical shear step joining two terraces.

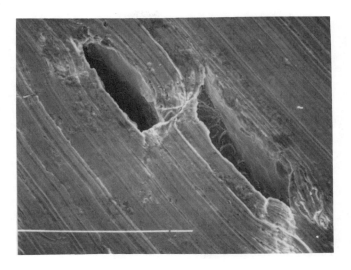

Fig. 15 - Lamellar tears showing deformation and inclusion fragments.

DISCUSSION

The step-like cracks running parallel to the plate surface either inside (Figures 4, 5, and 7) or outside (Figures 6 and 8) of the heat-affected zone are typical of lamellar tearing[3]. Lamellar tearing occurs during or shortly after welding, normally not during subsequent service. For lamellar tearing to occur, a susceptible material (poor ductility in the through-thickness direction) must be welded with a procedure and joint configuration which produces a high strain in the short transverse (through-thickness, Z-) direction. These three aspects of lamellar tearing: material, joint configuration, short transverse strain are discussed more fully below.

MATERIAL - No single grade of steel appears to be markedly more susceptible to lamellar tearing than another and it has been suggested[1-5], that the mechanism of lamellar tearing is too complex to relate to a given steel grade, inclusion type, or steel making process. Sulphides, silicates, alumina, and duplex inclusions in rolled carbon and low alloyed steels, whether semi- or fully killed have all been involved in lamellar tearing problems. Normalised, quenched and tempered, as-rolled, controlled rolled, fine and coarse grained steels have all experienced lamellar tearing.

That the killed carbon steel used for the hull plate is susceptible to lamellar tearing is evident from several micrographs presented in this report, and this should not be surprising when one remembers that the steel was made over 20 years ago. Further confirmation is obtained from the slice bend tests which started tearing after only about 2% surface strain. The planar Type II MnS inclusions shown in Figure 2 and more graphically in two dimensions in Figure 13 reduce the through-thickness ductility enough to cause lamellar tearing when the steel is strained in this direction. The fern-like pattern of the inclusions shows that simple cross-sections like that in Figure 2 may not adequately represent the extent of the inclusions in a steel and simple inclusion counting may not reflect the relative susceptibility of a given steel to lamellar tearing.

It is worth noting that ship classification societies such as Det Norske Veritas[6,7], Lloyds Register[8], American Bureau of Shippings[9], as well as the American Welding Society[10] recognize the lamellar tearing problem and suggest that special steel be considered for weld configurations that are susceptible to lamellar tearing.

JOINT CONFIGURATION – The joint configurations which have been found to be the most susceptible to lamellar tearing are tee (fillet or butt) and corner joints[3,4]. The cruciform joint is a more severe form of the T-joint because bending of the plate is restricted by the first weld. Very large components impose restraint due to their own weight whereas small components may be equally restrained if they have been stiffened.

Most of the cracking was found in the skewed partial penetration T-butt joint (Figures 5 and 6) between the bilge keel and hull plate. The cracks probably initiated at the partially penetrated root because the strain produced by the weld metal shrinkage hinges about the root of the weld. Contraction across a weld is a function of numerous factors including, number of weld beads and weld size. Figure 5 shows a weld with excess weld metal which does not contribute to the static or dynamic strength of the joint but develops greater shrinkage and thus more strain at the root of the weld.

The lamellar tearing runs for only about 75 mm at the intersection of the bilge keel and bulkhead welds on the hull plate. This suggests that the cruciform type joint configuration provides the necessary restraint to initiate lamellar tearing. The bulkhead welds stiffened and restrained the hull plate which cracked when the bilge keel welds were made on the highly restrained joint. Poor fit-up of the bilge keel, caused by careless flame cutting or distortion from previous welds also contributed to the overall restraint.

The crack under the bulkhead weld (Figures 3 and 4) did not extend beyond the immediate region of the cruciform joint configuration, about 30 mm in total length. These fillet welds are balanced (Figure 3) and were welded before the hull plate was stiffened. A patch which was welded into the bulkhead years after the original fabrication may have caused the lamellar tearing beneath the bulkhead. In this case, the bilge keel would have stiffened the hull plate.

Most fabricators have found lamellar tearing in steel plates in the range of 12 – 60 mm[3,4] in thickness but most felt that there were few problems below 25 mm in thickness. Indeed most problems have been reported in large thick steel structures such as bridges[11], boilers[12], ships[13,14], offshore drilling rigs and platforms[14,15], and skyscrapers[15,17]. However, lamellar tearing has also been reported in rigid cruciform type weld joints in steel as thin as 3 mm[18,19].

SHORT TRANSVERSE STRAIN – For lamellar tearing to occur, the welding procedure must develop strain in the short transverse, (through-thickness, Z-) direction of the plate. Initiation may be expected when the strain exceeds the short transverse ductility. The overall strain in the weldment depends on the amount of joint contraction, on the restraint by other members of the structure, and on a more localised scale on the interaction of the weld thermal gradients and the restraint of the surrounding metal.

Normally one would not expect lamellar tearing in 8 mm thick hull plate because the thin plate would distort and high strains would not develop in the short transverse direction. The present results support this because no lamellar tears were found except in the small region where the bilge keel and bulkhead welds cross i.e. where the hull plate is restrained by the cruciform joint configuration.

MECHANISM OF LAMELLAR TEARING – It has been suggested[3,4,5] that lamellar tearing takes place in a number of stages. Corrosion of the samples investigated prevented detailed examination of the lamellar tears but the sequence of tearing can be followed on the slice bend test microsections.

1. Numerous elongated inclusions as seen in Figures 2 and 13, make a steel susceptible to lamellar tearing.
2. The initial decohesion of the inclusion/metal interface can be seen at some of the inclusions in Figure 12. Figure 15 shows the decohesion at a more advanced stage.
3. Plastic deformation at the ends of the inclusions can be seen in Figures 12 and 15.
4. Tears at inclusions on the same plane linked by ductile tearing accompanied by plastic deformation, Figure 12.
5. Inclusions on different planes linked by shear fracture and the formation of shear walls, as seen in Figures 7, 12, 14, and 15.

The degree of through-thickness strain needed to initiate lamellar tearing depends on the through-thickness ductility which in turn depends on the inclusion type, shape, and distribution, as well as the properties of the metal matrix.

OTHER TYPES OF CRACKING – Hydrogen-induced and corrosion fatigue cracking are two other mechanisms which may have contributed to this failure. The low carbon equivalent ($\simeq 0.28\%CE$) in this thin hull plate is not expected to produce a microstructure in the heat-affected zone which is susceptible to hydrogen-induced cracking. Indeed, the maximum hardness found in the heat-affected zones examined was 270HV. No evidence of hydrogen-induced cracking was seen in the heat-affected zones of any of the welds examined.

Figures 9 and 10 show some signs of corrosion fatigue but there is no evidence to suggest that it played more than a minor role in the overall failure. Figure 6 shows that general corrosion contributed to the final failure. Indeed considerable general corrosion was seen on the boiler room side of the hull plate in the region of the bilge keel. No evidence of corrosion fatigue was found at the toes or roots of welds remote from the lamellar tears.

SIGNIFICANCE OF THE CRACKS - The first mention of lamellar tearing in what is now a quite extensive literature is dated 1961[13]. It follows that no precautions against lamellar tearing were taken by either steel manufacturers or ship builders when these ships were built. All steel plates and all locations within a plate will not have short transverse ductility low enough to be susceptible to lamellar tearing. However, it is probable that other lamellar tears are present where cruciform type weld joints are formed at the intersection of bilge keel and bulkhead weldments. Indeed, this is supported by a recent observation on another ship aboard which the forward water-tight section (~10 metres) of the port bilge keel was damaged and removed for repairs. The bilge keel was carefully removed so that no damage was done to the hull plate. Examination of the hull plate revealed several lamellar tears at the intersection of the bilge keel and internal vertical members. It is noteworthy that the lamellar tears were seen at only some intersections. The tears extended only 50 - 75 mm, had not penetrated the hull plate, and were not covered with corrosion. This shows that the cracks initiated from the root of the weld.

Lamellar tears are caused in part by the large thermal strains developed as highly restrained welds cool. Short transverse strains of such magnitude are not developed by service conditions, thus the cracks have been present since initial fabrication. The cracks may have propagated somewhat by corrosion fatigue but clearly this was not a major contributor to the overall failure. The final extension of the cracks through the hull plate was assisted by corrosion on the boiler room side of the hull plate.

There was no evidence of fatigue cracking from either the toes or roots of any of the welds examined; not even from the severe stress concentrations at the toes or roots of welds with poor profiles. In the present case, the presence of lamellar tears are not likely to affect the fatigue life. This is also supported by the small amount of work done on the sections removed from two other ships.

The small cracks are unlikely to present a brittle fracture problem in the thin hull plate.

CONCLUSIONS

1. The cracks found in the hull plate at the intersection of the bilge keel and the bulkhead were identified as lamellar tears.
2. Lamellar tears do not initiate during service but are already present after fabrication is completed. Clearly, in 20 years of service, the tears have not propagated far.
3. There is some evidence of corrosion fatigue but this had little effect on the overall failure. General corrosion on the boiler room side of the hull plate helped extend the crack path to the inside of the hull plate.
4. It is likely that other lamellar tears are present in the hull plate at other intersections of the bilge keel and bulkhead weldments. In some cases these tears may propagate through the hull thickness and become a nuisance by causing a minor leak. However, the metallographical results in this report and 20 years successful service suggest that this is not a major problem which would affect the hull integrity.
5. Lamellar tearing is a well known welding problem which is prevented by choice of materials or welding procedures by competent steel fabricators. The major ship classification societies recognize the problem and include provisions for the purchase of special steel grades which have sufficient short transverse (Z-) direction ductility.

REFERENCES

1. H.L. Drury, J.E.M. Jubb, "Lamellar Tearing and the Slice Bend Test", Welding Journal, Volume 52, Number 2, pp. 88s-98s, February 1973.
2. J.C.M. Farrer, "The Relationship Between Metallurgical Factors and Susceptiblity to Lamellar Tearing", in Investigations into Lamellar Tearing, a Compendium of Reports, The Welding Institute, Cambridge, March 1975.
3. J.C.M. Farrer, R.E. Dolby, "Lamellar Tearing in Welded Steel Fabrication", The Welding Institute, Cambridge, 1972.
4. "Control of Lamellar Tearing", Technical Note 6, Australian Welding Research Association, April 1976.
5. S. Ganesh, R.D. Stout, "Material Variables Affecting Lamellar Tearing Susceptibility in Steels", Welding Journal, Volume 55, Number 12, pp. 341s-355s, December 1976.
6. "Rules for Classification of Steel Ships, Materials and Welding", Det Norske Veritas, 1980.
7. "Rules for Design, Construction, and Inspection of Fixed Offshore Structures", Det Norske Veritas, 1974.
8. "Rules and Regulations for the Classification of Ships", Lloyds Register of Shipping, 1978.
9. "Rules for Building and Classing Steel Vessels", American Bureau of Shipping, 1981.
10. "Guide for Steel Hull Welding", Document D35-76, American Welding Society, 1976.
11. F.M. Burdekin, "Lamellar Tearing in Bridge Girders - A Case History", Metal Construction and British Welding Journal, Volume 3, Number 5, pp. 205-209, May 1971.
12. H. Wormington, "Lamellar Tearing in Silicon-Killed Boiler Plate", Welding and Metal Fabrication, Volume 30, Number 9, pp.

370-373, September 1967.

13. M. Watanabe, "The Pull-Out Type Fracture in Rolled Steel Plate", Symposium on Welding in Ship Building, The Institute of Welding, pp. 219-225, London 1961.

14. Y. Takeshi, "Lamellar Tearing and Marine Structures", Welding and Metal Fabrication, Volume 38, Number 12, pp. 740-746, December 1975.

15. J.M. Arrowsmith, D.C. Shenton, "Steel Plate for Offshore Structures", Metal Construction and British Welding Journal, Volume 8, Number 9, pp. 396-400, September 1976.

16. "Lamellar Tearing of Steel Worries Designers", Engineering News Record, p. 53, September 21, 1972.

17. "Steel Cracking Causes Design Change, Delays Job", Ibid, p. 12, January 29, 1976.

18. J. Vrbensky, "Some Weldability Problems in Thin Mn-V-N Type Steel Plates from the Crackability Point of View", Metal Construction and British Welding Journal, Volume 1, Number 2, pp. 44-49, February 1969.

19. J. Lombardini, "Cracking as a Criterion of Weldability", Ibid, p.40-43.

FAILURE ANALYSIS OF
A LIQUID PROPANE GAS CYLINDER

K. Mogami, S. Saito, H. Makishita
National Research Institute of Police Science
Sanban-cho, Chiyoda-ku, Japan

K. Ando, N. Ogura
Faculty of Engineering
Yokohama National University
Tokiwadai, Hodogaya-ku, Yokohama, Japan

Abstract

The failure analysis of a new developed liquid propane gas (LPG) cylinder was done. Deep drawing flaws were observed in the longitudinal direction on the inside of the LPG cylinder. It became clear that the flaws about 1.3 mm deep, steps, and the chevron pattern were observed on the fractured surface. Cleavage facets were also observed with scanning electron microscope (SEM). Owing to insufficient annealing, the hardness was relatively high and the elongated microstructure was observed near the fracture surface. Toughness was also deteriorated. The value of stress intensity factor K_I caluculated from the value of the internal pressure was lower than that estimated by the fracture toughness test. The residual stress was not removed completely and the hot straining embrittlement appeared at the bottom of the deep drawing flaws. Therefore it was proved that a crack at the bottom of the deep drawing flaw grew by some causes and the brittle fracture was produced.

FIVE NEW DEVELOPED LIQUID PROPANE GAS (LPG) CYLINDERS were failed in the daytime after using for about one year. The fracture of each LPG cylinder was $40 \sim 50$ cm long in the longitudial direction. In the case of one of these failed LPG cylinder, the material tests, observation of fractured surface with SEM, and fracture toughness test were carried out.
LPG cylinder failure arising from manufacturing fault was studied in this paper.

The outline of the failed LPG cylinder

(1) The manufacturing process of the LPG cylinder

The procedures of two deep drawing, annealing and the final drawing are employed to a round sheet cutted out from steel plate. Drawing ratio is 1.7. A screw mount for a safty valve is welded to a cup produced by deep drawing, a bottom plate is welded to the another cup, and these cups are welded to each other at the edges of these cups. Then, the cylinder is annealed again after welding, and pressure proof test is carried out. Then the cylinder is painted and LPG cylinder is manufactured.

(2) The structure of the LPG cylinder

The fractured LPG cylinder is about 128 cm high, its outer diameter is about 366 mm, and its wall thickness is about 3.3 mm. The internal capacity is 118 l, the self-mass is 43.4 kg, and the testing proof pressure is 3.1 MPa.

Examination and material test

(1) Appearance examination of fractured part

As shown in Fig.1, the LPG cylinder was fractured up to about 40 cm from the central welded part. The central part of the fracture was about 20 mm wide at the maximum value, being extruded about 25 mm high. Dividing the LPG cylinder into two parts along the central welding line and observing the internal surface, many flaws about 40~45 cm long were found in the longitudinal direction as shown in Fig.2. The schematic illustration and the appearance of the fractured surface are shown in Fig.3 and Fig.4. Flaws about 1.3 mm deep assumed to have been produced at the process of deep drawing was observed on the fractured surface. The fractured surface at the outside edge was inclined at a 45-degree angle. On the fractured surface above the most widely place, curved steps were observed. The fractured surface was smooth and flat below it, and the chevron pattern was observed near the bottom end of the fractured surface. Therefore it was proved that the crack, advancing from the bottom of flaw caused by the deep drawing, extended longitudially upward from the central welding part.

(2) SEM observation

On the smooth and flat fractured surface, the cleavage facets and quasi-cleavage facets were observed with SEM as shown in Fig.5. No intergranular crack was observed. Dimples were observed at the bottom of the deep drawing flaw, near the end of the fractured surface, and at the shear lip as shown in Fig.6. No striation was found.

(3) Microstructure

Six specimens were cut out at an equal circumferential interval at a distance of 200 mm from the central welded part. They were polished, etched by nital, and observed with a microscope. The microstructure near the fracture consisted of the ferrite and pearlite which were somewhat elongated as shown in Fig.7 (A). The remained plastic deformation was observed owing to insufficient annealing. As the distance from the fracture increased, the network of the ferrite and pearlite was observed in the other five specimens. The microstructure of specimens taken from the opposite side of the fracture indicated sufficient annealing as shown in Fig.7 (B). As shown in Fig.8, the flaws produced by the wrinkles, being pressed down at the deep drawing, were obvserved. Relatively sharp cracks were observed at the bottom of the these flaws.

(4) Hardness

Micro vickers hardness was measured at the positions where the microstructure was investigated. The part near the factured surface offered the maximum hardness of $H_V \doteqdot 170$ as shown in Fig.9. As the distance from the fracture increased, the hardness decreased to reach its minimum value of $H_V \doteqdot 90$ at the opposite side of the fracture.

(5) Chemical analysis

The chemical analysis of C,Si,Mn,P and S was done near the fracture. whose result is given in Table 1. All the element content satisfied the required standard.

Fig.1 - The photogragh of the fractured LPG cylinder.

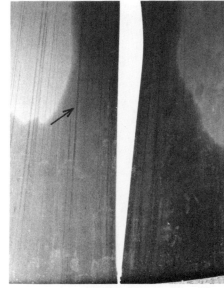

Fig.2 - The photogragh of the deep drawing flaws on the inside of the LPG cylinder.

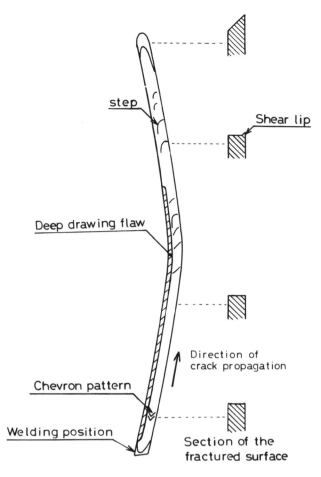

Fig.3 - Schematic illustration of the fractured surface of the LPG cylinder.

Fig.4 - The photogragh of the appearance of the fractured surface.
    A : deep drawing flaw
    B : brittle fractured surface
    C : shear lip

Fig.5 - SEM fractograph of the brittle fractured surface. Cleavage facets are visible at B in Fig.4.

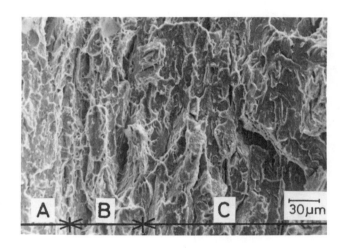

Fig.6 - SEM fractograph of the fractured surface near the bottom of the deep drawing flaw.
    A : deep drawing flaw
    B : dimples
    C : cleavage facets

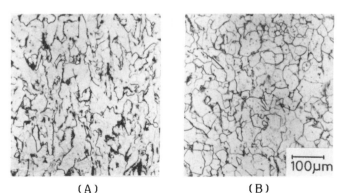

(A)   (B)

Fig.7 - Metallographical structure of the fractured LPG cylinder, near the fracture (A), the opposite side of the fractured position (B).

(6) Tensile test

Specimens for tensile test were cut out near the fracture, flattened, and prepared according to JIS Z 2201. Tensile tests were carried out for these specimens, whose result is given in Table 2. Their yield strength and tensile strength satisfied the specification of JIS SG 26, but their elongation was approximately 18~25 %. This value did not satisfy the specification requiring more than 28 %.

(7) Fracture toughness test

Three fracture toughness test specimens of the center crack type were prepared from the place near the fracture as shown in Fig.10. The fatigue crack of about 2 mm length was made on the conditions of ASTM E399. The fracture toughness test was carried out at a velocity of about 2 mm/min. The material was evaluated by the value of stress intensity factor K_I caluculated from the equation(1).

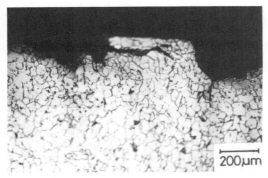

Fig.8 - Metallographic section of the flaw near the fracture.

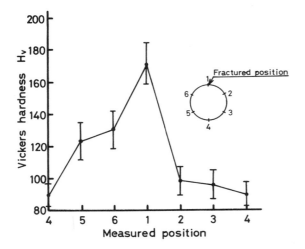

Fig.9 - Distribution of vickers hardness near the fractured surface. In the symbol $\mathbf{I}$, the middle black point is the mean value of vickers hardness and standard deviation is indicated by the bar enclosed lines.

$$K_I = \sigma\sqrt{\pi a}\ F(\zeta) \qquad \zeta = 2a/W \quad ----- (1)$$

$$F(\zeta) = (1 - 0.025\zeta^2 + 0.065\zeta^4)\sqrt{\sec(\pi\zeta/2)}$$

σ : gross stress (MPa)
a : crack length (mm)
W : width of specimen (mm)

As a result of the experiment, one of three specimens failed on the condition of brittle fracture and its K_I value was 49.6 MPa$\sqrt{m}$. The K_I values of another two specimens were not obtained, because the specimens showed the ductile fracture on the condition of big scale yield accompanied by deformation.

(8) Safety valve test

A work test of the safety valve of the failed LPG cylinder was carried out. The starting blow pressure was 2.19 MPa. This pressure value is lower than 3.1 MPa of the testing proof pressure. Therefore the failure was occured on the condition of the pressure below the setting pressure of the safety valve.

Table 1 - Chemical composition of the LPG cylinder.

in weight %

Elements	C	Si	Mn	P	S
Sample	0.13	0.010	0.42	0.004	0.018
JIS SG26	Max. 0.20		Max. 0.30	Max. 0.040	Max. 0.040

Table 2 - Mechanical properties of the LPG cylinder.

	Yield strength MPa	Ultimate tensile strength MPa
Sample 1	326	534
Sample 2	345	450
JIS SG26	Min.260	Min.410

Elongation %
25.4
18.0
Min.28.0

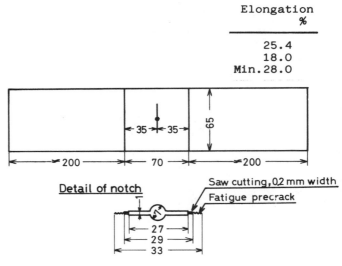

Fig.10 - Dimension of center crack tension specimen for plane-strain fracture toughness test, dimensions in mm.

Discussions

(1) Stress intensity factor caused by the internal pressure

Flaws owing to deep drawing existed on the inside of the LPG cylinder and the fracture started at the bottom of the flaw. Therefore the posibility of generation of the fracture from the flaw was discussed by fracture mechanics. In the present case, the cause of failure was discussed by assuming brittle fracture on the condition of small-scale yield.

Assuming an LPG cylinder to be a cylindrical container, tensile stress in the direction of the circumference caused by the internal pressure is calculated from equation (2).

$$\sigma = \frac{PD + 12Pt}{200t} \qquad ----------(2)$$

P : internal pressure (MPa)
The vapor pressure of propane gas at 40 $^\circ$C is 1.38 MPa.
D : inner diameter (mm)
t : wall thickness (mm)

From the equation (2), the tensile stress was calculated to be approximately 78 MPa.

The crack which advances into inside from the surface can often be approximated by the model of semi-elliptical crack as shown in Fig.11 (A). In this case, if d<<c, the stress intensity factor K_I can be calculated from equation (3) as the K_I of simple tension of sheet with crack on one side as shown in Fig.11 (B).

$$K_I = \sigma \sqrt{\pi d} \ F(\zeta) \qquad \zeta = d/t \ -------(3)$$

$$F(\zeta) = 1.12 - 0.23\zeta + 10.55\zeta^2 - 21.72\zeta^3 + 30.395\zeta^4$$

therefore $K_I = 10.7$ MPa$\sqrt{m}$

Assuming the flaws which existed on the inside surface of the LPG cylinder to be cracks, the value of K_I owing to the internal pressure was calculated to be $K_I \fallingdotseq 11$ MPa$\sqrt{m}$

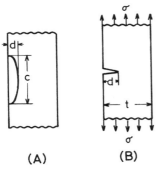

(A)   (B)

Fig.11 - Diagrams of the deep drawing flaw.
    d : flaw depth for surface flaw
    c : major axis of ellipse circumscribing the deep drawing flaw

(2) Cause of the failure of the LPG cylinder

The caluculated value of K_I is considerably lower than that obtained by fracture toughness test. Therefore cracks could not generate the fracture by taking their shape into consideration. However, the visual examination and SEM observation of the fractured surface suggested that the failure of the LPG cylinder had been caused by brittle fracture. Consequently it was proved that the following factors were affected to promote the extension of the cracks.

① In the vicinity of the fracture the hardness was relatively high, the microstructure was extended, and the elongation was low. Therefore the residual stress may be remained at the bottom of the deep drawing flaw because of insufficient annealing. So the real value of K_I at the bottom end of crack was higher than the calculated value by reason of the residual stress.

② It is well known that steel with plastic deformation of about 5 % at a temperature between 200 and 300 $^\circ$C shows so much deteriorated mechanical property and the possibility of generation of the fracture becomes relatively high at low stress. Such ductility deterioration is generally called "hot straining embrittlement" (HSE). The failed LPG cylinder was heated locally at 200~300 $^\circ$C at welding, and the HSE appeared at the insufficiently annealed bottom of the deep drawing flaw. Therefore it could be pointed out that the bottom of deep drawing flaw. its ductility being deteriorated. was in a state where the brittle fracture could be easily produced.

Results

The fractograghy of the fractured surface of a LPG cylinder were carried out and the characteristics of the material were investigated in order to find out the cause of the failure. The obtained results are as follows :

(1) On the internal surface of the LPG cylinder flaws produced at its production process of deep drawing were observed in the longitudial direction on the circumference. Cracks started at one of these flaws.

(2) The hardness was relatively high and the elongated microstructure was shown near the fracture. Therefore it was proved that annealing was insufficient near the fracture and the residual stress at the bottom of the flaws was not completely removed.

(3) Toughness was deteriorated owing to insufficient annealing. Therefore it was

proved that the crack at the bottom of the deep drawing flaw grew by some causes, such as the residual stress or the HSE, and the brittle fracture was produced.

References

1) JIS G 3116, Steel Sheets Plates and Strip for Gas Cylinders.
2) JIS Z 2201, Test Pieces for Tensile Test for Metallic Materials.
3) ASME E399, Standard Test Method for Plane-Strain Fracture Toughness of Metallic Materials.
4) H.Tada, P.C.Paris and G.Irwin, "The Stress Analysis of Cracks Handbook", p2.1, Del Research Corporation, Hellertown, Pennsylvania(1973)
5) JIS B 8243, Construction of Pressure Vessels

TURBINE BLADE FAILURES, WHO PAYS?

Jack R. Missimer
Tennessee Valley Authority
Engineering Laboratory
Norris, Tennessee, USA

Abstract

The assignment of financial liability for turbine blade failures rests on the ability to determine the damage mechanism or mechanisms responsible for the failure. A discussion is presented outlining various items to look for in a post-turbine blade failure investigation. The discussion centers around the question of how to determine whether the failure was a fatigue induced failure, occurring in accordance with normal life cycle estimates, or whether outside influences could have initiated or hastened the failure.

TURBINE BLADE FAILURE was defined, for the purposes of an Electric Power Research Institute (EPRI) survey on the subject [1], as: "a situation where inspection of the turbine revealed that one or more blades had lost the ability to perform their prescribed functions in a safe and reliable manner, and that corrective action was required prior to restarting the unit".

Over the years, EPRI has been actively involved in the documentation of turbine blade failures and the search for improved blade performance and survival in the harsh environment of a steam turbine (see, for example, References 2,3,4,5,6,7,8,9,10,11). The results of surveys conducted by EPRI have indicated that a predominant number of turbine blade failures occur in low pressure (LP) turbines (see Table 1) and that, even though the average cost of an LP blade repair is greater than it is for a high pressure (HP) or intermediate pressure (IP) turbine blade repair, the HP and IP turbine outages are generally of longer duration and thus end up costing more in lost power revenues.

One EPRI survey subdivided 314 blade failures by yearly increments of service and found that within the first two years of service life, one-third of the recorded failures had occurred. Within the first six years of service life, two-thirds of the recorded failures had occurred. The majority of the recorded failures also occurred within two years of the last inspection. To examine the success of the corrective actions taken to repair the turbine blades EPRI examined the turbine maintenance records and documented the turbine and blade row in which each failure occurred. Failures similar in origin to a previous failure occurring in a specific turbine and blade row were termed repeated failures. Table 2 summarizes the findings of the survey and reveals that in 40 percent of the cases where new blades of the same design were installed in a LP turbine, a repeat failure occurred.

Table 1. Summary of Average Downtime and Costs for Turbine Blade Failures [1]

	LP Blades	IP Blades	HP Blades
Number of Outages	345	47	54
Avg. Time (Hrs/Outage)	428	671	654
Avg. Repair Cost/Outage	226,100	117,500	73,000
Est. Lost Power/Outage	$2,307,700	$3,750,200	$3,411,400

Table 2. Reported Effectiveness of Corrective Action
in Turbine Blade Repairs [1].

	LP Repeat Failure Rate	IP Repeat Failure Rate	HP Repeat Failure Rate
Replaced w/ Same Design	40%	42%	40%
Replaced w/ New Design	38%	15%	12%

For cases where a new blade design was installed in LP turbines the repeat failure rate was only slightly better with 38 percent repeat failures. It was concluded that unless a blade failure is correctly diagnosed so that proper repairs can be made, the failures will continue to reoccur.

Financial liability for turbine blade failures lies with one of three entities: 1) The turbine manufacturer, whose obligations are normally spelled out in a parts warranty or in a maintenance contract; 2) The insurance company, who has insured the turbine against damage due to the malfunction of some other plant component; or 3) The owner/operator. This paper focuses on the cases that involve the malfunction of plant components other than the turbine. These cases are the ones in which some corrective action should be taken by the plant to prevent reoccurrence of the failure and in which financial compensation can be obtained from the insurance company.

TURBINE BLADE DAMAGE INSPECTION

Both steady-state and time varying stresses act on turbine blades. The steady-state stresses compose a base level of stress whose magnitude is a function of the operating load. These stresses vary somewhat from blade to blade but vary even more from one position in the turbine to another. These stresses are primarily the result of centrifugal forces, residual manufacturing stresses, and the steady-state operating temperature gradients. Time varying stresses are superimposed upon the base level stresses and are primarily the result of the rotor-stator interaction with the steam flow path and various randomly applied stresses such as those created by particle impacts on the blades.

"Failure due to fatigue" has become the popular catch-all phrase for many failures experienced by turbine blades. In the assessment of financial liability, however, the relationship of fatigue inducing, time varying stresses to other damage mechanisms that are present must be determined.

Subsequent to a blade failure, an internal examination of a steam turbine should be made to attempt to piece together observations at various locations in the turbine into a composite picture. Care should be taken that conclusions drawn from observations at various locations are consistent. The steam path through the turbine is the thread that ties the pieces together. The thermodynamics and gas dynamics describing the flow of the steam should agree with the evidence found on the blades.

An important part of an internal examination is a metallurgical examination. This can be considered on three scales: macroscopic, microscopic, and electron fractographic [12]. Combining observations made with the naked eye and with the help of a magnifying glass the macroscopic examination has two main thrusts. As an initial examination it serves to define items for further evaluation in the microscopic and the electron fractographic examinations. Catastrophic failures (severed blades, etc.), cracks, pits, rubbed and deformed areas are items to look for in the initial examination. A final macroscopic examination serves to correlate details found in the microscopic and electron fractographic examinations. The microscopic and electron fractographic examinations concentrate on the character of the microstructure, searching for changes in the microstructure which are characteristic of one or more of the various damage mechanisms.

The first and most obvious item to look for in a macroscopic examination of a turbine is catastrophic failure. This is the separation of blades from the rotor and/or casing. In a turbine, a separated blade will normally impact other blades, damaging them and possibly causing them to separate. The damage will progress downstream from the initial point of failure.

Damage mechanisms which can act on the blades include fatigue, corrosion, erosion and thermal shock. There are also combinations of these damage mechanisms, such as corrosion fatigue, which possess unique characteristics that are different from the characteristics of the two main damage mechanisms (in this case corrosion and fatigue). Depending on the location of the failure, the presence or absence of a specific damage mechanism may often determine financial liability for the failure.

As stated previously, a premature blade failure is not as likely to occur in the HP or IP turbines as it is in the LP turbine. For this reason, perhaps, if a blade failure should occur in the HP or IP turbines, it should be carefully diagnosed.

During a macroscopic examination, evidence of erosion, if it exists, should be visible on the first few stages following the main steam nozzles in an HP turbine, the reheat steam nozzles in an IP turbine, or the extraction points.

The boiler is the primary source of the solid particles that find their way to the turbine. The solid particles are primarily composed of iron oxides which are present in the form of scale on the inner walls of the boiler. Particles of scale break loose and are carried into the turbine by the steam. The scale is normally a fine powder by the time it reaches the turbine where it impinges on the turbine blades causing erosion [14,15].

The pits created by solid particles have more irregular edges than those created by liquid droplets. This is sometimes more easily seen with a magnifying glass or low magnification microscope. The damage done to the subsequent blade rows will diminish as the distance from the admission point increases. This will not necessarily be the case with liquid droplet erosion as will be discussed later. The solid particle erosion causes a blade material loss and, therefore, lessens the structural strength of the blade. The erosion pits also act as pockets for condensed steam to collect and for corrosion to occur. Both of these mechanisms serve to lower the endurance limit of the blade resulting in premature fatigue failure of the blade.

Liquid droplet erosion can occur in either the HP or IP turbines due to the entrainment of liquid droplets by the main steam or reheat steam, respectively. Often the high pressure heaters are located above the turbine. In such cases, a second mechanism that may cause liquid droplet erosion in the HP or IP turbines is the failure of the water level control mechanism on one of these heaters. This will allow water to drain back into the turbine. Liquid droplet erosion pits on the HP or IP turbine blades are evidence of water induction since condensation does not normally occur in these turbines. As mentioned previously, the pits formed by liquid droplets are generally smoother than those caused by solid particle erosion. Another technique for differentiation of solid particle erosion and liquid droplet erosion involves the progression of the damage downstream of the point of entry. The solid particle erosion will normally lessen as the distance downstream of the point of entry increases. Liquid droplet erosion will usually be more severe on the blade rows, two or three rows removed from the point of entry. After entry into the turbine and initial impact with a blade, the droplets that impacted a rotating blade will coalesce and eventually be slung from the blade with a velocity approaching that of the blade. The impact of these large, high velocity droplets on trailing blades is responsible for a significant portion of the erosion damage [13].

Further evidence of a severe water induction is the failure of the thrust bearing from excessive movement of the rotor in the direction of the steam flow. When a severe water induction incident occurs, the water blocks the blade passages and thus the steam path. The resulting pressure differential across the blocked section creates large axial loads on the thrust bearing and can result in bearing failure.

Major corrosive deposits found on turbine blades include [13]: sodium hydroxide, sodium chloride, hydrochloric acid, sodium silicate, sodium carbonate, sodium bicarbonate, ammonium chloride and potassium chloride. As the steam expands through the turbine, the solubility of the corrosive chemicals decreases. When the solubility limit is reached the chemicals precipitate into droplets of concentrated solutions which are in equilibrium with the steam. Chemicals such as sodium hydroxide are in a class of corrosives that can exist as concentrated aqueous solutions in superheated steam. Concentrations as high as 30 to 90 percent can exist over the entire superheat expansion line of a 2400 psi turbine. The solutions do not cease being corrosive until they become dilute in the wet region of the LP turbine. Deposits containing these chemicals can therefore occur in the HP or IP turbine and should be a point of investigation. In deciding whether excessive levels of a deposit are present, care must be taken in using physical and chemical equilibrium theory as a basis for the estimate. The steam velocities in a turbine are transonic resulting in transit times for the steam through the turbine on the order of one-tenth of a second. This time scale is small in comparison to the time scale of the water droplet nucleation and growth; therefore, the steam will become supersaturated to a theoretical moisture level of 3 to 4 percent above normal saturation before condensation occurs on the Wilson Line [10]. Chemicals such as sodium chloride are in a second class of corrosives which exist in their most concentrated form near the saturated vapor line in the slightly superheated region. Corrosion due to these chemicals should therefore occur primarily in the latter stages of the LP turbine. The presence of sodium chloride deposits in the HP or IP turbine is an indication of water carry-over.

The corrosive environment present in a steam turbine reduces the fatigue life of the blade material and can effectively eliminate the endurance limit [16]. Failures can therefore occur at stresses much below the normal endurance limits. Turbine blades subjected to both corrosion and fatigue characteristically fail more quickly than they would from either of the two damage mechanisms acting separately [17]. For turbine blades in the presence of steam laden with corrosive chemicals, the fatigue strength of the blade may be reduced by 20 percent or more [3]. Premature corrosion induced failures often point to improper operation of such plant components as the boiler or feedwater heaters.

Evidence of corrosion is found either in the form of a film, much like a rust film, or as pits. The presence of a film is not in itself a reason for alarm; in fact, the internals of a turbine will typically develop a film when the turbine is unbuttoned and exposed to the atmosphere. This film actually acts to protect the surface metal from further corrosion. Pits, however, are reason for alarm because they tend to initiate cracking. The pits act as a point of stress concentration causing the blade to see higher effective stresses (both steady-state and time varying) at the pit, and also act as stagnant cells where the corrosive solution can collect and react with the metal. Both solid particle and liquid droplet erosion can act to create pits that can subsequently become corrosion cells.

Corrosion fatigue cracks are predominantly transgranular as are fatigue cracks, and it is, therefore, often difficult to tell by fractographic or metallographic techniques if corrosion played a role in a fatigue failure. Corrosion fatigue cracks, unlike fatigue cracks however, may also be intergranular depending on the stress level and the specific corrodent [5]. It is therefore possible in some cases to find a branched crack or a mixed intergranular-transgranular fracture which indicates that corrosion fatigue is a likely cause. The existence of corrosion pits on a blade should immediately alert the inspector to the possibility of corrosion fatigue since the very existence of the corrosive environment means the fatigue strength of the blade may have been significantly reduced. Conclusive evidence of corrosion fatigue is considered to be a crack that starts from a corrosion pit [13].

Two other forms of corrosion that should be looked for are intergranular corrosion and stress corrosion. Intergranular corrosion occurs along grain boundaries due to the fact that the grain boundaries contain material which is more susceptible to the corrosive medium than is the material in the center of the grain [21]. Since typical fatigue cracks are transgranular, identification of intergranular cracking points to the presence of corrosion. Stress corrosion is the term given to corrosion that occurs in the presence of static stresses in which the damage exceeds the cumulative damage expected from the two mechanisms acting separately. Stress corrosion cracks may be either transgranular or intergranular [12]. The bulk of experimental evidence indicates that if the initial localized corrosion fissures are intergranular, the subsequent cracking is predominantly intergranular. If the initial fissures are within the body of the grain, the subsequent

cracking will be transgranular. In cracks that propagate rapidly, however, mixed modes of cracking may occur [16]. Again, it needs to be determined whether the cracks are purely transgranular and, therefore, indicative of fatigue failure or whether the cracks are intergranular and/or mixed mode indicating the influence of corrosion.

The HP and IP turbines are also susceptible to thermal shock from the entry of relatively cool liquid or vapor. Portions of the rotating or stationary components are rapidly cooled or quenched [18] creating large surface stresses which can exceed the elastic limit of the material [19]. In steam turbines, severe cases of thermal shock occur when the rotor is rapidly cooled and thus contracts, causing the shaft and shaft blading to rub against the turbine casing and casing blading. In some instances the rotor can contract to the point that the turning gear will not rotate the shaft.

Most turbine blade failures occur in the last few rows of the LP turbine. It is in this region that the steam begins to condense. The initiation of condensation (the Wilson Line) has been likened to a cascade of condensed droplets. The steam becomes supersaturated with a moisture level of approximately 3 to 4 percent above normal saturation moisture levels before condensation occurs. The location of the initial condensation may shift with changes in operating conditions and as it shifts, regions of the turbine may alternate between being in the wet and dry regions. Droplets of corrosive chemicals that were deposited on the blades while the blade was in the wet region can become more concentrated as the water evaporates under the influence of the dry region.

The damage caused by the long term bombardment of liquid droplets does not typically proceed at a uniform rate even if the liquid impingement is uniform with time [17]. There is an initial phase during which there is usually some deformation of the surface but no detectable loss of blade material. This is followed by a second phase during which the material damage rate increases to a maximum. During the third, and last phase, the damage rate begins to decrease, gradually leveling out at a much lower rate. Honneger [20] attributes this trend to a roughening of the surface, which by retaining water, cushions the material against further impacts. The severe erosion during the second phase, however, will normally reduce the blade efficiency to an unacceptable level.

Detection of liquid droplet erosion typically involves examination of the shape of the pits and the pattern that a group of pits make. If the impact of the droplet was at a large angle to the normal, the shape of the pit should be skewed in the direction of the particle impact indicating this. In some instances several droplets will impact the surface forming a recognizable pattern aligned with the steam flow path. When premature condensation occurs in the turbine, evidence of corrosion may appear upstream of any significant droplet erosion. Again, this is due to the fact that the major erosion damage results from the impact of the larger, coalesced drops.

Conclusions

The assignment of financial liability for a turbine blade failure rests on the ability to determine the damage mechanism or mechanisms responsible for the failure. The optimum investigation incorporates a three prong approach including macroscopic, microscopic and fractographic examinations, as appropriate. Post-turbine blade failure investigations normally focus on the question of whether the failure was a fatigue-induced failure or, if it was not, whether the blades in the vicinity of the failure were approaching the limit of their life expectancy in a fatigue failure mode. To answer these questions requires an estimate on the manufacturer's part of the expected life of the blades in question. Second, the post-failure investigation must determine if outside influences could have initiated the failure or even hastened it. Outside influences could have an immediate effect such as the entry of solid particles or liquid droplets into the turbine or a long term effect such as the presence of an excessively corrosive steam chemistry over an extended period of time. If evidence of an outside influence exists, the effect of the influence on the failure must be quantified.

A key to the investigation of a particular incident is the determination of the proximity of the failure to various turbine inlet and extraction points. Additionally the thermodynamics and gas dynamics associated with the particular turbine under investigation must be used to determine if observations of as-found conditions are to be expected or are indicative of an abnormality that existed.

The final determination of the damage mechanisms responsible for the failure should be documented in such a way that financial liability for the failure can be clearly identified. Attempting to determine the scenario that must have occurred by eliminating possible scenarios until only one remains is not a recommended technique in this case. It must be kept in mind that the default position in the assignment of financial liability is normally the owner/operator.

REFERENCES

1. Dewey, R. P., and N. F. Rieger, "Survey of Steam Turbine Blade Failures," EPRI Report CS-3891, March 1985.

2. Kantola, R. A., "Condensation in Steam Turbines," EPRI Report CS-2528, August 1982.

3. Bates, R. C., F. J. Heymann, V. P. Swaminathan, and J. W. Cunningham, "Steam Turbine Blades: Considerations in Design and a Survey of Blade Failures," EPRI Report CS-1967, August 1981.

4. Benedict, R. P., J. W. Wonn, M. C. Luongo, and J. S. Wyler, "Detection of Water Induction in Steam Turbines," EPRI Report CS-1604, November 1980.

5. Rettig, T., K. Kinsman, and R. Richman, "Workshop Proceedings: Low-Pressure Steam Turbine Blade Failures," EPRI Report WS-78-114, July 1980.

6. Benedict, R. P., J. W. Wonn, and M. J. Hurwitz, "Detection of Water Induction in Steam Turbines, Phase II: Field Evaluation," EPRI Report CS-3135, June 1983.

7. Bates, R. C., J. W. Cunningham, N. E. Dowling, F. J. Heymann, O. Jonas, L. D. Kunsman, A. R. Pebler, V. P. Swaminathan, L. E. Willertz, and T. M. Rust, "Corrosion Fatigue of Steam Turbine-Blading Alloys in Operational Environments," EPRI Report CS-2932, September 1984.

8. Reinhart, E. R., and J. P. Porter, "Nondestructive Examination of Steam Turbine Blades: An Assessment," EPRI Report CS-3675, November 1984.

9. Allmon, W. E., et al., "Deposition of Corrosive Salts from Steam," EPRI Report NP-3002, April 1983.

10. Jonas, O., A. Pebler, and R. C. Bates, "Characterization of Operational Environment for Steam Turbine-Blading Alloys," EPRI Report CS-2931, August 1984.

11. Kratz, J. L., L. D. Kramer, and R. J. Ortolano, "Development of Low-Pressure Turbine Coatings Resistant to Steam-Borne Corrodents, Volume 2: Detailed Studies," EPRI Report CS-3139, Volume 2, June 1983.

12. Colangelo, V. J., and F. A. Heiser, Analysis of Metallurgical Failures, Wiley Interscience Publications, 1974.

13. Speidel, M. O., "Introduction: Corrosion Fatigue Failures," Proceedings of EPRI Symposium on Corrosion Fatigue of Steam Turbine Blade Materials.

14. Brown, E. G., and J. F. Quilliam, "Solid Particle Erosion of Utility Steam Turbines - 1980 EPRI/ASME Workshop," EPRI Report CS-3178, July 1983.

15. Buden, K.G.R., "Turbine Damage Due to Solid Particle Erosion," Presented at the ASME-IEEE Joint Power Generation Conference, September 1976.

16. Harwood, J., "The Influence of Mechanical Factors on Corrosion," Proceedings of the Symposium on Corrosion Fundamentals, University of Tennessee Press, 1956.

17. Juvinall, R., Stress, Strain, and Strength, McGraw Hill Book Co., 1967.

18. Dickson, J. D., and E. E. McKinley, "Turbine Effects from Steam and Water Abnormal Cooling," Proceeding of the American Power Conference, Vol. 32, pp 376-389, 1970.

19. Burger, C. P., "Thermal Modeling," Experimental Mechanics, pp 430-442, November 1975.

20. Honegger, E., Brown Boveri Rev., Vol. 14, p 94.

21. Jastrzebski, Z. D., Engineering Materials, John Wiley & Sons, Inc., 1959.

ANALYSIS OF A HELICOPTER BLADE
FATIGUE FRACTURE BY
DIGITAL FRACTOGRAPHIC IMAGING ANALYSIS

R. H. McSwain
Naval Air Rework Facility
Pensacola, Florida, USA

R. W. Gould
University of Florida
Gainesville, Florida, USA

ABSTRACT

A helicopter main rotor blade spar fracture has been analyzed by conventional and advanced computerized fractographic techniques. Digital Fractographic Imaging Analysis of theoretical and actual fracture surfaces was applied for automatic detection of fatigue striation spacing. The approach offers a means of quantification of fracture features, providing for objective fractography.

THE CRITICAL NATURE of the main rotor blade to aircraft flight safety is unquestionable; the loss of a main rotor blade in flight is catastrophic. The rare occurrence of a partial failure of the main rotor blade spar provides the opportunity to examine the mechanism of failure without the associated loss of aircraft and life. For the failure described here the aircraft was hovering at approximately 10 feet above a ship when one spar section failed. The failure was accompanied by an explosive noise and sudden violent buffeting of the aircraft. The aircraft commander responded immediately by dropping the aircraft back to the ship deck. Visual inspection revealed that a crack had progressed through one member of a dual spar plate assembly at a fold pin lug hole as shown in Figure 1. The remaining spar plate carried the blade load until the aircraft was landed.

Laboratory analysis involved the determination of the mode of failure, evaluation of material properties, and calculation of the crack growth rate from the fracture surface features. The crack growth rate was analyzed using conventional striation counting techniques (1,2,3), ground-air-ground analysis techniques, and Digital Fractographic Imaging Analysis Techniques (4).

MATERIALS ANALYSIS

The lower spar plate was cracked chordwise through both sides of the blade fold pin hole, as shown in Figure 1. The trailing edge fracture showed beach marks typical of fatigue, as shown in Figure 2, with a crack initiating at the center of the spar shown at the arrow, beneath the blade fold pin bushing. Beach marks extended 19.5 mm out from the origin along the centerline of the spar. The remainder of the fracture (20 mm) was slant fracture typical of overload. The entire fracture surface was uncorroded, indicating a relatively fast propagating crack. Chemical analysis and hardness analysis of all materials related to the failure indicated proper compositions and hardnesses.

The bushing from the fold pin hold in the fractured spar section was measured and found to have an improper outside diameter of 72.42 mm, identical to the size of the hole in the spar. Specifications for the bushing required an oversized diameter which would have produced an interference fit of approximately 0.43 mm in the hole. The upper bushing had an outside diameter which very closely approximated its specified range. The lower (undersized) bushing from the failed spar plate had a new appearance with no oxidation present. The upper (proper sized) bushing was heavily oxidized indicating it had been in service for a considerably greater length of time. It also had no identification markings remaining. Records investigation revealed that the undersized bushing had been installed during blade rework approximately 620.7 hours prior to the failure. The undersized bushing was probably the contributing factor to the fatigue failure.

Macroscopic examination of the fracture surfaces revealed distinct beach markings originated at a shot peening cavity in the fold pin hole. The origin of the fatigue beach marks was located in the canter of the plate thickness, beneath the undersized bushing. The surface of the fold pin hole in the spar had been coarsely shot peened and tops of the shot peening asperities had been removed as per drawing and process specification provisions. The residual stress in the shot peened area was evaluated by x-ray diffraction techniques and found to be 165 MPa (24,000 psi), indicating the shot peening was effective in producing a residual compressive stress at the surface of the fold pin hole. Further examination of the fracture surfaces revealed an initial smooth area of beach markings which extended approximately 5.6 mm from the origin as shown in Figure 2, representative of slow fatigue crack growth.

Fractographic analysis of the fractures by scanning electron microscopy revealed striations typical of fatigue, as shown in Figure 3. Each striation usually corresponds to a single load cycle (1,2,3). The striation pattern indicated random loading with additional distinct spike load indications superimposed in the fracture pattern. The primary load cycle was one cycle per blade revolution. The aircraft had four main rotor blades which produced additional lower intensity load cycles. Superimposed on those loadings were flight loads due to flight maneuver air turbulence. Though striation counting is useful for analyzing crack growth rates, the presence of random loading makes the process much more difficult (1,2,3).

The fatigue crack progressed approximately 5.6 mm in a uniform mode corresponding to the smooth semicircular crack zone as shown in Figure 2. Beyond that zone the crack began to propagate by a repeat cycle of overload followed by fatigue as indicated by the alternating light and dark bands of fatigue and overload. The bands are clearly shown at the edges of the spar in Figure 2 and extended to approximately 19.5 mm from the origin before complete overload occurred. The bands of overload and fatigue allowed a macroscopic analysis of the crack growth rate to be performed which could be compared to the microscopic striation analysis.

CRACK GROWTH RATE ANALYSIS

Though the striations were nonuniform across the fracture surface, a striation count was attempted to obtain an estimate of the crack growth rate. The goal was to determine if the time since installation of the bushing was sufficient for fatigue crack initiation and propagation. Typically four scanning electron fractographs were taken at periodic distances from the origin. Striations from at least six areas on each fractograph were counted. However, there was great difficulty in locating sufficient regions of uniform striations for counting. For the initial crack zone at 2.5 mm from the origin, the striation count was 14,000 striations per mm. At 5 mm the value was 11,000 striations per mm. In two outer fatigue bands separated by regions of overload, the value was 10,000 striations per mm at 7.5 mm and 8,850 striations per mm at 10 mm. In every fatigue

area examined, the fracture surface showed large superimposed loads and also random load cycles. The peak load cycles tend to slow the crack growth rate because of plastic deformation at the crack tip which results in residual compressive stresses in the region into which the crack must propagate (1). It has also been shown that an abrupt load change from high to low intensity may result in an area void of striations for a minimum of 10 cycles (2).

The striation data was further related to crack growth rate assuming a load frequency of one cycle per rotor head revolution, or 17,886 cycles per hour. Extrapolation of striation spacing to the initial fatigue zone (20,000 striations per mm) produced a crack growth rate of 0.889 mm per flight hour. This would have resulted in failure in less than 20 hours which seemed unreasonable. Again, the striation data was questionable.

Macroscopic crack growth analysis by monitoring crack length during testing and microscopic analysis by striation counting has been shown to differ due to striation irregularities caused by random loading (2). To further evaluate the crack growth rate, a macroscopic analysis was performed using ground-air-ground (GAG) cycle bands which were present beyond the initial uniform propagation zone. The width of three of the best defined GAG fatigue bands was 0.15 mm, 0.18 mm, and 0.25 mm. The bands were located at distances of 7.6, 10.2, and 12.7 mm from the origin. The mission of the aircraft under investigation involved approximately 2 flight hours per flight, resulting in crack propagation rates of 0.076, 0.090, and 0.125 mm per hour. These values were approximately a factor of 20 slower than the growth rate obtained for the striation counting and were more realistic. There were approximately 26 bands of fatigue in the GAG cycle zone extending from 5.5 mm to 19.5 mm from the origin. Assuming an average width of 0.25 mm resulted in 6.5 mm of actual fatigue crack propagation beyond the uniform crack propagation zone.

Using the GAG cycle analysis and assuming an average crack growth rate for the banded zone of 0.127 mm per hour, a time of 51 hours was obtained for propagation of the fatigue crack through the banded GAG cycle zone. By extrapolating the three growth rates of 0.076, 0.090, and 0.125 per hour to a slower growth rate of 0.025 mm per hour for the early cracking zone, a propagation time of 220 hours was obtained for slow crack growth. Thus the total estimated time required for propagation by fatigue in the spar was 271 hours. The time on the blade since bushing replacement was 620.7 hours, leaving approximately 348 hours for fatigue crack initiation. These values appear reasonable.

In order to obtain a more objective analysis of the striation spacing, Digital Fractographic Imaging Analysis was applied to the fracture surface features. A scanning electron micrograph of the fatigue striations was digitized and the two-dimensional Fourier transform was calculated. The transform contains the spatial frequency content of the input image, which in the case of fatigue striations is the striation repeat frequency. Computer generated models of fatigue striations were used to calibrate the technique. Figure 4(a) shows computer generated parallel lines with two repeat frequencies and 4(b) shows the resulting Fourier transform power spectrum. The two outer bright spots on the horizontal axis of the transform are located symmetrically about the center spot and their distance from the center spot corresponds to the spatial repeat frequency of the area between the fine lines in the input image. The smaller spots on the horizontal axis of the transform correspond to the coarser spaced areas between the groups of fine lines image. The smaller spots on the horizontal axis of the transform are repeated due to harmonic effects. The center spot corresponds to a repeat frequency of zero, or the background intensity in the input image. By analyzing a number of spacings for various input images the technique was calibrated for a range of input image feature spacings.

A high magnification scanning electron fractograph of an area located 1.25 mm from the origin is shown in Figure 5(a). The Fourier transform power spectrum from the fatigue striations in Figure 5(a) is shown in Figure 5(b). The figure shows small spots close to the center spot. Further analysis indicated the presence of six distinct spots. The average spot spacing corresponded to an input pattern spacing of approximately 2 mm. By considering the magnifications of the input image fractograph, an actual striation spacing of 20,250 striations per mm was obtained. That spacing was similar to the striation density obtained manually in the striation counting process. Thus, the manual counting techniques were accurate, but the striation density was altered by the random and spike loadings resulting in an apparent accelerated crack growth rate. Furthermore, the computerized technique was more objective than manual striation counting, was faster, and was much easier because the Fourier transform is extremely sensitive to repeat feature detection.

CONCLUSIONS

The analysis of a failed helicopter main rotor blade spar was performed to the extent that the failure mechanism was identified, the material properties were verified, and a crack growth rate was established. Manual counting of striations in scanning electron fractographs produced an unrealistically fast crack growth rate. Ground-air-ground analysis produced a more realistic crack growth rate. Digital Fractographic Imaging analysis of the fracture surface striations resulted in a similar fatigue striation spacing as produced by manual counting, indicating the complicated load pattern was responsible for the apparent high microscopic crack growth rate. The computerized technique offers the ability to quickly and accurately determine striation spacing and in the case of a more uniform loading environment to establish highly accurate fatigue crack growth rates.

ACKNOWLEDGEMENTS

This work was supported by the Naval Air Systems Command, Washington, D.C., under Contract No. N00014-83-K-2035.

REFERENCES

1. McMillan, J. C., and Hertzberg, R. W., "Application of Electron Fractography to Fatigue Studies," Electron Fractography, ASTM STP 436, 89-128 (1968).
2. Whiteson, B. V., Phillips, A., Kerlins, V., and Rawe, R. A., "Special Fractographic Techniques for Failure Analysis," Electron Fractography, ASTM STP 436, 151-178 (1968).
3. Abelkis, P. R., "Use of Microfractography in the Study of Fatigue Crack Propagation under Spectrum Loading," Fractography in Failure Analysis, ASTM STP 645, 213-34 (1978).
4. McSwain, R. H., "Digital Fractographic Imaging Analysis," Ph.D. Dissertation, University of Florida, (1985).

Figure 1. Failure of one member of a dual spar plate assembly at a fold pin hole. (0.22X approx.)

Figure 2. Fatigue fracture in spar plate which initiated beneath bushing at position shown by arrow. (1.84X approx.)

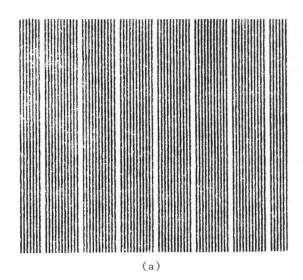

(a)

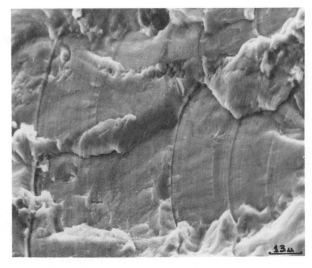

Figure 3. Fatigue striations from random loading on fracture surface of failed spar. (638X approx.)

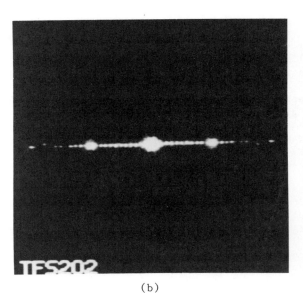

(b)

Figure 4. Theoretical fatigue input image/Fourier transform pair: (a) computer generated parallel lines with two repeat frequencies; (b) resulting Fourier transform power spectrum.

(a)

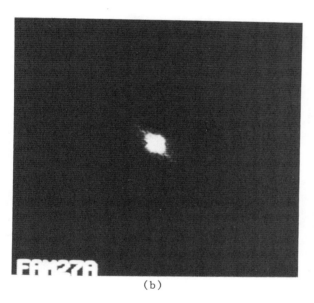

(b)

Figure 5. Fourier transform power spectrum of
fatigue striations: (a) scanning electron
fractograph of fatigue striations on failed
spar fracture surface (11,475X approx.) and
(b) resulting Fourier transform power spectrum.

THE INTERACTION BETWEEN STRUCTURAL FAILURE, PRODUCT LIABILITY AND TECHNICAL INSURANCE

H. P. Rossmanith
Technical University Vienna
Vienna, Austria

ABSTRACT

Today failure analysis and fracture mechanics technology provide a powerful tool in the prevention and analysis of structural failures. In addition the basic principles of product liability are now firmly integrated into the national and international laws. International co-operative programs have confirmed the need for an interdisciplinary approach to failure prevention and it is mandatory for the enhancement of safe and reliable performance of engineering products that experts from the technical, legal and insurance fields intensify their cooperation and mutual understanding.

This contribution deals with the interaction between failure preventive design and manufacturing and failure occurrences, quality control, legal liability and technical insurance.

INTRODUCTION

At the First International Conference on "Structural Failure, Product Liability and Technical Insurance" in Vienna, Austria, in 1983 /1/, all participants shared the conviction that there is an urgent need for constructive cooperation and interaction between engineers, technicians, lawyers, managers and technical insurers. The participating experts stressed on recent advances in failure research and fracture mechanics as well as the impact of this technology on product liability with a firm stronghold on failure preventive design and manufacturing technology, and legal and insurance aspects of product liability.

Today failure research and fracture mechanics technology have advanced to the stage where they are of direct engineering value for the prevention of sudden fracture in structures. This new technology enforces an interdisciplinary system-type approach to failure prevention and it is mandatory that the technology stresses on the interaction between material properties, design, fabrication, inspection and operational requirements so as to enhance safe and reliable performance of engineering products.

The rapid rate at which modern technology is being developed has accelerated the manufacturing of engineering products and construction of large-size structures and components. With respect to possible damage of the structure the supplier and user should be aware of the current state of the art so that fracture mechanics can be fully utilized to produce a better product.

Ignorance of modern development in fracture research and structural failure development may cause severe damage of constructions associated with possible loss of life leading to legal problems concerning guarantee, payment of damages, product liability and insurance. Advancement and results of structural failure research offers new measures and guidelines for the conception and the extent of damage, the methods of damage assessment, the repartition of loss, the avoidance of damages and the insurance against losses. They will also influence the liability of the producer for the safety of his products.

In 1973, the American Society for Testing and Materials (ASTM) in its March issue of the *Standardization News*, entitled *Product Liability* /2/, calls the reader's attention to the (relatively) new field of *product liability*. The merging of the computerized precise world of engineering with the polymorphic social world results in the emerge of a new engineering ethic, and the impact of product liability on engineering is impressively demonstrated by the fact that 27 out of some 70 pages of the March issue have been devoted to this important subject. Subsequently, short courses on 'Fracture Mechanics, Failure Analysis and Product Liability' have been offered at Lehigh University and elsewhere. These short courses were designed for engineers and managers "...who have concern for the fracture of engineering components and their resulting legal consequences"/3/.

Recent developments and changes in the US products liability concept, the various approaches to the problem in various European countries and the recent uniform guideline on products liability of the Commission of the European Community have given the subject of product liability a key position in the technical-economical-legal interaction.

STRUCTURAL FAILURE

The very idea that all constructions contain imperfections and defects has for long been repugnant to many in the engineering world and this attitude is still maintained by some in spite of recognition that acceptance of alternative approaches such as e.g. Well's *Fitness-For-Purpose-Concept* /4/ entails added responsibility.

Structural failure of a component or structure implies reduced level of operation and performance. A fracture, e.g. induces a locus of reduced load-carrying capability in a structure or in a structural component. Brittle fracture is characterized by sudden and unexpected occurrence of a crack the extremely high-speed extension of which quite often leads to catastrophic failure of the structure. The number of brittle failures in failure statistics is low, but when they do occur they may be far more costly in terms of loss of life and material.

The cumulative occurrence of large-scale structural failures during the first half of this century has been calling for a means to prevent and/or predict damage by failure. In several particular catastrophical failures the primary cause of failure could be traced to the existence of stress raisers such as cracks and sharp corners. Fracture mechanics evolved from scientific investigation and research associated with the many various facets of failure analysis. Extensive research on brittle fracture in structures of all types has revealed that numerous factors can contribute to brittle fracture but there are essentially three primary factors: *material toughness*, *crack* or *defect size* and *stress level*, that control the susceptibility of a structure to brittle fracture. If a particular combination of stress loading and flaw size in a structure reaches the toughness level fracture may occur.

Many structural components do not fail due to single overloads but as a result of repeated cyclic load application even at stress levels far below the critical level for onset of brittle failure. Results of innumerable tests have provided a sound basis for the evaluation and prediction of the fatigue strength of structural components.

Although previous studies have indicated that design frailty and fragility are two of the principal causes of product failure the greatest improvements can usually be achieved in the design area rather than in testing and quality control. If product testing and quality control operate at appropriate levels the flexibility necessary for overall cost reduction is almost exclusively restricted to the design stage of a product. Costs incurring in fracture avoidance by over-designing figure up to billions of whatever currency one may choose. Hence, during the past decade a new design philosophy and methodology has emerged: *failure preventive design*.

In structural engineering the term design refers to the synthesis of various disciplines to create a technical structure that suffices the requirements imposed. Within the framework of fracture mechanics the word design is intimately associated with the prevention of brittle fracture, i.e. with the elimination of structural details that can be potential fracture-initiation sites and appropriate stress levels. Complete elimination of stress raising details may not be possible to achieve in large-scale structures and appropriate material selection and design stress level compensation for non-optimal structural brittle fracture design. In many low-to-medium-strength structural materials plane stress conditions and/or elasto-plastic behaviour are encountered under slow loading and at normal service temperatures. Then, linear-elastic fracture mechanics is no longer applicable and extensions into plane stress and elasto-plastic fracture mechanics such as R-curve analysis, crack-opening-displacement and J-integral should be employed.

Conventional design of structural components subjected to fluctuating loads base on a design fatigue curve as obtained from unnotched specimen fatigue testing. Although these design fatigue curves have been utilized to analyze the fatigue behaviour of cracked components such as weldments the concepts of fracture mechanics offer a more rational analysis of fatigue behaviour. Moreover, the assumption of the presence of an initial flaw-type defect represents a conservative approach to design to prevent fatigue failure. In addition to fatigue growth, flaws or cracks can also grow by stress-corrosion during the life-time of the structure or, furthermore, they can grow by the possibility of joint operation of both mechanisms. Thus, knowledge of the fatigue, corrosion-fatigue and stress-corrosion behavior of the structural materials is required to enable *failure preventive design, failure preventive manufacturing* and *failure preventive engineering*. Criteria selection is the most important part in this respect. Questions of *'how much toughness is necessary?'* and *'how much is the designer willing to pay for materials with superior toughness?'* should be answered only after a careful study of the particular requirements for a structure.

Regarding the predictable costs of failure occurrence it is obvious that manufacturers and engineers devote considerable efforts and expenses to failure prevention and avoidance. The three widely practised approaches to achieve cost effectiveness while still maintaining maximum safety and reliability are /4/:

- *overdesign* on the basis of generous safety factors to account for a variety of uncertainties,
- *product testing*, and
- *production quality control*.

The objective in structural design of complex structures such as planes, bridges, pressure vessels, pipe-lines, ships, trains, spacecraft etc is to produce a structure that will perform the desired operating functions efficiently, economically and safely. Optimization is the key to modern business and this is already reflected in the initial phases of the development of a structural project, i.e. structural design. To achieve these objectives methods and techniques of failure preventive design, failure preventive manufacturing and failure preventive engineering are at the engineer's service.

When the general guidelines are integrated into the requirements of a particular structure a *fracture control plan* has been established. A fracture control plan consists of a set of recommendations developed for a particular structure and, hence, should not thoughtlessly be applied to other structures. The development of fracture control plans for large technical structures is very difficult and an expensive task /4,5/. In addition, a fracture control plan is not a once-for-ever task but with increasing knowledge and experience during the service of the structure adjustments may be made in terms of refinements or recommendations and guidelines. On the other hand service-based adjustments may also be made in the design, manufacturing, inspection and operating conditions which in turn tend to establish adequate fracture safety. The comprehensive plan will contain information pertaining to design, manufacturing, materials, inspection, quality control and service operation. Efficient operation and execution of a fracture control plan requires a great deal of interdisciplinary and inter-group coordination and cooperation. Complete avoidance of structural failure by fracture can only be assured when the elements of the fracture control plan supplied by the different divisions or groups are coordinated by an especially established fracture control group.

Today - and this has been already realized in many branches of industry - the development of a new product is dictated by a generalized form of optimization process where economy and safety represent two major decisive and most often opposing and mutually exclusive criteria. A necessary condition for a marketing product is highest possible efficiency through maximum utilization of materials, fabrication techniques etc for a calculated desired life-span at maximum reliability and safety against loss. In real life, in order to remain competitive on the market, the manufacturer must constantly weigh the cost of fracture prevention against that of fracture occurrence. The costs to the manufacturer involved in fracture avoidance as well as in fracture occurrence can be very high /6-9/.

PRODUCTS LIABILITY

Products liability is defined as legal liability in respect of bodily injury or property damage caused by products manufactured or supplied by the injured if such injury or damage occurs away from the premises of the insured after he has delivered or transferred the product to others by relinquishing their physical possession. Thus, claims resulting from damage occured to the products manufactured or supplied by the insured is not covered /9/.

Product liability, therefore, is a subject which today affects us all whether we are manufacturers, consumers, attorneys or insurers.

Product liability flows from four sources of defects:

- *design defect*,
- *construction defect*,
- *operating defect*, where the product is safe in design and construction but defective because adequate instructions have not been provided, and
- *warning defect*, where a technically adequate product became defective because post-manufacture duty to warn was not adequately fulfilled.

A crucial question in the products liability concept therefore is concerned with the characterisation and definition of a defect. Two thorough investigations may be found in Refs /10/ and /11/, where the first reference refers to a more technical definition whereas the second reference provides a juridical characterization.

One statutory definition for a product goes back to the Second Restatement of Torts, 1965 /10,11/:

"...who sells any product in a defective condition which is unreasonably dangerous is subject to liability for physical harm caused to the ultimate user if the seller is engaged in the business and the product reaches the user without substantial change in manufactured condition."

Since, on the other hand, any company concerned with the manufacture of components or the production of systems and system components aims at optimizing the manufacturing process in terms of cost and efficiency quality assurance comprising quality planning and quality control offers a potential menas for both, industrial safety - in preve ting injuries to or death of people - and the prevention of damage to materials and equipment. Quality planning generally begins with planning of the quality of a product parallel to the planning and/or design phase of the structural system or component.

The guideline issued by the European Commission on Product Liability Law provides for strict liability. The determining factor is thus the defect of the product. Obviously, consumer protection-based reasoning has had a major influence on the formulation of the definition and European exporters fear additional cost for industry involved in any increase in liability. The EEC Products Liability Law provides the individual member states with the possibility of establishing national rulings and regulations on liability and to exclude producer's liability for the disputed development risks.

TECHNICAL INSURANCE

Classical product liability insurance concentrates on the damage suffered by the final consumer. The European Commission is also concerned exclusively with questions of liability as these affect the final consumer. But the model for product liability insurance developed by the Casualty Insurers Association implies abandonment of the classical relation between final manufacturer and final consumer in favor of the so-called *delivery chain* /9/.

duction and promotion.

The calculation of a product liability insurance premium is the conversion of the legal and social environment into calculable costs /9/. The calculation of a product liability premium must be supported and preceded by a thorough risk evaluation which is determined by product related and insurance related factors. Therefore, a product liability premium is composed of ponderable and imponderable factors. Because of its intimate relation to competitiveness the question of additional premium requirement is a very difficult one to handle for the insurance industry. Several different viewpoints of European insurance companies on this matter may be found in Ref./1/.

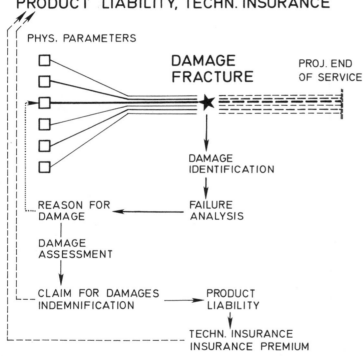

Figure 1: The interrelation between structural failure, failure analysis, damage assessment, product liability and technical insurance in the field of failure preventive engineering /1/.

Technical development, a growing awareness of claims for compensation, increasing pressure exerted by customers and nowadays environmental protection as well as the trend in court decisions and legislation towards an extension of liability form only a few reasons for increasing importance of products liability insurance. Internationalization of products liability and product liability insurance is receiving importance through highly increasing export activity of industry, plant construction abroad and the set-up of subsidiaries abroad for sales, pro-

CONCLUSION

The interaction of the three disciplines, *engineering, law and technical insurance* is sketched in Figure 1. A complex structure that involves a number of possible failure-inducing parameters is designed, manufactured, controled and planned to operate efficiently, reliably and safely over its expected projected and desired span of life, i.e. at least up to the projected end of service point. Disregard of proper execution of only one of the steps in the failure preventive

design, manufacturing and operation philosophy may lead to premature failure of the structural component or, even worse, of the entire structure.

Failure research closes the chain via *failure identification, failure analysis* and *reason for failure* aiming for the goal of detecting the 'lost' physical parameter. In a causal extension of the flow diagram *failure and damage assessment, claim for damages, indemnification, product liability, technical insurance* and *insurance premium* enter the field of view of the engineers.

The preceding investigation has confirmed the need for an interdisciplinary approach to failure prevention and has revealed that it is mandatory for the enhancement of safe and reliable performance of engineering products that experts from the technical, legal and insurance fields intensify their cooperation and mutual understanding.

REFERENCES

/1/ Rossmanith, H.P.(Editor): *Structural Failure, Product Liability and Technical Insurance.* Proc.1st Int.Conf., Vienna, Sept.1983, North-Holland Publ.Comp., 1984.

/2/ *Product Liability.* ASTM Standardization News, Vol.1, Np.3, March, 1973.

/3/ Lehigh University Short Course on *Fracture Mechanics, Failure Analysis and Product Liability*, R.Hertzberg (Organizer), 1978-1985.

/4/ Rolfe, S.T. and Barsoum, J.M.: *Fracture and Fatigue-Control in Structures; Applications of Fracture Mechanics*, Prentice-Hall Inc. 1977.

/5/ Irwin, G.R.: Private Communications (1977-1985).

/6/ *Handbook of Loss Prevention.* Allianz-Vers. GmbH, Berlin-München, 1978.

/7/ *The Economic Effects of Fracture in the United States.* NBS Special Publication 647-2, 1983.

/8/ *Understanding Fracture.* ASTM Standardization News, Vol.7, No.4, April 1979.

/9/ Schubring-Giese, F.: *Product liability insurance: viewpoint of an international insurer.* In Ref./1/, 255-264, 1984.

/10/ Ross, B.: *What is a design defect?* In Ref. /1/, 23-72,1984.

/11/ Noel, D.W. and Phillips, J.J.: *Products Liability in a Nutshell.* 2nd Edition, West Publishing Co., St.Paul Minn., 1981.

THE EXPERT WITNESS AND THE ATTORNEY: VS OR VIS-A-VIS

John William Juechter
Consultech Inc.
East Greenwich, Rhode Island, USA

ABSTRACT

The expert witness in a product liability suit has a need for access to technical and general information in the initial stages of investigation; a need to be kept continually abreast of developments in the proceedings; a need for counselling prior to depositions and trial; and a need to be present at, or at least accurately apprised of, unfolding developments during the trial.

The needs and views of the in-house expert and the forensic expert differ markedly with regard both to their goals and short-comings. Each type of expert exhibits peculiar strengths which should be used to their best advantage.

The examples used in the development of these ideas are drawn from 13 years of forensic experience in product liability. The ideas presented are intended to assist the untried in-house expert and the attorney new to product liability suits to prepare effectively for their roles individually and as a team. Experienced forensic experts and product liability attorneys may also find items of interest to use as a checklist.

MUCH has been published about the role of the expert witness—as seen by the trial attorney. Nothing has been written about the role of the attorney—as seen by the expert witness.

If there is an opportunity for justice to be served in a product liability case, be it property damage, person injury, or even death; key rolls in the case will probably be played by one or more technically qualified (but legally ignorant) experts, usually engineers or scientists. Such experts will generally fall in one of two categories: those working in the field of forensics, full or part time (forensic expert); and those in actual daily work with the product (in-house expert). The latter experts usually are encountered defending the product, while the former may be found on either the plaintiff's or the defendant's side.

We can expect the forensic expert to have a broader knowledge of the methods employed in investigation and preparing a case for potential trial; the in-house expert to have a greater in-depth knowledge of the product.

In working with counsel, both experts have many needs in common. They will also have differing individual needs.

At least in theory, the forensic expert is not in the case to "win." His role should be that of the "friend of the court", constrained to telling the facts and drawing his conclusions without bias to the side which has engaged his services.

The in-house expert, while he may be equally ethical in his profession, has a vested interest in the group with whom he associates on a daily basis and may be expected to view "winning" as far more necessary.

FORMING THE TEAM

In all of the following discussions, it is assumed that a suit, once initiated, is not settled and the case continues on to trial. Actually, settlement can occur at any point in the process, at which time, all further effort is usually aborted. Thus, the descriptions furnished below are generally applicable up to the point of settlement or until the case is otherwise disposed of.

DEFINITIONS - For purposes of this paper we have elected to make the following definitions:

Forensic Expert - a graduate engineer or scientist, or a technically trained specialist in the field who has no specific ties to any of the litigants in the subject case other than having been engaged to investigate and to render an opinion regarding the causal aspects of the matter at hand. Thus, such a forensic expert is considered (and expected to be) objective in his approach and in enunciating his conclusions.

In-House Expert — a technically trained individual, possibly with the same background as a Forensic Expert, but with the added qualification of being employed by one of the litigants in the subject case, and considered by that litigant as the best qualified of his employees to speak in defense of the employer. An in-house expert is generally considered to have a bias toward his employer's viewpoint.

BECOMING INVOLVED — In most product liability cases there is the beginning, pre-suit period when one or both sides are trying to take a definitive view of the facts of the case to determine whether to seek a settlement or to prepare to litigate through a trial if necessary. For purposes of this paper, we have chosen to consider several scenarios wherein the experts are introduced to the case in various ways.

The Plaintiff-Oriented Case — This type of product liability case originates when an injury to body or pocketbook occurs and the injured party (or his estate) engages the services of a forensic expert to determine if a viable potential defendant exists. In such a situation, the services of a forensic expert may be engaged by an insurance company, or by an attorney on behalf of his client. Under those circumstances, the forensic expert will usually be the first expert on the case, involved in conducting a pure investigation, but with the potential plaintiff's goals in mind.

Many such cases originate when the worker's compensation insurance carrier engages the services of a forensic expert to examine the circumstances surrounding an industrial accident to determine, in as much detail as possible, how an accident happened; whether the accident is attributable to the configuration of a piece of equipment or structure; whether such a configuration is unsafe in the light of the state-of-the-art for such equipment at the time it came into being; and, if so, who is responsible for the unsafe configuration. Other cases initiate when the injured party engages his or her own attorney to evaluate the possibility of a suit, in which case, the attorney generally engages the services of a forensic expert.

For example, a worker is injured on a piece of manufacturing equipment. The first step is to determine if the configuration of the machine is as-sold or if it was changed by the employee, the employer, or the intermediate owners and/or sellers, if any. Sometimes a guard which would have prevented the accident has been removed by a former owner or the employer.

In one such case, an injection molding machine was modified so that jewelry beads on a string could be molded. The manner chosen was to bypass the hydraulic and pneumatic safeties provided by the manufacturer so that the operator could hold the end of the string on which the beads had been molded. The operator's hand was crushed when the die closed unexpectedly. There was no cause to fault the manufacturer of the machine, although such a case had been initiated by the employee's attorney before consulting an expert. There could have been a case made against the designer and builder of the die, but the statute of limitations had run before the expert was consulted.

In another case, a house caught fire and was a total loss, even though the house had been equipped with a central station fire and burglar alarm system and no alarm had been received by the central station. The fire insurance carrier engaged the services of a forensic expert to determine why no signal was received. Examination of the system after the fire revealed that when the system was serviced a few days before the fire, the serviceman mis-wired the system and apparently did not check its performance thereafter.

The Anticipatory-Defendant Case — This type of product liability case occurs when a supplier of products or services learns of an accident involving that which he has supplied and, recognizing the potential for a suit, decides to make an immediate investigation while the elements of the case are still fresh.

If the potential exposure in the case appears limited and the loss occurred some distance from the nearest technical base, the supplier's insurance company may engage the services of a forensic expert to investigate for the defense.

If the potential exposure appears large, the initial investigation may be undertaken by the in-house expert (for the defense).

A typical example of the first category of anticipatory-defendant cases is when a new machine causes a permanent, but non-crippling injury to an operator and the employer contacts the manufacturer of dealer to inquire about the possibilities and probabilities of how the accident may have occurred or might have been prevented. A visit to the scene by the manufacturer's most qualified employee may not be justified, and so the liability carrier may engage the services of a local forensic expert.

Examples of the second category of anticipatory-defendant cases occur frequently when an serious automobile accident occurs which involves severe injury or is fatal to one or more persons. If the car is a new one, the dealer may learn of the accident shortly after it happens, and the automobile manufacturer will frequently send its own representatives to check out the vehicle(s) involved.

The Reflective-Defendant Case — This is the type of product liability case which may be thought of as that which occurs when the forensic engineer in a plaintiff-oriented case has found that which he considers just cause and has identified a potential defendant from whom to request financial redress of the experienced injury. A letter is sent to the identified potential defendant, usually with a request that the letter be forwarded to the potential defendant's insurance company. At that point, the potential defendant's in-house expert will probably be consulted as to the feasibility of defense.

If the potential defendant is a large company with considerable experience in defending its product or services, the in-house expert will probably make the initial investigation. At some later date, possibly years later, the services of a forensic expert may still be engaged because of the generally accepted belief that the in-house expert must be biased.

If the potential defendant is a small company, its insurance company (or, sometimes, its attorney) will probably engage the services of a forensic engineer to make the initial investigation.

An example of the first category of the reflective-defendant case is a large mechanical power press manufacturer who is a frequent potential defendant of suits originating from injuries which occur from time to time on their machinery in the field. Two aspects of power presses which result in their being involved in injuries are their longevity and the fact that they are universal machines. [By universal we mean that they can be used for an infinite variety of tasks and operations, the scope of which is only limited by the physical dimensions of the machine and the ingenuity of the tool engineer. Since the location, size, shape, and required access to the point-of-operation is not defined until the tooling is complete, it is impossible for the machine manufacturer to provide guarding for the point-of-operation.] They are generally well built and survive decades of use in a variety of applications, frequently having a series of owners. The presses are frequently modified by their owners and are used without proper guards and maintenance. Because such injuries are frequently serious, and because the original configuration of such old machinery is difficult to determine without previous access to the manufacturer's records, initial investigation by the in-house expert is considered by many such manufacturers as the only practical approach.

An example of the second category of reflective-defendant cases is a manufacturer of bicycle components whose product was alleged to have caused a serious bicycle accident. Since the product was a low cost item and only one of several components which may have caused such an accident, the manufacturer could not afford to make his own investigation; so the method chosen was for the investigation to be made by a forensic expert who had the background and experience needed to make a reconstruction of the accident.

Another example of the second category of reflective-defendant cases is a manufacturer of aluminum ladders. One characteristic of metal ladders is that the cause of the accident can usually be determined from close examination of the ladder and of the site of the accident. We have yet to find a metal ladder which has failed, even though the condition of the ladder after the accident may look, to the uninitiated eye, as if the ladder had to have failed because of the distortion of the ladder and because of breaks in parts of the ladder. Under such conditions, an experienced expert can make the initial examination of the ladder and generally provide a report

clearly explaining how the ladder was misused.

There are exceptions to all of the above scenarios. Sometimes suits are brought to a successful conclusion for either the plaintiff or the defendant when only one or neither of them has used a forensic expert.

One example of this is a suit where we were asked by the plaintiff's attorney to examine an aluminum stepladder on behalf of the plaintiff. The right rear leg had broken off a stepladder which had been used outside in a garden. It was obvious that the leg had been broken off because it sank into soft ground and the ladder fell over to the right with the plaintiff on it. We advised the attorney that there was nothing wrong with the ladder, but he had invested so much of his time in the case that he chose to go forward with it. The attorney for the ladder manufacturer did not have an expert to defend his client since there was none on the plaintiff's side, and the jury found for the plaintiff.

However, as can be seen from the foregoing, in most product liability cases one or more forensic experts are involved early in the process, and usually at least one in-house expert plays an important role, since without the investigations of one or more experts, the cases would probably not go forward.

THE ADVANTAGES - While the courts generally require that notice of a potential suit be given to the identified potential defendant as soon thereafter as possible, for a number of reasons, the potential defendant frequently does not learn of the subject loss until long after the evidence has disappeared, or aged, or been altered by testing, etc.; or until the participants have died or become otherwise uncommunicative.

Typical examples of such disappearances of the evidence are automobiles which may be junked shortly after the plaintiff's expert has finished with them; old machines which may have been rendered uneconomic to repair by a failure of a part which caused the subject accident, and buildings damaged by fire which may be either demolished or renovated before the defendant's expert has the opportunity to make a firsthand examination of them. Many such disappearances also are attributable to the long delays typically encountered between the time a suit is filed and the case is finally tried or settled.

Under such circumstances, the plaintiff's forensic expert has an advantage, that of probably being the only technically trained investigator to examine the evidence shortly after the loss. He also generally enjoys a geographical advantage in that his office is usually in the area where the loss occurred, and the trial, if one is held, will generally be held in the area in which the loss occurred. Thus the plaintiff's expert's lines of communication to the plaintiff's attorney are usually short.

The defendant's forensic expert is probably at a disadvantage in not having seen the evidence that was available to the plaintiff, and in not having the long-term experience with the product or service that the in-house expert enjoys. He, the defense expert, too, may enjoy the advantage of a local assignment if the potential of the case is small or if the potential defendant is not used to defending itself.

However, if the potential defendant is a large company or a frequent potential defendant, they may have had experience with a forensic expert who is not resident near the area of the loss, and in that case, the defendant's forensic expert may not enjoy the local contact with the defense attorney, although he probably will have a good, proven working relationship with the in-house expert.

On the other hand, if the product is complicated, and if the probable cause of the loss can be verified only through expensive, specialized testing; the plaintiff's forensic expert may be at a disadvantage and the defendant's forensic expert, backed up by the in-house expert and the facilities and experience in the product or service that are immediately available to the in-house expert, may well have the advantage.

One example of this latter circumstance is a case wherein a restaurant deep fat fryer fire was blamed on an improper installation of the automatic dry extinguisher located over the fryer. The installed location of the nozzle over the fryer was not quite positioned as required by Code. With the assistance of the in-house expert from the manufacturer, we were able to set up an actual test of an identical fryer and extinguisher installation, also deviating identically from Code; and to operate the fryer until the shortening in it caught fire. The extinguisher worked properly and extinguished the fire, as all concerned expected that it would. However, as in the actual case, the power to the fryer was not cut off by the extinguisher's operation (despite the manufacturer's requirement of the installation) and the shortening re-ignited after having been successfully extinguished. Since the test required the use of a fire-proof building, a mockup of the original installation, the use of a portable generator, and sundry other items, it was practical only with the effort of the in-house expert.

Thus, the relative advantage each expert enjoys may vary from case to case.

DEVELOPING THE DATA — In developing a case, the data needed is accumulated from a number of sources:

When a loss occurs, there are several common elements which usually exist:

the location of the loss;
the injured party or his representative;
the device which caused the injury;
witnesses, before or after the fact; and
historical, narrative data regarding pre-loss conditions.

Typically, this data is gathered by a claims adjuster or by an attorney who, on the basis of the accumulated preliminary data, makes a decision to, or not to, engage the services of a forensic expert.

INVESTIGATION FOR CAUSE — The forensic ex-pert has the task of taking the preliminary data and making an investigation, generally of the physical items and site involved, to determine:

the identification of the article(s) invol-ved [any and all identifying names and numbers and the unequivocal identification of the actual item involved in the loss];

if applicable, a more detailed history of the article(s) involved [when and where it was purchased, what changes, including service were made or done to the object];

the most probable manner(s) in which the subject loss occurred [generally, is it possible and logical that the accident or loss occurred as is reported or as is assumed];

the most probable cause of the loss [the manner in which the subject loss occurred and the item or items which most probably caused the loss];

the identification of the party(ies) respon-sible for the loss [the complainant, the employ-er, the owner, the seller, the manufacturer, the builder, the architect, etc.]; and

if a successful suit appears likely under the particular circumstances.

The plaintiff's forensic expert generally brings together, in his report, the data directly applicable to the physical loss of which he is aware as of that time. He may also, at that time, make suggestions as to what data should be sought through discovery to reinforce or clarify his conclusions.

Initial Discovery — At this point, a letter is generally sent to the potential defendant(s) informing each that litigation may be initiated because of a particular loss, the responsibility for which has been assigned to them. An attorney usually takes over the case at this point, if one has not been involved before, and discovery be-gins covering:

interrogatories;
statements from witnesses;
costs stemming from the loss; and
searches for similar cases.

It is at this point that the vis-a-vis rela-tionship between the forensic expert and the person responsible for the case frequently starts to deteriorate to one of benign neglect. The forensic expert has no immediate direct role in the case, and, may not be consulted further until the morning of the trial.

In many instances, the monetary value of the suit is relatively low (less than $50,000) and the attorney who will try the case may not be advised that the case is coming to trial until a day or two before the trial (or even the night before). This deplorable condition is attribut-able to slipshod law firm practices and to the confused working of the legal calendar. At that point there is no opportunity for a working con-ference between expert and attorney. A brief meeting to discuss the case may be held the morning of the trial and then the trial proceeds. Such practices are unfair to all the parties involved, particularly to the client.

<u>Discovery & Case Development</u> — During this stage of development, the following may be expected to occur:

the compliant is written, filed, and served;

depositions of witnesses and officials or technical specialists of the potential defendants are taken (probably including the defense forensic expert;

documents relating to the design and specific life history of the article(s) involved in the loss are requested; and

additional interrogatories may be posed and answered.

During this stage, also, the forensic expert may be ignored or recognized as a continuing member of the "litigation team".

PREPARATION FOR TRIAL — This is the stage wherein:

the data collected is cataloged, reviewed, and evaluated;

the weaknesses and strengths of the case are determined; and

the proposed scenario of the trial is laid out.

While the Preparation for Trial stage generally occurs in the life cycle of each case which gets to the court house steps, the stage is sometimes reduced to a night-before-trial reading of the case by the trial attorney. Such a practice ignores the fact that proper preparation usually spells success.

The foregoing developing of data has been written from the point of view of the plaintiff. A very similar path may be followed by the defense team except that many of the cases only come to the attention of the potential defendant when the first 2 stages: Preliminary Data and Investigation for Cause have been completed on the plaintiff's behalf. Thereafter, a similar process will probably be followed by the defendant (potential defendant).

VIVE LA DIFFERENCE

One of the first difficulties encountered in the relationship between an attorney and an ethical forensic expert is the difference in their focus or role.

The attorney is, traditionally, involved in projecting his client's case in a light most favorable to his client. His role is to prove that his client's side or view is the correct one. That which is unfavorable will be veiled, if not obscured; and that which is favorable will be promoted.

The forensic expert traditionally has had the role of amicus curiae (friend of the court) wherein he is expected to testify to the truth of the matter without regard to the interest of the litigant that engaged his services. This means that the forensic expert has the duty to report his findings and conclusions to his client whether the results be favorable to the client or unfavorable. The forensic expert cannot do his job if he starts out to prove something as does the attorney whose goal is it to win the case.

The expert must examine the evidence and base his conclusions solely on the evidence, regardless of the effect that has on his client's position.

This is a role that unfortunately is misplayed by a number of forensic experts who are willing to take the advocate role, trading ethics for notoriety. Among their peers they are known as the "guns for hire" or the "technical whores".

The attorney new to product liability practice or the attorney who has encountered a number of advocating forensic experts is likely to have trouble, at least initially, working with the "traditional" (amicus curiae) forensic expert. Indeed, in some cases, a working relationship can never be established.

One striking example of this was an aluminum ladder case wherein the defense attorney apparently never believed in the veracity of his experts. The aluminum extension ladder was badly deformed, and the plaintiff had a touching story to tell of the ladder's failure. Both the forensic expert and the in-house expert agreed in their analysis of the ladder and how it was misused; but the defense attorney was intimidated first by his own lack of understanding of how the misuse of the ladder could be deduced from the battered remains, and by the imposing mien of the plaintiff. Even the plaintiff's attorney evidenced the desparation of defeat. Needless to say, a plaintiff's verdict was awarded.

We do not intend to suggest that the forensic expert has no interest in his client's case or in the outcome thereof. With few exceptions, an expert should not be the expert of record in a case unless he believes in the validity of his client's case. The expert also needs to distinguish concern or pity from the ethical considerations of the facts as they exist.

Plaintiff's attorneys tend to embrace one of two theories of retribution. They may favor the "deep pockets" theory wherein, if a severe injury has occurred, some financially responsible entity (usually, a corporation) should be held liable, regardless of their actual responsibility for the accident. Other attorneys follow the "shotgun" approach of suing everyone possibly attached to the loss on the theory that discovery may uncover the involvement of one or more of the potential defendants or that they will, in trying to defend themselves, help to incriminate each other. Antithetically, the forensic expert must evaluate the case on its technical merits and draw his conclusions accordingly.

Occasionally, even following the dictates of his conscience, a forensic expert may find himself as the named expert in a case and yet may be opposed to the direction in which the case is headed.

The entity identified by the forensic expert as the one (or one of those) responsible for the loss may be or become immune to suit or prove to have no significant assets and the plaintiff's attorney, following the "deep pocket" of "shotgun" approach, may decide to sue someone else. In such a case, the forensic expert may have to testify against a potential defendant whom he feels is not responsible. Since he probably has a moral obligation to the plaintiff not to refuse to testify on the plaintiff's behalf, the forensic expert can testify to the facts he is aware of, drawing only those conclusions which, in his mind, are justified.

On one occasion, we were involved in a case wherein the driver of a piece of off-the-road machinery was injured to the extent that he became a quadriplegic. Our investigation indicated that the manufacturer of the machinery had not provided a safe internal environment for the operator and was, therefore, responsible for the injury. Unfortunately, due to some legal complications (and medical complications of the plaintiff), the case against the manufacturer was settled for a very low figure and, at a later date, the case was heard against the dealer who sold the machine. Since we believed the dealer's role was minor compared to the manufacturer's we were not in agreement with his bearing the brunt of the potential award; but, because of our previous involvement, we testified to the facts as we understood them, and the jury found against the dealer; a result we could not agree with but had to accept.

Another area of potential misunderstanding between attorney and associated expert is cooperation with experts representing the opposite side in the suit. Strangely enough, attorneys seem to be able to work together, although on opposing sides, without fear of conflict of interest, yet when one expert cooperates or agrees with another as to various phases of an investigation, the associated attorney generally seems to fear collusion is taking place. This stems from the attorney's bias toward always presenting his client in the best possible light, and in his not understanding the proper role of the expert.

Since the expert's role is supposed to be to seek and reveal the truth, two opposing, ethical experts, working together in a joint investigation, can become a unique fact-finding team. Their varied backgrounds will provide slightly different approaches to the problems encountered. The conclusions which they draw from the investigation will be kept confidential and, again because of their different points of view, may vary considerably. However, such a team will generally agree on the facts, if not the conclusions.

ON TECHNICAL IMPORTANCE

In a case being prepared for trial, there is no question that the attorney is the main figure in the event. He determines who the witnesses will be, in which order they will testify, what the scope of their testimony will be, is the most visible member of the team, and coordinates the entire presentation of the case.

However, the individual who set the potential suit in motion is the forensic expert who conducted the original investigation and wrote first concluded that a potential suit existed.

While it appears that the laws of evidence are more rigid and complex than the laws of nature thus far discoveredd, participants who can interpret each set of laws are equally necessary to bring the trial to a successful conclusion. Thus, both the expert and the attorney have equally important roles to play in the courtroom.

PREPARING THE CASE

INITIAL INVESTIGATION – As noted earlier, most product liability cases depend on a forensic expert to interpret a set of physical facts and abstract concepts to determine that a justifiable suit can be initiated against an identified potential defendant. While this decision may be made shortly after the loss occurs with the evidence fresh; it is frequently made with much of the overall piecture still missing. Until a reason for pursuing such a case has been recognized, there is little cause to expend the time and monies needed to complete the discovery program.

Thus, the forensic expert makes his first major decision, the one which starts the wheels of justice grinding, with only a portion of the knowledge available regarding the loss situation.

This is a situation which is not recognized by many attorneys and insurance company employees. We are not condoning the forensic expert's guessing what the facts are; we are recognizing that initial decisions must frequently be made with many of the pertinent facts missing. In most cases, if the forensic expert has been able to make a reasonable investigation, the missing facts, when known, will not change the overall conclusions, but may, at most, only modify the expert's conclusions.

The facts which generally are the ones most likely to be unknown at the time of the plaintiff's expert's investigation are exactly how the injured party will describe the events leading up to the loss, and what the detailed history is of the objects involved in the loss.

If the forensic expert attributes an automobile accident to a defect existing in the vehicle at the time of the accident, he may not be able to determine, at that time, whether the defect is a factory defect, a dealer-induced defect, or a defect arising from work done on the vehicle by someone else. His conclusion may be that the car was defective, probably attributable to one or more of the entities in the distribution chain. Discovery may later allow him to pinpoint the responsible party and eliminate the others.

Sometimes, the forensic expert's investigation points to there being but one logical way for the particular accident to have happened. He draws his conclusion accordingly. At a later date, the plaintiff (who may have been under doctor's care while the expert was preparing his report) may claim that the accident happened in a different way. The difference may be minor, or it may have a considerable bearing on the case.

In one instance, an operator was injured by a pair of nip rolls located deep within a textile machine. The evidence available to the expert indicated that he was trying to thread the machine at the time of his injury, in which case, the manufacturer of the machine would have been responsible for the injury since he should have provided some safe way to thread the machine, a foreseeable event. However, the plaintiff insisted that he had slipped and fallen, and that, in trying to catch his balance, his hand went into the machine through a tortuous path to the nip rolls. Under those circumstances, the manufacturer was probably not responsible.

INTERROGATORIES & DEPOSITIONS — The next opportunity to develop technical data that may be of interest to the forensic expert is when interrogatories are formulated, and when answers to interrogatories are written. Posing questions for interrogatories offers an opportunity to ask specific technical questions, and since frequently, there are technical areas where the known data is not comprehensive, the forensic expert's assistance should be sought. Typical data needed is the meaning of model and serial numbers on equipment, a copy of the manufacturer's operating instructions for the equipment in question to learn more about how the manufacturer regarded his machine, and drawings or other technical data to define the probable condition of the product "as-sold."

When a good working relationship exists between expert and attorney there are 2 time tested ways of utilizing the available technical talent:

One approach is for the attorney to request the expert to make a list of questions covering the technical (and other) data he (the expert) feels is desirable to advance his portion of the case. After receipt of the list from the expert the attorney may edit and rephrase the questions he deems appropriate and have the expert review it to ascertain that the meaning of the questions has not been changed.

Another approach is for the attorney to formulate the entire set of questions and then ask the expert to review the list, commenting on whatever questions he believes may require modification.

The second approach will probably entail less of the expert's time, but the first approach may result in a greater diversity of questions being asked as the expert's imagination is given freer rein than in the second case.

The assistance and advice of the forensic expert should also be enlisted in formulating the answers to technical questions in answers to interrogatories. Again, either of the 2 methods previously explained may be used with good results.

The approach that should <u>not</u> be used is for the attorney, using the forensic expert's report, to formulate the answers to questions which request an advance description of what the expert is expected to testify to. There is the definite possibility that a misunderstanding of the expert's total position will result in the expert being placed in a very embarrassing situation.

TECHNICAL FALLOUT — As the case develops, discovery will add to the store of knowledge about the loss and its associated facts and events. Depositions of the plaintiff (who is frequently unavailable to the forensic expert during his investigation), of witnesses, and of opposing experts will be taken in which a vast amount of data will be revealed, some of which will probably be contradictory to the position being held by either or both forensic experts.

Such information should be conveyed to the associated (on the same side) forensic expert as soon as it becomes available. There are occasions when the plaintiff's view of the loss is quite different to that which witnesses conveyed to the plaintiff's expert. In most cases, the differences are minor and many times help the expert to understand more clearly how the accident or other loss occurred. In some cases, however, the plaintiff describes a totally impossible situation, one that is at odds with the experts findings. Immediate resolution of such differences is necessary if the case is continue to be viable.

Another situation that sometimes develops is when the opposing expert or the plaintiff changes his view as to how the loss occurred or why the loss occurred. Such a shift in emphasis or direction may require a complete reconsideration of the position previously taken by the associated forensic expert, so any such change should be brought to his attention as soon as possible. Such information should not be kept a secret by the associated attorney, to be passed on to his expert if and when he (the attorney) deems appropriate.

Many times the in-house expert has not been involved in litigation before. He may have no knowledge of what to expect if and when he testifies. He may even be concerned that the associated forensic expert, without the background in the product that he (the in-house expert) has, cannot understand the intricacies and nuances of the product and will be unable to provide adequate testimony. The associated attorney has to define the scope of each expert's testimony, their order of appearance, and coordinate their probably significant areas. It is essential that this be done and the information disseminated well before the trial takes place.

<u>Preparation for Trial</u> — At some point a tentative trial date is set, a fact that should be conveyed immediately to all concerned. While the trial will probably be postponed again and again, it is much better for all to be aware of the posted date and of any potential conflicts than to be taken unaware with unchangeable other commitments.

If the associated attorney has done a proper job of keeping the forensic and in-house experts apprised of the developing situation, then no surprises should occur as the final stage of preparation for trial takes place. All associated experts should have been provided with copies of all of the depositions taken, of the interrogatories and answers to interrogatories on nical data which has been discovered. The 3-fold object of such a policy is:

First, for the associated attorney to receive advice from the associated experts as needed as the data is received;

Second, for the associated experts to be kept aware of the overall makeup of the case; and

Third, for all to be aware that the team effort is continuing.

THE TRIAL

There appear to be 2 extreme ways of preparing for trial. In one method, the experts are qualified by the associated attorney by arranging to have them operate the particular equipment in question so that they are specifically familiar with the item, then kept abreast of developments in the case, and are part of the final review and strategy meeting before the trial begins. Since rules of evidence differ from state to state and even from court to court within a state, it is important that the experts understand just what is applicable and what is not admissible. No expert should ever be allowed to be tricked into talking about his insurance company clients!

In the second method, the associated expert is notified of the trial a day or so before the trial begins and the review meeting takes place in the courthouse the morning the trial begins, or the morning of the day on which the expert is called to testify.

Needless to say, the former approach is far more likely to produce positive results than in the latter.

One refreshing example of an attorney's preparing his expert for trial involved a meat grinder case where the plaintiff's attorney insisted that his expert use the grinder, grinding meat in it so that he could testify to firsthand knowledge of the process.

In most cases, there are unanticipated questions which arise as the discovery of the case unfolds. Some of the questions may not be answerable, practically; but many problems can be simple, if they are known about in sufficient time and when there are adequate facilities available with which to work on the problems. Even the simplest question may be unanswerable at the courthouse on the morning of the trial.

Many attorney's consider that, other than for his testimony, an expert's work is complete once the trial starts. For some attorneys, this is true. But many attorneys will continue to make use of the expert's training and expertise throughout the trial.

The associated experts should be present in court, as a minimum, when the plaintiff testifies and when the opposing expert(s) testifies, regardless of whether the subject expert has been engaged by the plaintiff or by the defense. It is probably easier to understand why an expert should be present when his counterpart (and associate(s)) testifies than it is to understand the need to be present when the plaintiff testifies. However, the plaintiff's view of the loss on the day of his testimony is frequently at odds with his earlier concept of it and the additions or omissions he makes may have quite a bearing on the testimony the expert will give. Also, the cross-examination of the plaintiff may provide further insight into how the facts are to be represented. Much of this cannot be passed along to the expert secondhand. His presence is essential for complete coverage and understanding.

With the expert present there is also an opportunity for the associated attorney to obtain immediate input from the expert as to the meaning of certain facts related by the plaintiff (or opposing expert) and in the form of suggestions as to additional questions to ask.

Conversely, there is generally little reason to have an expert present during the presentation of medical and economic testimony. These are rarely areas of interest to the expert, nor is it likely that he will be able either to be better prepared by having heard the testimony or to offer significant suggestions regarding the testimony.

Another benefit which may result from the forensic expert's hearing all of the pertinent testimony is that he also has the opportunity to hear how it was introduced and, since he has more freedom of expression than do most witnesses, he has the opportunity, under cross-examination, to correct or re-dress statements of supposed truth made to him by the opposing attorney.

In a recent trial involving a plaintiff driving under the influence on the wrong side of the road, the plaintiff's attorney, cross-examining the nervous, non-English-speaking defendant driver, badgered the driver into accepting his (the attorney's) suggested values for a number of possible answers relating to the moments before and of the accident. When he later tried to introduce those answers as evidence during his cross-examination of the defendant's expert (who had been present during the entire trial), the expert was able to point out the manner in which the answers were obtained, and that they were not truly the defendant's answers. Thus, the fact that those answers conflicted with the expert's testimony was not significant. This could not have been achieved had the expert not been present during the entire trial.

Before the trial begins, the associated attorney should determine, as closely as possible, when he wishes the expert to be present; possibly, where he wishes the expert to sit; and when he wishes the expert to leave the courtroom (immediately after his testimony, after the other expert's testimony, etc.). It is also advisable to explain to the expert the reason for such decisions.

AFTER THE TRIAL - While it seems almost axiomatic that an attorney will inform his associated expert(s) as to the outcome of the trail, many attorneys fail to follow through with that which is at least common courtesy, and which may have additional importance if the trial has been declared a mistrial, as sometimes happens. It of the twists and turns of the trial and to critique his performance with an eye to improvement.

CONSIDERATION AND RELATIONSHIP

We have already addressed the difference in the overall outlook between attorneys and experts as they relate to duty toward the client and to truth. There generally is also another difference of great importance—how they are to be recompensed for their labor.

Since a forensic expert is expected to testify to the truth of the matter and not to only that which best suits his client, it follows that he must be paid for his efforts, as each stage is completed, without regard to the subsequent events of the case, i.e., its success or failure. Were an expert to accept a contingency fee arrangement in a case, his objectivity would immediately become questionable. Therefore, he should operate on a pay-as-you-go plan.

When a potential plaintiff's attorney asks a forensic expert to make an initial investigation in a case, the attorney should be prepared to meet the expert's request for payment in advance of conducting the investigation. If a case does not warrant either the plaintiff or his attorney advancing the requested funds, it is probable that the merits of the case are questionable. If the expert is required to wait to be paid until after his initial report is submitted, or, even worse, until the case comes to trial, the expert has little reason to believe that he will be paid at all if he files a negative report, and, again, his objectivity is placed in jeopardy.

Since, as noted, an expert cannot work on a contingency basis, it is also unfair to ask him to be an unwilling financial backer of a case, as some attorneys are wont to do. The expert expects to be paid only on the basis of his billing rate and time spent, and not as a percentage of the award in the case as the attorney is frequently paid. While it is true that a positive initial report can probably assure an expert of additional work on the case and, possibly, testifying at a trial; he probably also has other clients who need his services during the same period.

There seems to be a belief among some attorneys that experts like to prolong their stays in court because of the generally higher billing rates charged during trials and because much of the time spent is not physically or mentally taxing. Actually, since most experts are highly self-motivated, the enforced inactivity of dallying in the courthouse is distasteful and debilitating. It also creates problems trying to schedule work with other clients since a trial, once started, generally takes precedence over other work. Trial work, for many experts, is a necessary evil. To testify and then face the opposing attorney's concerted effort to discredit one and one's testimony is not a pleasant way to pass the time.

It should be noted also, that the scheduling of the expert's time in court is generally determined by the attorney and not by the expert.

SUMMATION AND CONCLUSIONS

The relationship between attorneys and forensic experts working together on cases in the field of product liability is frequently poor, with the forensic expert's talents being underutilized. In most such cases, the client suffers as a result.

In general, the forensic expert and the trial attorney have totally diverse fields of inclination, different educational backgrounds, distinctive experience and training, dissimilar goals regarding the client's distress, and contradistinctive methods of recompense. However, to pursue a law suit to its successful conclusion, these 2 entities must work together as part of a team.

Since the attorney is unquestionably in charge, he is responsible for coordinating all effort and for making the best use of the particular talents of the members of his team.

As we have tried to demonstrate in this paper, based on hundreds of personal experiences, the ideal is rarely reached and the inadequate is too often encountered.

In the forensic expert and in the in-house expert, the associated attorneys have a resource of technical expertise that is available and usable. A forensic expert probably wrote the report that concluded that a suit appeared feasible. The understanding, by both the plaintiff and the defendant in the suit, of the product or service involved and of its ramifications, was probably brought about through the experts' (forensic and in-house) efforts.

The definition of the data needed to clarify both the plaintiff's and the defendant's positions was probably provided by the experts. Some of the questions posed in both the interrogatories and the various depositions taken were probably posed by the experts also. And the experts will probably be called upon to explain the differences in opinion between the plaintiff's and the defendant's experts and to suggest which differences are significant and which are probably meaningless.

When the case goes to trial, the experts will be called upon to explain to the jury and to the judge; and to face cross-examination by a hostile attorney.

Since so much of the case (both in preparation and presentation) is dependent upon the talents of the expert(s), it is essential that they be treated always as vital members of the litigation team. Such utilization, cooperation, and coordination, when achieved, will generally work to the benefit of all involved: client, attorney, and expert.

INFLUENCE OF FAILURE ANALYSES
ON MATERIALS TECHNOLOGY
AND DESIGN

Ubbo Gramberg
Mobay Chemical Corporation
Pittsburgh, Pennsylvania, USA

Abstract

Failure analysis is a most important field of materials engineering since the deterioration of components subjected to specific working conditions provides concise information about application limits of materials. Primarily, failure investigations are conducted to find the cause of specific damages and to derive measures for failure prevention. In addition, the results of failure investigations have initiated the development of advanced materials and have stimulated improved design of equipment. Furthermore, advanced test methods have also been developed for evaluating and predicting the expected performance of a material, based on failure analysis information.

DESIGN AND CONSTRUCTION of equipment for chemical plants involve many experts from different technical fields. If process-related demands such as environment, flow rate, and heat transfer are known, the basic design normally follows conventional lines in the system service conditions - construction - materials. The selection of materials and their processing is just one part of the task of design and construction.

Experience shows, that this task can be decisive for the availability of the equipment.

Service conditions and construction methods form the base of materials selection. Laboratory testing of materials performance can play a vital role, but frequently the choice is based to a large degree on experiences with materials under similar service conditions. Failure analysis of equipment damaged during service, therefore, is an important approach getting data about materials behavior if subjected to complex working conditions. Failures are usually unexpected; nevertheless, they occur and if analysed can provide excellent information regarding the recognition of types of hazards. In addition, they offer options to avoid these hazardous conditions and thus prevent in the future similar kinds of failure in the equipment.

Since in chemical plants the superimposition of mechanical straining and corrosive environments can lead to a materials response often not predictable, only the analysis of failures provides the necessary information about materials performance within the equipment.

This is valid in particular in cases where local corrosion causes a deterioration of materials. This paper will show that the results of failure investigations are not only important for detection and prevention of specific failures, but that they can provide the basis for improvements in materials and equipment design. In addition, there is a considerable influence on test methods.

MATERIALS DEVELOPMENT

Chemical equipment can be subjected to quite complex service conditions, including mechanical straining, elevated temperatures, aggressive environments, and velocity-enhanced types of attack such as erosion or cavitation. Therefore, materials development has been quite active in this field trying to meet the ever-growing process-side demands. New chemical processes often call for improved materials, since available materials may be strained to their limits or even beyond, thus causing failures of equipment with sometimes very serious consequences. Thus, the improvement and development of materials is a continuing process in chemical engineering, mainly based on failures in the laboratory and in service.

The following examples, therefore, are typical for the processing industries. First the initial problem is outlined, then the materials solution is shown:

A well known initial problem with austenitic stainless steel grades was their tendency to fail due to intergranular attack along heat affected zones in weldments, fig. 1. Investigations revealed that this deterioration was caused by a precipitation of chromium carbides at grain boundaries and resulting in chromium depletion in zones close to the grain boundaries. The local decrease of chromium led to an impaired corrosion resistance and thus susceptibility to local attack in specific environments. The solution to this problem was attained in two different ways. First, through the addition of so-called stabilizing elements such as titanium and niobium (columbium) forming carbides with higher bonding energies than chromium carbides. This prevents the precipitation of chromium carbides during heat input. The carbon content and consequently the stability of the austenitic structure are not affected. Typical modifications are the Ti-grades of the 300 series stainless steels.

The other method of improvement was a considerable reduction of the carbon content in order to keep the remaining carbon in solid solution even after heat input. The L-grades of the 300 SS are examples of this development. But since carbon contributes to the stability of the austenitic structure suitable measures were required to maintain the single-phase structure. This was provided by an increase in the nickel content, or alloying with nitrogen or a combination of both. High pressure metallurgy allows to dissolve up to 0.25% nitrogen in austenitic SS (LN-grades of 300 series) which brought about two important advantages. First, nitrogen increased the yield strength without impairing the

ductility. Second, the increased stability of the structure made an addition of elevated molybdenum contents possible. Higher Mo-alloyed SS performed considerably better in pitting initiating environments, as shown in fig. 2 (a comparison of the current density - potential curves of various steel grades in seawater at ambient temperature). Basically, these grades can be considered the origin of high-alloy extra low carbon, nitrogen bearing steels which have made many inroads into chemical equipment materials engineering with regard to an improved resistance against pitting and crevice corrosion.

A basically similar development took place for Ni-Mo-Cr and Ni-Mo alloys (Hastelloy C and Hastelloy B grades). These alloys are used to handle highly corrosive environments. However, it was found opposite to stainless steels, that intermetallic compounds of complex composition were responsible for a sensitivity of these materials to intergranular attack. Therefore, an expensive research program had to be conducted to explain the influence of various alloying elements and to provide the base for the development of grades not sensitive to a heat-input during processing.

Another development of materials had to be initiated to meet the product side demands of the ammonia synthesis. High-pressure, high-temperature, hydrogen containing environments caused a reaction between hydrogen and the iron-carbide phase in the carbon steels, used in the beginning for the construction of reactors. As a result of this reaction, methane was formed in the microstructure thus decarburizing the steel. Consequently the strength level of the material decreased. Further stages of the failure process were characterized by the effect of methane trapped in the microstructure: Blistering of areas close to the surface, disbonding of grain boundaries, and finally intergranular cracking. The failures had devastating consequences.

Bosch, the inventor of the ammonia synthesis, found a genial solution for the problem. A product-size Armco-iron shell was reinforced by a medium-strength constructional steel having tightly-spaced blow holes to allow the diffusing hydrogen to pass through without causing damage. Due to the high costs of manufacture, this solution was only interesting as long as no less expensive solution based on a suitable material was available. A review of the experiences gained during the first years and the results of additional laboratory testwork showed an important influence by alloying elements such as chromium and molybdenum. The activation of reactions between hydrogen and chromium or molybdenum carbides requires higher pressures and temperatures if compared to iron carbides due to their higher stability. The results were compiled by G. A. Nelson in his highly informative diagram, fig. 3. This diagram explains the deterioration of carbon steels. Plain carbon steels can only be used in contact with hydrogen containing environments at partial hydrogen pressures below 50 bar (700 psig) at temperatures below 220°C (430°F). This is an important criteria for the safe performance of low-alloy Cr-Mo steels in boilers and superheaters since the reaction between superheated steam and iron can cause the formation of hydrogen in these high-pressure units.

A last example for the impact of failures on materials development is provided by the handling of hot concentrated nitric acid which is used for various important processes in the chemical industry.

Initially pure aluminum was used for manufacture of such equipment. Due to its low strength properties, equipment design required reinforcement with aluminum alloys. Welding of aluminum caused continuing problems due to a selective attack on the heat affected zones as well as on the weld metal in service. Various types of stainless steel were also used, but the conventional grades, austenitic as well as superferritic ones, show a steep increase of attack above concentrations of 70 to 90%. Based on the experiences with high-silicon cast iron, the development of an austenitic steel with high amounts of silicon was initiated. Its resistance in boiling nitric acid of concentrations above 60% is shown in fig. 4. In contrast to other SS grades, the resistance increases above concentrations of 80% nitric acid due to the formation of a stable silicon oxide surface layer. This grade of stainless steel containing 4.5% silicon has shown excellent performance in contact with concentrated nitric acids for some years now. However, a drawback has to be mentioned. Similar to molybdenum, silicon also has a tendency to segregate and to form carbides as well as intermetallic phase precipitates. Segregation leads to selective attack parallel to rolled surfaces, precipitates formed during welding or insufficient heat treatment result in

intergranular corrosion. Therefore, the metallurgy as well as the welding of this steel requires high skill and considerable experience, making this steel an expensive solution for welded constructions.

DESIGN

Though hardly ever recognized, experiences derived from failures are the main base for design improvements. The failures do not have to occur in service; many of them are observed in a preliminary stage during laboratory investigations. A good example is the optimization of welded constructions which are subjected to severe fatigue straining during service. The design calculation of such equipment requires considerable effort related to the application of modern techniques. But even methods such as Finite Element Analysis Calculations are not able to take into account residual stresses of unknown magnitude and orientation or the impact of weld defects. Therefore, testing and failure evaluation is necessary in many technical fields.

A simple example will demonstrate the utilization of failures for design and manufacturing. Due to a rigid construction, a bypass used to take samples from a pipeline failed frequently caused by fatigue fractures initiated at the weldment between bypass and pipe, fig. 5. To avoid cracking two separate measures were taken. To keep the operating stress level low a more flexible construction as well as an increased wall thickness of the bypass-line adjacent to the pipe was chosen. In addition, better penetration of the weld was suggested in order to avoid internal notches acting as stress risers. Of course no further failure was observed in the system after incorporating the design improvements.

A more sophisticated example of a similar problem and its quantitative evaluation is shown on fig. 6. Welded type A manifolds in a pipe system subjected to varying internal pressures failed after low numbers of pressure changes. Shut downs, cleaning, and repair operations caused considerable maintenance costs. Therefore an investigation of slightly different designs was conducted in the laboratory. The test system was pressurized cyclically between 7 and 140 bar (100-2000 psi), the number of cycles until fatigue crack initiation is shown for three samples each. Failure evaluation of type A provided important ways for improvement. Type C shows, in comparison with type A, that small variations in design and manufacture can lead to a considerable improvement in the performance. The increase of the initial numbers of cycles necessary for fatigue crack initiation to values four times higher was sufficient to provide an economical and reliable performing system.

TEST METHODS

It has to be expected that failure evaluations will have a considerable impact on test methods. Only if the failure cause and the failure type are known, can similar future breakdowns be avoided. To prevent on-site damage, the materials response to specific service conditions has to be tested in the laboratory first. Since total simulation is not possible for each single system, some general lines have to be developed for appropriate testing in order to get as much specific information as possible.

Many examples are available illustrating the important contributions of failures to the development of materials testing. In contrast to former times with the more empirical approach such as boiling tests in corrosion engineering, or impact testing of small notched specimens, modern techniques try to evaluate the materials response on a more sophisticated basis. The failure type is investigated, its mechanism is evaluated and then transferred to the test situation on a scientific and reproducible basis.

Pitting corrosion is a common failure mechanism in the chemical industry, see fig. 7. In order to fight pitting, the fundamental interrelations between damaged materials and the environment had to be investigated. This led to the introduction of electrochemical testing into corrosion engineering, since corrosion in aqueous solutions always can be explained by electrochemical reactions. It was found that pitting is correlated to a specific electropotential influenced by the environment-metal system.

Developed to explain failure mechanisms this test method can now be utilized for the development of materials with increased resistance against pitting. Experience is now gained in the laboratory and not under on-site conditions. There is always a considerable interaction between failure cases, advanced test methods, and modification or development of materials.

The previously mentioned sensitization phenomena are another important example for a failure-originated improvement of testing. Appropriate environments had to be developed to identify a material's tendency to this type of failure in order to form a base for materials evaluation and improvement in the laboratory. Appropriate test methods are standardized in ASTM A262, A708, G28, and G67 for example.

The mechanical deterioration of complex designed constructions was considered an important problem as soon as catastrophic failures occurred with severe consequences. The Fracture Mechanics Approach as a quantitative explanation of the mechanical performance of equipment design containing inevitable defects can be related directly to failures of units such as ships, bridges, and pressure vessels. In contrast to the aforementioned corrosion test, fracture mechanics is not used for everyday problems due to a considerable amount of calculations and testing required. The fracture mechanics approach can be applied for insights into failures, for materials development, and to define requirements regarding location and indentification of stress rising defects in materials.

DATA STORAGE

The experience gained during failure investigations should be available to all concerned experts. Thus, it is necessary to store the information in a data base. Optimal applicability of such a system requires a suitable data key system as well as simple storage and retrieval operations.The best system for processing more than approx. 100 failure investigations per year is the computerized data storage. Fig. 8 shows an extract from a coding system used in chemical engineering.

SUMMARY

Failures of technical equipment will continue to occur in spite of the cooperation of experts in the fields of design, fabrication, supervision, and materials engineering. Since the deterioration of equipment components yields important facts about application limits of materials and their processing, especially in systems with complex conditions, failures should be investigated regarding their cause and additional information about factors contributing to the initiation and propagation of the damage. Proper

preventive measures can only be utilized if
the cause of failure is known. But apart
from the importance of failure
investigations for the specific cause,
results can stimulate an improvement of
design and initiate materials developments
with the aim to meet service side demands.
Materials testing also benefits from the
knowledge gained during failure analyses in
order to shift an investigation of the
materials performance from plant to the
laboratory.

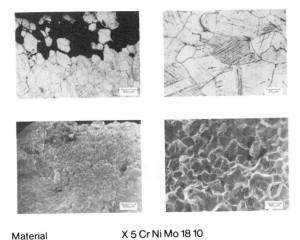

Material   X 5 Cr Ni Mo 18 10
Life time   14 days
Conditions   electrostatic gas cleaning

Fig. 1 - Intergranular Corrosion of Electrode Wires Sensitized due to an
         Improper Heat Treatment

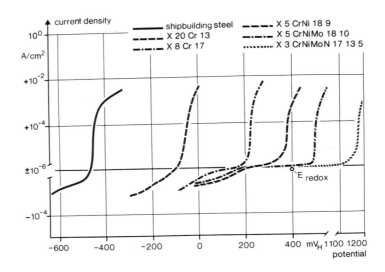

Fig. 2 - Current Density - Potential Curves of Various Steels in Seawater
at Ambient Temperature

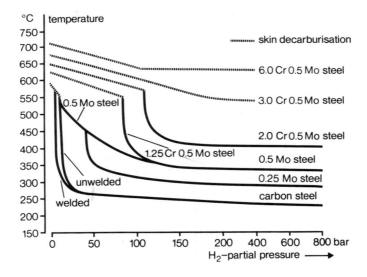

Fig. 3 - Service Limits for Carbon Steel and Low-Alloy Steel in Hydrogen
Containing Environments According to G. A. Nelson

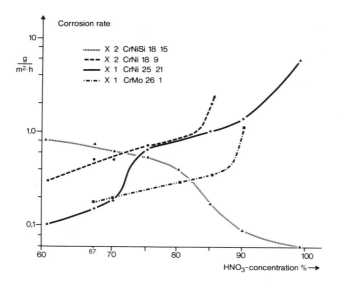

Fig. 4 – Corrosion Rates of Some Stainless Steels in Boiling
Nitric Acid

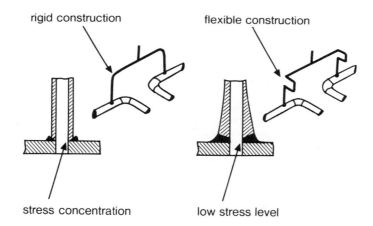

Fig. 5 – Measures Derived from the Failure of a Sampling Bypass Line

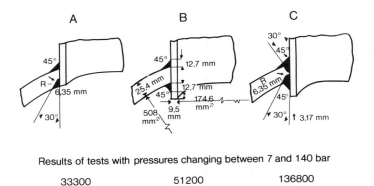

Results of tests with pressures changing between 7 and 140 bar

33300	51200	136800
32800	54800	180700
40800	114000	203100

load cycles until frature

cit. Lane

Fig. 6 – Influence of Design and Weld Geometry on the Fatigue Performance
of Manifolds

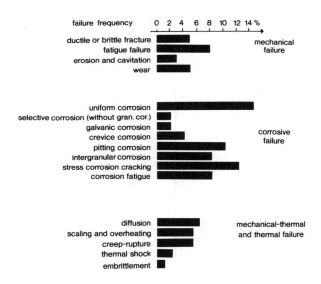

Fig. 7 – Statistical Evaluation of Failure Causes in Chemical Engineering

1. investigation purpose	1.1 failure analysis 1.2 materials selection 1.3 .... ⋮	7. pressure during service or testing	7.1 less than 1 bar 7.2 1 bar 7.3 .... ⋮
2. types and sources of failure	2.1 ductile fracture 2.2 brittle fracture 2.3 .... ⋮	8. period of service or testing	8.1 less than 1 h 8.2 1 to 10 h 8.3 .... ⋮
3. materials	3.1 structural steel 3.2 steel for tubes 3.3 .... ⋮	9. form of products, parts of construction	9.1 strip, plates 9.2 tubes, semi tubes 9.3 .... ⋮
4. protection against corrosion and wear	4.1 cathodic protection 4.2 anodic protection 4.3 .... ⋮	10. chemical engineering	10.1 sublimation 10.2 crystallization 10.3 .... ⋮
5. corrosive medium	5.1 atmosphere 5.2 oxygen 5.3 .... ⋮	11. conducted tests	11.1 strain measurements 11.2 tensile test 11.3 .... ⋮
6. temperature during service or testing	6.1 below −120°C 6.2 −120 to +10°C 6.3 .... ⋮	12. commissioner	12.1 department X 12.2 department Y 12.3 .... ⋮

Fig. 8 – Coding System for Failure Analysis Data Storage in Chemical Engineering

METHODOLOGY FOR ANALYSIS OF GUN TUBE FAILURES CAUSED BY AN IN-BORE EXPLOSION OF A PROJECTILE

James G. Faller
U.S. Army Combat Systems Test Activity
Aberdeen Proving Ground, Maryland, USA

ABSTRACT

A methodology for analyzing an accidental projectile explosion in a large caliber gun tube is presented along with two examples where it was applied at Aberdeen Proving Ground (APG). These involved an accidental explosion of a 152-mm gun and a deliberately induced explosion of a 105-mm gun. In the latter case, the results of a metallurgical examination were compared with those generated by simulation of the explosion using a 2D hydrodynamic computer code. The computer flexibility was exploited by modeling a variety of conditions for two sizes of projectiles placed in and partly out of the gun tube at the time of the explosion and by initiating the detonation at either the base or nose of the projectile. In these results the computer was shown to provide a limited insight into the extent and severity of damage of the gun tube and to serve as a useful complementary tool to metallurgical techniques.

ALTHOUGH IT OCCURS rarely in the firing of ammunition and weapons, it does on occasion happen that a projectile containing high explosives will detonate before emerging from the gun tube. Referred to as a projectile premature, the explosion usually destroys the gun and releases blast and fragments which may injure on kill personnel manning the weapon. In test facilities such as APG personnel are not endangered, since experimental weapons and ammunition are always fired from behind barricades.

An investigation of the weapon premature has as its principal aim to determine "what caused the failure". The question can be further broken down into such component elements as "Was the tube or ammunition at fault?" or "Was it a high-order explosion (i.e., a true detonation) or a low-order explosion (i.e., a violent burning)?" In this connection it is necessary to determine the exact location in the gun tube at which the explosion occurred and where it initiated inside the explosive filler of the projectile. The "signature" of the exploding round is made of various scars and markings from fragments, shock pulses, and hot gases, with indentations in the tube and projectile caused by the geometry of the interacting parts.

Detailed metallurgical procedures have been developed at APG for the analysis of accidental prematures. In the past it has always been the practice to validate the causal hypothesis of failure by comparing an accidental premature with one that was deliberately induced. This step, imposed by the many uncertainties and frequent lack of good data from the field, has proven both time consuming and expensive. The development of hydrodynamic computer codes held out the promise of analyzing a premature failure at considerable savings by greatly reducing or eliminating the necessity for simulating the incident by field destruction of a full scale weapon.

The paper presents the results of a study to evaluate a typical hydrodynamic computer code by comparing the results generated by the program with those derived from the metallurgical examination of a deliberately induced premature. To gain a perspective on the controlled experiment an accidental gun tube explosion incident is also described.

METALLURGICAL TECHNIQUES

The following steps should be considered in the analysis of an in-bore projectile explosion:

1. Identify all pertinent firing conditions and data for deviations from normal procedures. Any available nondestructive test data on the tube and projectile should be examined.

2. Examine tube and projectile fragments and reconstruct both as completely as possible. An intact projectile placed along side the assembly is useful in matching markings on tube fragments caused by projectile body parts.

3. Trace chevron patterns on fracture surfaces to the point of origin. This helps to determine whether the tube or the ammunition caused the failure.

4. Look for the burnished circumferential area that locates the rotating band of the projectile and the nose fragment ring that locates the ogive of the projectile. Smears of copper (from the shaped-charge liner that forms into a jet) on portions of the steel barrel are usually downbore from HEAT projectile. Aluminum smears are left by fuze parts and tail and boom assemblies.

5. Observe the fracture mode in the detonation area. Heavy shear is usually prevalent, but the presence of a fracture in the brittle mode indicates a lower order of projectile functioning.

6. Note fragment size, both tube and projectile. Smaller sizes indicate higher orders. The symmetry of the pattern should be noted. A central initiation (such as from the fuze) will cause a symmetrical pattern.

7. Observe any spallation of the tube, indicative of high order functioning.

8. Examine tube fragments from the detonation area for evidence of transfer of toolmarks from the outside surface of the projectile and for indications of land flattening which is indicative of a high order functioning.

9. Examine projectile fragments for rifling engraving; higher orders of functioning cause more extensive engraving. Examine the inside surface of projectile fragments for pitting, caused by "jetting" of high explosive during high order functioning.

10. Examine all fracture surfaces for evidence of prior fatigue cracks. Such evidence, especially when accompanied by a nonsymmetrical breakup in a tube, may indicate the explosion resulted from the initial failure of the tube.

11. Mount microspecimens of tube from the detonation area and from a location distant from the detonation area for comparison of structure. Martensite streaks caused by adiabatic shear indicate intense shear loading but cannot be used to differentiate between high and low order detonations. The presence of an unusually large quantity of microvoids or microcracks is considered evidence of shock loading from a high order detonation. A "transformed" severely altered surface layer on both tube and shell fragments is another unique feature which cannot be used to distinguish between high and low order detonations.

12. Conduct microhardness surveys of tube fragments from both the detonation and nondetonation zones. Some investigators have reported that high order functions cause an increase in Rockwell C hardness of four points or more in detonation area fragments. Low order events cause a lesser increase. A similar treatment of shell fragments may indicate in a longitudinal survey the direction of propagation of the explosion, if these fragments can be assembled.

13. Check to see that material meets specification requirements through a chemical and mechanical property check of tube and projectile. In the case of an exploded projectile this may not be possible.

After collecting all the preceding information it will be necessary to arrive at the point of being able to deduce a) the exact location of the projectile at the time of detonation, b) the order of functioning, c) where detonation initiated (i.e., base or nose of projectile), d) description of the event, and e) the known or probable cause.

ACCIDENTAL PREMATURE IN 152-MM GUN TUBE

The failure occurred in the XM81E12 tube upon firing tube round No. 986, an XM409E4 HEAT round. Magnaglo and black-light inspections from the breech end had detected only minor cracking along the fillets of the missile guide slot and at the corners of the detent holes, upon examination after 914 rounds. This extent of cracking was considered to be normal for a tube of this type, with its firing history.

The macroexamination showed the tube to have exhibited a considerable amount of ductility prior to rupture. No evidence of prior fatigue cracking was noted on any fracture surface. Fracture rearward of the projectile function area was restrained by the recoil sleeve. Figures 1 and 2, Appendix A, clearly shows these effects and a symmetrical-shaped damage pattern. Little real land flattening was observed, with damage chiefly being caused by fragment impact. The most distinctive bore-signature feature was the nose fragment ring, which can be seen in Figure 3. Figure 4 is a view of a large tube fragment, showing the nose fragment ring and a fracture origin clearly shown at the black indicator arrow. At the rate of loading during this low-order functioning, the brittle mode was initiated at the primary rupture, rather than extending out of the detonation zone from a region of massive slant shear, as is usually the case.

Figure 5 shows a close-up of the nose fragment ring. A transverse microspecimen removed to identify the foreign material noted on the surface of the impacted depressions disclosed that it was wrought pearlite and ferrite of the type of steel used for the windshield of the round. This agrees with the hypothesis that the nose ring is caused by windshield fragments being knocked out at an angle by jet particles striking from the rear.

Mechanical property data showed that the tube met all the specification requirements. The energy V-notch transition temperature was about -20°F.

Considering all these factors, it was concluded that the failure did not result from any shortcoming in the tube and it was recommended that the projectile or fuze be suspect. It was further noted that it would be advantageous to examine both tube-and-projectile fragments together, in a better hope of arriving at a correct analysis.

DELIBERATE PREMATURE IN 105-MM GUN TUBE

TEST SETUP - The tube which was fired without a carriage was supported on heavy armor mounts and clamped at its center of gravity about 200 cm from the breech face. Several layers of 1.2 by 2.4 m (4 by 8 ft) sheets of Cellutex were placed in front of a concrete barricade behind the gun to absorb the recoil. Armor shielding was used over the gun to contain fragments.

The 105-mm M456A1 high explosive anti-tank (HEAT-T) projectile with M509A1 PIBD fuze was modified for prearming after chambering in the tube. The modification consisted of removing some of the normal fuze safeties and making the fuze arm when one pull wire (coming out of the front part of the projectile and through the muzzle) was removed. A wooden beam was placed in the center of the gun tube at the point related to the desired in-bore functioning. When the round was fired and the wooden obstruction was impacted by the nose of the projectile, an electric charge was generated in the piezoelectric crystal at the nose and applied to the detonator functioning the round.

RESULTS - Failure Description - When the tube was reassembled from the recovered fragments (fig. 6), it was divided into five zones for purposes of analysis (fig. 7). Zone 1 consisted of the intact muzzle end, 40 to 50 cm long and weighing 31.0 kg. Zone 2 consisted of six pieces, all of which were recovered and fitted together with the fragments of zone 1. Zone 2 extended from 40 to 75 cm from the muzzle surface. The fragments weighed a total of 12.2 kg with masses ranging from 0.73 kg to 3.0 kg. The herringbone "V's" of the fracture surfaces pointed toward the detonation zone indicating the cracks were growing toward the front of the tube at this location. The rear of the fragments were bent out radially, in a petalled manner.

Zones 3/4 contained the area of the detonation stretching from 70 cm to 95 cm. Zone 3 included small pieces which had the machine markings from the OD of the tube but included none of the ID. The 41 recovered pieces ranged in mass from 0.9 g to 71.2 g and totalled 794 g. Zone 4 included small spalled pieces which had tube ID but no tube OD. Twenty of these pieces were recovered ranging in mass from 5.7 g to 119.9 g and totalling 635.0 g. The remaining recovered fragments from zones 3/4 numbered 7 and contained both OD and ID surfaces. These ranged in mass from 44.2 g to 396 g and had a total mass of 1180 g. Zone 5, behind the area of detonation, was fragmented into three long petals and one short petal.

The ratio of fracture area to mass and the percentage of mass recovered for each zone are graphically illustrated in Figures 1 and 2, Appendix B. The bar graphs emphasize that the large number of individual fragments belonging to zones 3/4 constituted only about 18% of the total mass of that section of the gun tube.

A very light circumferential indentation band encircled the bore fragments between 85.3 cm and 87.1 cm from the muzzle. The indentation, shown in Figure 8, was the width of the obturator seal and was used to locate the projectile position at the instant of detonation. The tip of the spike when the projectile detonated was 50 cm from the muzzle face, 21 cm forward of the intended detonation point. A ring of indentations was also noted on fragments from zones 1 and 2, 44 to 50 cm from the muzzle face (fig. 9). The ring extended from 45° to 90° from the longitudinal axis at the spike tip.

The most severe land flattening occurred near the center of zone 4. Land flattening was observed visually throughout zone 4, all of the rear of zone 2, and at the front of zone 5.

Several fragments from zones 2 and 3/4 displayed silvery smears on the bore surface (fig. 9). The smears resembled aluminum metal, the probable source for which was the tail boom of the projectile. No copper smears were found in any of the fragments, implying that the detonation caused a normal jet which travelled out of the tube.

Metallurgy of Gun Tube – Metallographic examination of the tube was done on samples taken from the bore fragments of zones 2, 4, and 5 and from the origin of rifling. The most distinctive observation made in these metallurgical sections was that the surface of the rifling grooves had structurally transformed from the bulk material to depths of up to 0.08 mm. The structural change was revealed by the presence of two different layers, a so-called white layer (fig. 10),

which was porous and non-etching, and an untempered martensite layer, which etched lightly and tested to a hardness in the range of HRC 56-62. The white layer, identified by X-ray microbeam analysis as being of tube steel composition, was best developed in zone 4.

Spall fragments from zones 3/4 exhibited a second-type of structural transformation, which penetrated the bulk in a narrow line and etched white in contrast to the surrounding quenched and tempered structure (fig. 11). This product was identified as an adiabatic shear layer consisting of untempered martensite.

The fracture mechanism of herringbone (chevron) and spall fractures was in both cases microvoid coalescence or ductile dimple rupture.

Tube hardness on an average was HRC 41.4 and did not vary between the detonation zone and elsewhere in the tube. It was noted, however, that the standard deviation in measurements made in the detonation zone was six times greater than in areas removed (table 1, app C).

The mechanical properties of the tube were determined on fragments of zone 2 and on specimens machined from the breech area. All properties met the specification requirements.

Metallurgy of Projectile – The 34 fragments recovered from the steel section of the projectile (body and standoff spike) ranged in mass from 1.0 g to 49.9 g and totalled 217 g. The 37 fragments recovered from the aluminum fin and boom assembly ranged in mass from 0.5 g to 22.5 g and totalled 230 g. The recovered fragments represented 5% of the projectile weight. The Rockwell hardness of a steel fragment from the mid-body of the projectile was 7 points higher than the hardness of an unfired projectile (HRC 36.4 versus 29.4).

COMPUTER SIMULATION

DESCRIPTION OF CODE - The HEMP2D (ref 1) is a numerical code useful in solving a large class of problems which include those of projectile impact, high explosive detonations, and low energy stress strain. As input the user submits information which includes an abbreviated description of the part geometry from which a detailed computational mesh is generated, material parameters for the stress-strain relations and the equations of state, and appropriate boundary and initial conditions.

The program integrates in time the conservation equations for mass, momentum, and energy, to which are coupled equations of state for hydrostatic pressure and consitutive relations for stress deviations, velocity strains, total stress, and the von Mises yield criterion. Additionally, a failure criterion is incorporated. The conservation equations and the constitutive relations are solved by finite differences.

The default constitutive model in the code for the metals assumes an elastic perfectly plastic response (von Mises) and a criterion of failure based on the hydrostatic pressure. Failure occurs instantaneously when the hydrostatic pressure reaches a critical tensile value. Further expansion occurs at the fixed value of pressure. In the event of a recompression, compressive pressures are again allowed. This failure model has no basis in micromechanics and, therefore, has no possible way of bringing into coincidence results from experiment and computations.

PROBLEM - The 105-mm, M456A1 projectile detonated in the induced in-bore premature and a comparison projectile, the 120-mm, M356 with larger size and different explosive filler distribution, were treated in the study (figs. 12 and 13). Formal drawings were used to obtain the dimensions of gun tubes and projectiles. The starting HEMP2D computational mesh for these projectiles and gun tubes is shown in Figures 14 and 19. The projectiles were detonated in positions fully in-bore and partly out-bore at the muzzle. Their geometry is fully described by a half-section plot above the horizontal axis of symmetry. Some geometrical features of the projectiles were neglected including the spike, fin and boom assembly of the M456A1, and the rotating band, base plug, and nose plug of the M356. Experience has shown that these simplifications reduced computational difficulties and time without affecting the results. Parameters for the constitutive relations and the equations of state entered in the program are listed in tables 2 and 3 of Appendix C.

COMPUTATIONS - The changes taking place in the gun tube geometry from the start of the explosion to elapsed times varying between 25 μs and 55 μs for the M456A1 projectile and up to 60 μs for the M356 projectile are plotted in Figures 15-18 and 20-21. These changes in the gun tube are manifested by:

a. Expansion and deformation.

b. Zones of hydrostatic compression exceeding 100 kb of pressure.

c. Zones of hydrostatic tension equalling -15 kb of pressure.

The initial shock wave passing into the gun tube is compressive and the progress of the shock front may be followed by the distribution of bow tie symbols. Shock waves reflected from the gun tube wall become tensile in character and their progress is traced by the triangle symbols, which also represent failure in the material by the tensile pressure cutoff criterion.

The results bring out the following salient characteristics about an in-bore explosion:

a. Deformation, expansion and failure occur mainly to a section of the gun tube equal to the span of the explosive filler.

b. When the explosion starts at the base end of the projectile explosive the mechanical shock wave travels from the area of the gun tube just above the initiation site to that area of the gun tube just above the nose end of the projectile explosive; the opposite occurs when the explosion starts at the nose end of the projectile explosive. The mechanical shock wave in the gun tube follows the detonation wave in the explosive.

c. The gun tube and projectile expand radially under the force of the explosive.

d. The gun tube section around the explosive filler always starts to expand on the end closer to the explosion initiation site.

e. Differences in gun tube deformation between nose and base initiated explosions become less pronounced with elapsed time from the start of detonation.

f. There is ample energy in the explosive filler to cause massive damage to the gun tube.

g. The difference in gun tube deformation and failure signatures between base and nose initiated detonations is enhanced by a longer axial space and a more uniform distribution of explosive filler.

The compact geometry of the M456A1 projectile coupled with a nonuniform distribution of explosive filler caused the peak bulge of the gun tube to occur at about the same place for both base and nose initiated detonations (figs. 15-18). In contrast, the extended length and uniform distribution of explosive filler of the M356 projectile caused the peak to be separated by about a third of the projectile's length, holding out the real possibility that base and nose initiated detonations can be differentiated in a premature involving this projectile.

Since the HEMP2D allows for sliding but not for sliding with friction, sliding if allowed meant that no transfer of shear across the interface was possible. This was not considered a realistic boundary condition, and it was felt the neccessity of having shear transfer took precedence over having the projectile moving at the time of detonation. As it turned out, there was no noticeable difference between the various conditions of sliding; hence, the results presented were limited to the cases felt to be most correct.

Although the nose detonations for the M456A1 projectile are shown as point detonations in Figures 16 and 18, they are actually treated as ring detonations in the problem. The imposition of axial symmetry assures that this will happen for any point not placed on the axis of symmetry.

COMMENT - The principal difficulty with HEMP2D and other similar numerical codes arises from the inadequacy of the models in describing material failure (ref 2). Under conditions of dynamic loading, failure can occur by several mechanisms which depend on the type of material, state of stress, and rate of loading. Efforts to model failure in such a complex environment have not greatly advanced. Zukas (ref 2) projects that the problems of accounting for dynamic material fracture, though exceedingly difficult, are surmountable within the next ten years.

Finally, it should be stressed that HEMP2D calculations cannot be used to replace a metallurgical analysis and such was never intended to be the case. A metallurgical analysis of the failed parts from an actual premature must always be performed and will usually answer questions of shell location, order of function, and identification of tube or ammunition as the probable cause of failure. In some instances, the direction of explosion propagation can be determined. It is considered significant that HEMP2D calculations, and more importantly HEMP3D calculations which are not restricted by symmetry assumptions, can indicate for low order prematures (where destruction of the gun may not be severe) the origin of the explosion in the HE filler.

CONCLUSION

Since computer codes fail to accurately account for dynamic material behavior and fracture, results from these codes are best thought of as providing only complementary information to the metallurgical analysis of an in-bore premature.

ACKNOWLEDGEMENT

The failure analysis of in-bore prematures at APG owes its inspiration and direction to R. H. Huddleston, who retired as the Chief of the Physical Test Division. Many of the ideas used in this paper originated with him and M. J. Drabo, his capable assistant for many years.

REFERENCES

1. Giroux, E. D., HEMP User's Manual, Lawrence Livermore Laboratory, UCRL-51079, Revision 1 (1973).

2. Zukas, J. A., Nicholas T.; Swift, H. F.; Greszczuk, L. B.; and Curran, D. R., Impact Dynamics, John Wiley and Sons, New York 1982.

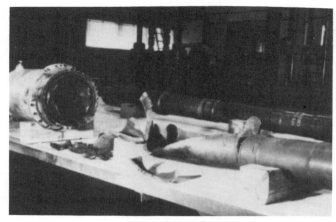

Fig. 1 - 152-mm M81 gun which failed by accidental functioning of XM409 projectile.

Fig. 4 - Extensively deformed fragment with a
failure origin of this segment of 152-mm gun.

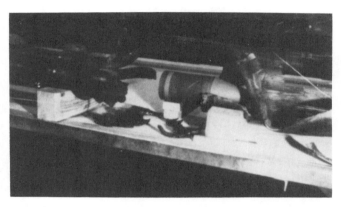

Fig. 2 - Suspected location of XM409 projectile
at instant of detonation. Recoil sleeve has
been removed.

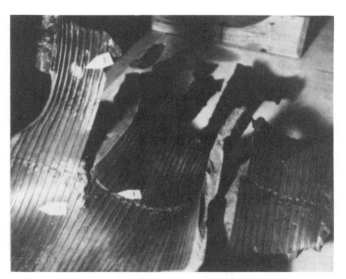

Fig. 3 - Area of thrust collar of 152-mm tube.
A circumferential ring of fragments (1), an
aluminum smudge (2), and an area lightly coated
with copper (3) are shown.

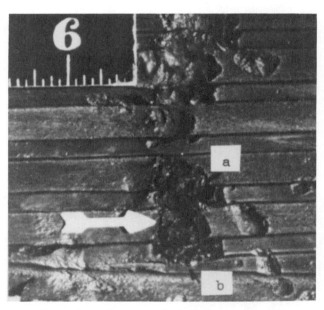

Fig. 5 - Circumferential fragment ring showing
location of two fragments (a and b) identified
as part of the projectile windshield.

App A

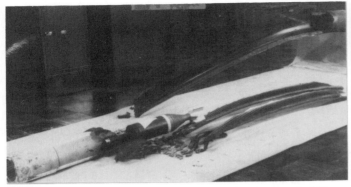

Fig. 6 - Reassembled 105-mm gun which was deliberately failed by detonating an M456A1 projectile in the bore.

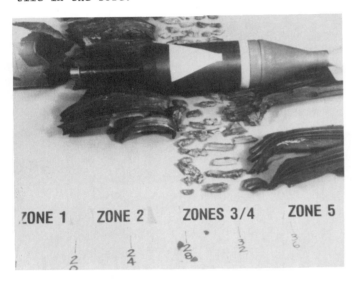

ZONE 1   ZONE 2   ZONES 3/4   ZONE 5

Fig. 7 - The five zones into which the failed 105-mm tube was divided. Location of round at instant of detonation is shown.

Fig. 8 - Fragment of zone 4, showing (a) cracks along prior tool marks on land, (b) craters in groove, (c) the indentation used to precisely locate the projectile at instant of detonation and (d) land flattening.

App A

Fig. 9 - Fragment of zone 2, showing projectile nose cap indentations and silvery smears from fin and boom assembly.   Magn. 0.6X

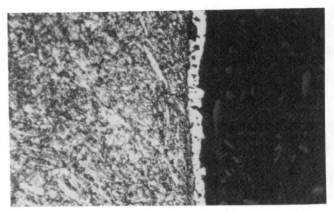

Fig. 10 - Structurally altered material, white layer, in groove of zone 4 fragment 33 in. from muzzle. Material overlays normal quenched and tempered structure of steel. 2% Nital; Magn. 310X

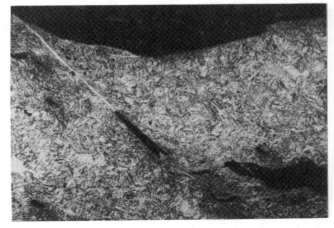

Fig. 11 - Adiabatic shear layer (white-etching strip) in spall fragment of zones 3/4. Picral; Magn. 160X

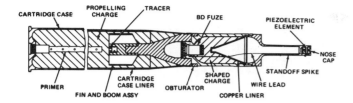

Fig. 12 - Sketch of 105-mm cartridge: HEAT-T M456 series.

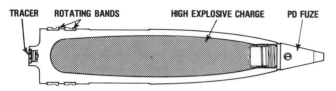

Sketch of 120-mm projectile modeled in HEMP2D calculations

Fig. 13 - Sketch of 120-mm projectile: HE-T M356 series.

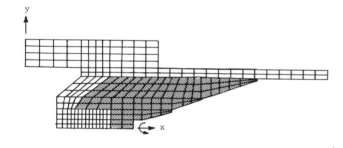

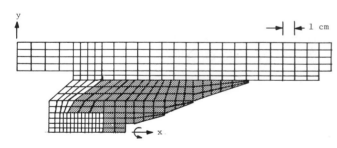

Fig. 14 - Computational meshes of 105-mm gun and M456A1 projectile for fully in-bore (bottom) and partly out-bore at muzzle (top). Space occupied by explosive filler is shaded.

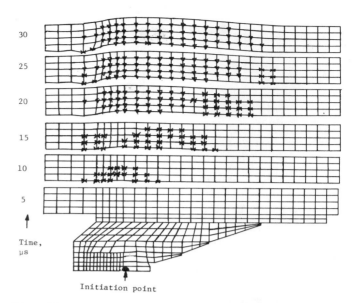

Fig. 15 - Computed changes in fully in-bore geometry of 105-mm gun for base initiated detonation. The bow ties flag the zones (above and right) with hydrostatic pressure above 100 kb compression; the triangles flag zones which have met hydrostatic pressure criterion of 15 kb tension.

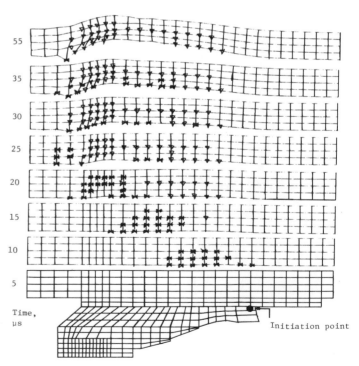

Fig. 16 - Computed changes in fully in-bore geometry of 105-mm gun for nose initiated detonation.

App A

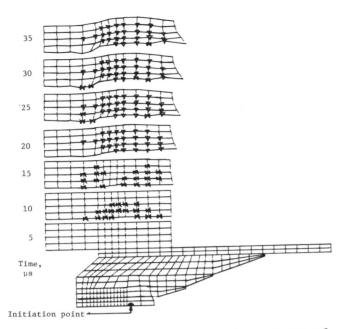

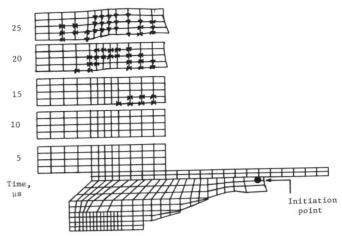

Fig. 17 – Computed changes in muzzle geometry of 105-mm gun for base initiated detonation.

Fig. 18 – Computed changes in muzzle geometry of 105-mm gun for nose initiated detonation.

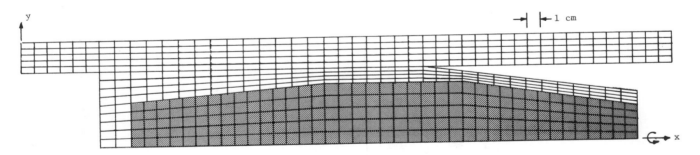

Fig. 19 – Computational mesh of 120-mm gun and M356 projectile. Space occupied by explosive filler is shaded.

App A

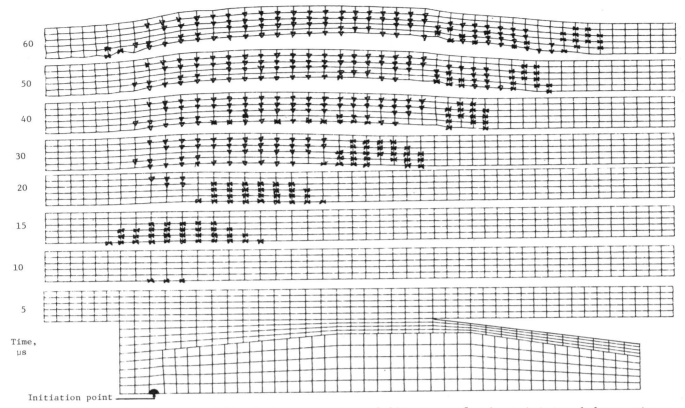

Fig. 20 – Computed changes in fully in-bore geometry of 120-mm gun for base initiated detonation.

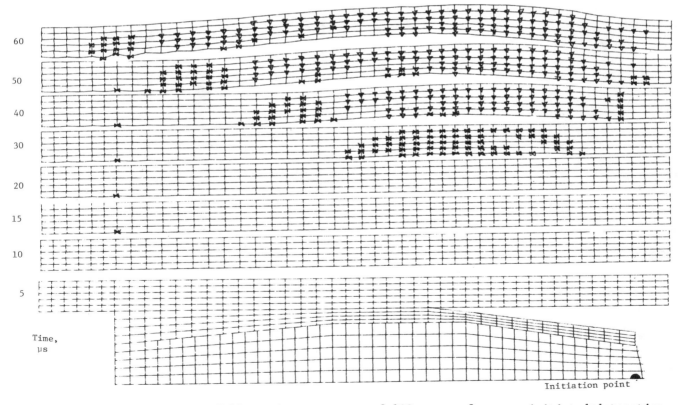

Fig. 21 – Computed changes in fully in-bore geometry of 120-mm gun for nose initiated detonation.

App A

TABLE 1. HARDNESS OF TUBE RELATIVE TO DETONATION SITE

Front of Detonation	At Detonation			Behind Detonation
41.5	41.8	39.5	39.0	41.1
41.1	43.4	40.6	42.2	41.6
42.2	41.1	42.5	41.2	41.2
41.2	40.1	41.3	42.3	41.4
	41.8			41.5

Avg 41.5 ± 0.5 SD 41.6 ± 1.2 SD 41.0 ± 1.3 SD 41.1 ± 1.5 SD 41.4 ± 0.2 SD

App C

TABLE 3. EQUATION OF STATE FOR COMPOSITION B EXPLOSIVE (JWL PARAMETER TABLE 3-36, HEMP USER'S MANUAL)

C_1 (mbar)	C_2 (-)	C_3 (-)	C_4 (mbar)	C_5 (-)	C_6 (-)	C_7 (mbar^{-2})	C_8 (-)	C_9 (-)	Detonation Velocity (cm/sec)	Energy (mbar-cm^3/ (cm^3)o)	Density (g/cm^3)
5.24	4.2	0.081	0.099	1.10	0.309	0.34	500	1.706	0.798	0.085	1.717

$$P = C_1(1-C_3/V)e^{-C_2 V} + C_4(1-C_6/V)e^{-C_5 V} + C_7 \eta \epsilon^*$$

where

$$\eta = \frac{1}{V} = \frac{1}{\rho_o/\rho}$$

TABLE 2. EQUATION OF STATE App C PROPERTIES FOR GUN TUBE AND PROJECTILE STEELS

Density (g/cm^3)	C_1 (mbar)	C_2 (mbar)	C_3 (mbar^{-2})
7.8	1.65	1.87	1.7

Tensile and Compressive Yield Stress (mbar)	Shear Modulus (mbar)	Minimum Pressure (mbar)
0.011	0.8	-0.015

$$P = C_1 \mu + C_2 \mu^2 + (C_3 + C_4) E^*$$
$$\mu = (\rho/\rho_o)-1$$
$$C_4 = 0$$

P = Pressure
E = Energy
ρ_o = Reference density
ρ = Actual density

*Form E in HEMP User's Manual, $C_4 = 0$.

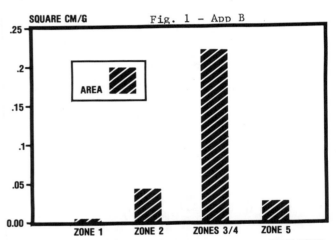

SQUARE CM/G   Fig. 1 - App B

AREA

ZONE 1   ZONE 2   ZONES 3/4   ZONE 5

THE RATIO OF FRACTURE AREA TO MASS IN EACH ZONE

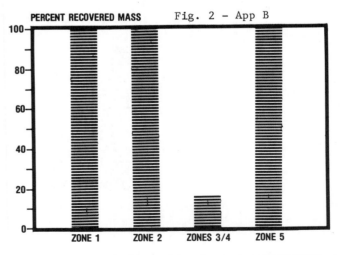

PERCENT RECOVERED MASS   Fig. 2 - App B

ZONE 1   ZONE 2   ZONES 3/4   ZONE 5

THE PERCENTAGE OF MASS RECOVERED IN EACH ZONE

ARTIFICIAL INTELLIGENCE AND DEEP KNOWLEDGE IN METALLURGICAL TESTING AND FAILURE ANALYSIS

Minas Ensanian
Electrotopograph Corporation
Ensanian Physicochemical Institute
Eldred, Pennsylvania, USA

ABSTRACT

The commercialization of the science of Artificial Intelligence also known simply as (AI), is taking place by means of so-called Expert Systems (ES) viz., computer programs or systems that in very specific and narrow technical or problem domains, can to varying degree exhibit some useful reasoning power and function as an expert advisor. There is every reason to believe that such systems will bring about a revolution as profound as that resulting from the introduction of the modern digital computer, and in almost every field of human endeavor. Without doubt its most immediate and important contribution centers upon the concept of knowledge and its manipulation. AI is closely associated with machine intelligence and machine decision-making. This paper provides a brief overview of AI as well as a new sensor science and technology known as Electrotopography (ETG),which combination provides the metallurgist with a new approach for the characterization of metals,and metal fatigue,and failure analysis in particular. The new tools permit the mapping of metallurgical phenomena onto higher dimensional manifolds or n-D spaces, thus providing important new insights and perspective.

ARTIFICIAL INTELLIGENCE IS PRESENTLY being considered or used in medical diagnosis, weather forecasting, military intelligence, mineral exploration, determining the structure of chemical compounds, research in pure mathematics, operations research, design engineering,monitoring and repair of machines etc. The purpose of this paper is to outline the methodology that will be required in order to use AI in real-time metallurgical problem - solving. There are three main aspects to this procedure , viz., knowledge acquisition, representation, and manipulation or processing. The architecture of an expert system is shown in Figure 1.

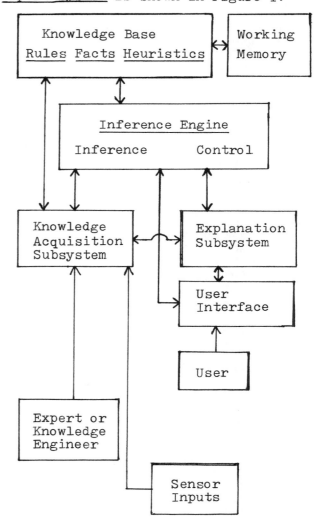

Fig. 1 - The architecture of an artificial intelligence, knowledge-based expert system

The power of an expert system resides in its Knowledge Base (KB). Over the years a new profession has been under development and it is known as Knowledge Engineering and such individuals are presently in great demand worldwide since it is usually not possible to build an expert system without them. It is the job of the knowledge engineer to interview and question a human expert in a specific knowledge or problem domain and attempt to "capture" both his knowledge as well as his particular personal methodology for problem solving or making educated or informed guesses.It can be a very difficult and time consuming process and it can become especially complex when an attempt is made to incorporate expertise from different experts into a single KB.

The knowledge base is the computer representation of the expert's problem domain knowledge and it contains facts, rules, and user-defined functions. The KB is to Data as the Inference Engine is to a conventional computer program,is one way of placing this all in perspective. The inference engine is that part of the computer program that knows how to reason using the facts and expert rules and heuristics etc., and on the basis of the information which is available it infers a solution to the problem.

Knowledge comes in in many varieties, for example, there is general knowledge, and there is domain specific knowledge, there is declarative knowledge and there is procedural knowledge (also classified as dual semantics- two perspectives on the same knowledge).

There are four methods of representing knowledge in knowledge bases: 1) semantic networks 2) rules (meta rules are rules about rules) 3) frames which contain slots (which contain rules, facts, hypotheses, actions and subprograms) and 4) formal logic. These are the four major methods for representing knowledge and relationships.

Frames are filled with slots that contain the attributes of the things which are under consideration. Expert systems (after M. Akhtar) have two special features. First, they are primarily concerned with making decisions. Second they are interactive computing systems that can learn, test and tutor, advise, categorize, consult, design, communicate, interpret, justify, manage, monitor,plan, present, retrieve, schedule and diagnose.

For both practical as well as perhaps philosophical reasons, scientists as well as information engineers consider a truly self-learning machine or computer system as the crowning achievement of AI in the future, if it is ever possible.

Learning however, is not to be considered as a mere accumulation of data,it should be more properly viewed as an accumulation of information. We seek structural properties in the set of data.

In view of the information explosion there is a great need for machines sharing human reasoning powers. Human experts are at their best,when they are able to cope with uncertain, incomplete(and rapidly changing situations) and even false information.

In a conventional computer program for example in BASIC, once the program is run, we can put into action the well known TRACE feature which prints out the program lines as they are executed. In AI we can request the program to explain its line of reasoning, that is, why did it reach a certain conclusion ? On the other hand, the program can just as easily inform the user that the knowledge base is incomplete and that additional information is required and even request or suggest that certain tests or experiments be performed. This can be particularly true in medical diagnosis. The AI program is interactive, that is, a two-way street.

Returning to Figure 1, let us briefly examine the inference engine. A heuristic in its most elementary terms, may be viewed as rules-of-thumb, e.g., if you want to drive to California, just follow the setting sun, or the shortest distance between two points is a straight line (not always true).

Sampling, is perhaps the oldest and the most widely practiced of all heuristic methods, since it may involve inferring some property of a population from the properties of a small, randomly selected, subset of elements. Estimation is another example of inference. There are heuristic definitions and heuristic rules and as previously indicated, the central AI problem of designing a system that developes its own heuristics is still unsolved.

To many a pedagogue, generalist, or epistemologist, artificial intelligence, that is, practical high-level machine decision-making not only represents a supreme challenge, but is also one of the most interdisciplinary subjects that are known. It almost bridges the full spectrum of the humanities, the physical, biological, mathematical sciences, and engineering.

Although an attempt was made in its very early days to develop AI from the viewpoint of a General Problem Solver (GPS), in present (general) practice it is domain oriented. Speaking abstractly, what is domain knowledge ? It consists of descriptions, relationships, and procedures. It is of the most critical importance to recognize that any physical system or problem is capable of being described at multiple levels of abstraction. For example, it will subsequently be shown here, that a metallurgical engineering component such as a jet-engine turbine blade, potentially possesses a near-infinite number of physicochemical (includes physicomechanical parameters as well) dimensions. It may be noted in passing that for a given product space (package of special product information) there may be multiple levels of abstraction, irrespective of the dimensionality of the system under study. Different mathematical operations (operators) on the same data will produce different viewpoints or perspectives. All manufactured products such as the above turbine blade, can be described as an information vector or matrix of any order. The dimensionality of the information vector or matrix simply depends on the number of elements or entries that it contains.

Problem solving via a knowledge base can be viewed from another perspective as shown in Figure 2.

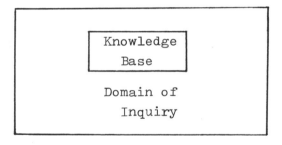

Fig. 2 - Problem solving via fixed and variable information

Let us assume that we are interested in the general subject of metal failure and wish to develop an expert system that will be concerned with the catastrophic failure of military aircraft jet engine turbine blades. What would go into the knowledge base would be so-called fixed information, that is, invariable data and information about metal failure in general. If we are subsequently confronted with the failure of such a component, this kind of knowledge would be referred to as variable information and it would not go into the knowledge base because it only was concerned with the failure problem at hand.

The Domain of Inquiry is the subject that the expert system is supposed to be all about. Except in an ideal world (after C. Naylor) the knowledge base would encompass the whole of the Domain of Enquiry. Therefore, we must show the knowledge base as being smaller than but existing within, the Domain of Enquiry.

If a knowledge base knows everything about a particular problem domain, we refer to it as the closed-world assumption. If on the other hand, we know nothing, then it is called the open-world assumption.

Returning to fixed and variable information, it is obvious that if an expert system is to grow and mature so-to-speak, then there cannot be a sharp line between so-called fixed and variable information, since this would preclude learning and allowing the system to adapt itself to change. So we must allow for the frequent alteration of so-called fixed information. All of this might remind the mathematician of things such as successive overrelaxation, recursion, or even simple iterative processes, or Bayes' Rule. At this juncture mention must be made of the concept of Monotonic Reasoning. A reasoning system based on the assumption that once a fact is determined it cannot be altered during the course of the reasoning process.

Finally, mention must be made of (O-A-V Triplets) i.e., Object-Attribute-Value Triplets, a method of representing factual knowledge. An object is an actual or conceptual entity in the domain of the expert, e.g., mechanical properties of metals and alloys. If an object is a test specimen, then the attributes are the associated properties (e.g., tensile, yield, and %R/A). Each attribute can then take different values.

PROBLEM SOLVING - Successful problem solving depends upon knowing the initial state. Our primary interest in this paper concerns critical metallurgical engineering components (CEC), that is, components whose catastrophic failure can lead to great loss of life, valuable property, and/or expensive downtime. Let us further assume that we are concerned with the design of an unattended (robots replace humans) manufacturing facility for the production of CEC's.

Problem solving (after P. Harmon and D. King) is a process in which one starts from an initial state and proceeds to search through a problem space in order to identify the sequence of operations or actions that will lead to a desired goal. A problem space is a conceptual or formal area defined by all of the possible states that could occur as a result of interactions between the elements and operators that are considered when a particular problem is being studied.

A problem space can become unbounded and exhaustive search impossible if there are a large number of elements and operators and they are poorly defined.

Let us consider the case where a mechanical test specimen passes through a regimen on a fatigue machine to failure, that is, actual mechanical separation. A state is a snapshot of a system (the actual test specimen) as it passes through a so-called problem space. If we could match any given state of the test sample with a state in the problem space, then we would know the present condition of the sample.

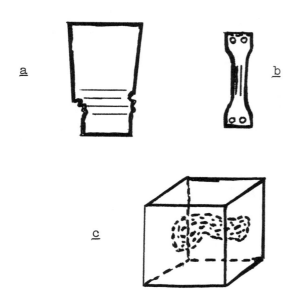

Fig. 3 - Examples of n-dimensional metallurgical systems. a turbine blade, b mechanical property test sepecimen, and c a solid cube of an alloy steel with some type of a physicochemical anomaly

A problem space should not be confined to linear parameters. In other words, nonlinear accumulative fatigue should also be included. If an engineer comes face-to-face with a problem in metal fatigue, then in terms of the modus operandi of AI he may be required to manufacture a population of samples that both define as well as bound the problem space, provided of course, a means exists to take snapshots as one travels through the problem space. The end result is that a box or even a room filled with metal specimens can represent a very powerful physical manifestation of the specific problem space. It could even be argued, depending upon one's viewpoint, that unless you can produce a box of specimens, that either the problem is ill-defined or even that you cannot prove that such a problem does exist .

A problem space is then in reality a collection or set of state spaces. Having now gone full circle, we return to the question or problem of the so-called

initial state of a given system.All things being equal operationally speaking, scatter in mechanical testing even for a homogeneous population (test specimens were all taken or cut from the same plate) must be traceable to differences in initial states. Does this therefore mean that if a metallurgical system such as shown in Fig. 3b were raised to a higher level of abstraction that eventually it would become possible to actually forecast such scatter diagrams ? The answer is yes.

Problem solving provided once again that we have a means of snapshoting state spaces, then reduces to an exercise in pattern matching. High-level patterns (after F. Hayes-Roth et al.) correspond to highly aggregated, abstracted, or condensed descriptions of what are actually highly detailed phenomena. The belief is that the higher the level of structure that can be matched or recognized, the more conceptual ground can be covered in one inferential leap.

THE PHYSICAL NATURE OF INFORMATION - The concept of an initial state leads to formidable conceptual difficulties for the materials scientist or engineer. How can such a state be defined or characterized ? In the new sensory science of ETG (Electrotopography) the different types of information that can roughly define an initial state, fall into six general categories as shown in Table 1.

Table 1 - Criteria for determination of Initial States

Type	Source	Example
1	global or volumetric	relative internal stored energy,degree of disorder,deformation state,distribution of elemental constituents degree of cold work or stress-relief, cooperative phenomena
2	surface	surface chemistry and physics
3	linear	physicochemical/mechanical uniformity in any direction
4	genetic	chemical composition and crystallographic state
5	boundary-geometric	geometry or shape of test specimen
6	experimenal regimen	tension,compression, vibration, thermal

What do we mean by the physical nature of information ? What is the true nature of this seemingly elusive or intangible quantity ? Information is as real (for all practical purposes) and as important as quantities such as mass or energy. Every manufactured metallurgical product such as those shown in Fig. 3 carry information. Metals have "memory" and they retain a record of everything that has been done to them, and as a matter of fact, it is virtually impossible to work on a metallic system without it retaining a substantial memory of the event.

This memory is not a hypothetical or abstract quantity, it is real and it has profound thermodynamic consequences as well as other physical attributes. We must remember that the manufacture of a steel for example, by classical or historic procedures is a stochastic process, that is, characterized by the presence of randomness. The alloy steel cube shown in Fig. 3c consists,or is made up of major, minor, and trace chemical (metallic and nonmetallic) elemental constituents, and all questions concerning such an entity must ultimately reduce to the question of how each and everyone of these same chemical elements are distributed throughout the three dimensional volume defined by product or object.

The only way that one can prove for example, that a steel bar is "clean" or "uniform" is by destructive testing. An indirect proof such as that offered by statistical sampling of a product population is unsatisfactory and not acceptable, in this context. This also applies to prolongations employed in the forging industry.

Every characteristic, attribute or property of a metallic system is determined by the distribution of its elemental constituents. The greater the randomness of these distributions the greater the amount of information carried or stored by the object. In other words, information in metallic systems is associated with the type of atomic species present, how much of it there is, and how it is distributed ? From a purely theoretical standpoint, one might state that a perfect crystal system carried little or no information (exhibits perfect order).In a population of steel specimens the amount of information provided by each of the elemental constituents can be determined experimentally and is a useful quantity.[1]

Operationally speaking, the amount of information that can be obtained from, for example,the cube of alloy steel shown is very large. In fact in terms of $8\frac{1}{2}$ x 11 pages, it can many times exceed the scientific output of the whole world.The volume of scientific literature can be estimated to be of the order of some 13 to 15 million pages.

All metallurgical products are unique and are originals and this is true to what ever level of abstraction they may be raised. The same is also true of metal working machinery. Every machine has its own unique eccentricities, for example, a machine that compacts metal powders leaves its individual signature on the product, and this signature can influence the products performance under dynamic or service conditions. Recent advances in sensor technology, now make it possible to read such signatures and thus identify individual machines, that is, match a metallic object with the machine that worked on it.

Metal testing is an act of collecting information, and as such it can be shown to represent a general communications system (after C. Shannon and N.Wiener). The theory of informational entropy and its relation to thermodynamical entropy has been discussed by Bosworth.[2]

Two products can be identical and have equal mass, but if one of them is of inferior manufacture (possesses a larger measure of informational entropy), this fact will manifest itself under service conditions, that is, informational entropy (also known as informational negentropy after Brillouin-1953) can so-to-speak be transformed into thermal entropy.[3] As Bosworth points out, a factory produces a large number of replications, in a somewhat distorted form, of the original information as conceived by the design engineer. In other words, it is not possible to make the same product twice ;refers to all micro and macroscopic details.

THE METALLURGICAL LIBRARY - In the mechanical testing laboratory, it is customary to discard broken test samples as soon as the test has been completed unless they must be filed for legal or insurance purposes (including product liability considerations);product may be a CEC.

Although it has not as yet been discussed, we shall assume that the technology exist,that will enable capturing the manufacturing history of a product.

Under these circumstances, every single product that comes down the production line is akin to a book or package of information. If such information can be obtained under real-time conditions it can be invaluable in both theoretical as well as economic terms. First of all it allows the design engineer to determine the replication index of each and every product. The index provides a useful measure of the degree of distortion and/or the fuzziness of the product.

Second, it allows for <u>evolutionary feedback</u> in real-time, that is, product information is immediately sent back upstream thus providing greater and tighter process controls, and third, it allows collection of data for the metallurgical library.

A <u>library</u> contains both information as well as every manner of reference material or product. <u>It is in essence the physical and informational manifestation of a total problem space.</u> It might contain products that have performed well in service for many years. Ever form of failed or defective product. Products from competitors or from other plants within the same industrial organization. It is not at all uncommon for a company to have several plants just a few miles apart and as far as anyone knows, both are doing the exact same thing, yet products from one of the plants is consistently better than that of the other. The functions of a library are outlined in Table 2.

Table 2 - Functions of a
Metallurgical
Library

1 A knowledge engineer can begin the development of a knowledge base for an AI system in the absence of a human expert. All of the records he needs are there and he can order tests to be made if necessary.

2 The library is a place to develop strategy for <u>reverse engineering,</u> how do you attempt to save your investment in a product which is marginal ? The common practice of metal reworking

3 All products have a finite life-cycle. Examining old records of this kind can provide critical insights for new product development.

4 Theoretical research for a better understanding of complex metallurgical phenomena(causal relationships)

5 Legal considerations, public relations. Development of intellectual property.

6 Failure analysis and prevention

METAL FATIGUE

STATE SPACE REPRESENTATION - A state can be now considered to be a finitely describable mathematical entity. The description of a state-space problem is the specification of three things: a set of possible stating states, a set of operators (computational procedures), and a set of desired states / goals. A given state space problem can be considered a collection of smaller state-space problems.

A solution to a state-space problem also known as a <u>solution path</u> is again the specification of three things viz., one of the many possible starting states, a set of operators (finite sequence)that transforms the starting state to the <u>desired state</u>. There may of course be a large number of solutions to a given problem. In general, (after P.C. Jackson, Jr.) if the elements of a given problem are partially specified, they are referred to as <u>situations</u> and <u>actions</u>. On the the other hand, if they are completely described, they are called <u>states</u> and <u>operators</u>.

A very useful and interesting tool to describe a start-state and a goal-state is the famous 15-Puzzle that is sold in all game stores. There are 16 spaces one of which is empty. In this situation there are 20,922,789,888,000 or 16! different states in the state space of the puzzle.

In fatigue testing, if we retest an unbroken specimen that was originally tested below the fatigue limit, this situation represents the new starting state; the <u>goal state</u> is mechanical failure and complete separation, ironically.

SYSTEMS ANALYSIS - A mechanical test specimen (Fig. 3b) is a <u>system</u>, and it is capable of converting <u>inputs</u> into <u>outputs</u>. The test specimen consits of numerous <u>subsystems</u>, viz., the major, minor, and trace elemental constituents of which it is composed. Now regardless of how these same elements are combined and/or interact, it is reasonable to assume that when such a sample is placed under a constraint, that all of the chemical elements will not behave in exactly the same way. They have different electrical compositions, charge, and atomic or ionic radii. In Figure 4 a mechanical test sample is considered analogous to a black box.

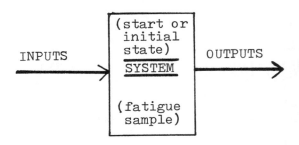

Fig. 4 - A black box representation of a mechanical property test specimen

It is obvious that the <u>output</u> for a given <u>input</u> will depend on the <u>state</u> of of the system (particular situation).

Returning to the 15-Puzzle, it should be noted however, in passing, there are trillions of states from which the <u>goal state</u> cannot be reached. This is a very important consideration with reference to the <u>theory of metal reworking</u>, where we may be in a state from which it is physically impossible within economic constraints, to reach a desired goal,i.e. a product with satisfactory or acceptable quality or features.

All of this discussion on state-space presumes that a means exists to generate the required information vectors or matrices. On the assumption that this type of information can be made available it is of interest to note that the possibility then exists for establishing totally <u>new criteria</u> for evaluating fatigue states and behavior. In other words, in sharp contrast to conventional hard evidence (morphological parameters) for defining fatigue condition, one can now employ <u>systems</u> characteristics. These are a much higher level of abstraction, and as subsequently be shown, are associated with a more fundamental characteristic of matter than its morphological, viz., its <u>electronic parameters</u>. Both cracking and diffusion are preceded by changes in <u>chemical bonds</u>.

ELECTROTOPOGRAPHY

<u>Chemical Amplification</u>[4] refers to the observation that minor deviations or additions can cause large changes in the behavior of a system. The phenomena is ubiquitous in all areas of technical practice, and the influence of trace impurities on the behavior of metals is well established.

If a manufacturer attempts,for example,to make two <u>nearly</u> identical metal electrodes, the subsequent voltage,when the two metals are brought into common contact with an electrolyte, is a reflection of their microscopic differences.

Even if the two electrodes were so nearly identical that it was impossible to separate them by conventional means, their minor differences would be chemically amplified under these conditions.In the elementary school science class when students are introduced to the "lemon" battery (strips of Cu and Zn are usually inserted into a lemon and a voltage is observed), the students are taught that the two metals must be different,and far apart on the Galvanic scale. The fact that the two metals, e.g., a pair of bright 10D common nails, can be identical is not well known, and seldom mentioned in textbooks on electrochemistry.

Such voltages can be substantial, often in excess of 1000 millivolts (mV), although they may be short-lived and at

times multiple polarity transitions may be observed. These voltages are a thermodynamic manifestation of the informational differences between the two coupled metals (electrodes). The theory of metals as well as the origins of galvanic potentials is not complete, notwithstanding advances brought about by quantum electrodynamics.

State-space representation of metals, has historically not been heretofore possible,since there was no methodology for generating the required vectors. In other words, there was no way to even roughly approximate the distributions (throughout the volume element defining the product) of the major, minor, and trace, elemental constituents, nondestructively, and in real-time.

Let us assume that the material that we are concerned with is composed of iron, carbon, manganese, silicon, nickel, chromium, sulfur, phosphorus, and molybdenum, only. The state-space matrix would appear as follows.

$$\begin{bmatrix} Fe & C & Mn \\ Si & Ni & Cr \\ S & P & Mo \end{bmatrix} \begin{matrix} STATE \\ SPACE \end{matrix}$$

Fig. 5 - A symbolic representation of a state-space for a metal alloy

Fig. 6 - A symbolic representation of a metal product - the metallic system is a mechanical property test specimen

By inspection,a state-space representation is a <u>multidimensional</u> abstraction of a real system, in this case, a mechanical test specimen as shown in Fig. 6. Each chemical element or symbol represents a <u>real</u> physicochemical dimension of the metal product. The second requirement for this kind of formalism, is that all quantities in the matrix refer to so-called <u>systems</u> or <u>volumetric</u> parameters.

The air pressure in a balloon is a volumetric or systems parameter since it can be,either determined or sampled from any position on the surface and is a measure of global property or is a global summary or approximation. In the case of a football, the air pressure is still a

GLOBAL PROPERTY even if there is a local deformation such as during kicking. The state-space representation leads to a number of conceptual difficulties viz., if we are only to deal with global properties, then this is only possible if all members of a given elemental species, act cooperatively, and heretofore which, there is no such evidence.

Many of the physical properties of metals are difficult to reconcile and reference is made to a so-called electron -gas. Returning to Figures 5 and 6, this would mean that the metallic continuum consisted of nine different electron-gases, since there may be no other way to resolve the metallic system (theory of informational decomposition in artificial intelligence) into nine global or state functions.

When the electrical resistivity of a metal is measured, no one would ask about the contributions that each of the elemental constituents, individually made to the final result. There may be no way to determine that.

When a manufacturer of bar products claims that his bars are all uniform with respect to all of the elemental constituents, the only way to prove it, is to cut the bar into pies, and then section each pie into 4 or 8 additional sections, all of the time, remembering the exact location of each piece, and then procede, by whatever means, to analyze the chemical composition of each unit and then map the same. The number of such maps would equal the number of the chemical elements being considered. This is obviously a very tedious and costly procedure.

MECHANICS OF ETG SENSORS - Global characterizations of metallurgical systems in real-time, have two fundamental requirements. The signals or waveforms must relate to volumetric parameters that are also chemically specific.

Electrotopography (ETG)[5-7] is a new robotic sensor technology that has been under development for more than several decades and is in its initial stages towards commercialization. Its primary application concerns state-space representations of metallurgical systems (CEC) and transformations. Although selected aspects of the subject are proprietary, the fundamental principles, operationally at least, are relatively simple and straightforward.

The ETG method of waveform generation is depicted in Fig. 7 . Returning to the lemon battery, we begin with two electrodes in common contact with an electrolyte. The latter may be in the form of a rubber-like (doped) polymer, a glass, or even a high temperature ceramic

material. A near-infinite number of such sensors is possible (to date more than 35,000 different sensors have been studied) by virtue of the combinatorial explosion that can be brought about when we consider even briefly, potential materials of construction. For example, pure metals, alloys, metalloids, semiconductors, metal matrix composites, polymers, glasses, ,organic, biochemical, biological materials, and hundreds of millions of chemical compounds as dopants. Likewise, other physical principles or agents such as heat, light, sound, pressure, mechanical vibration, can also be incorporated into this general scheme.

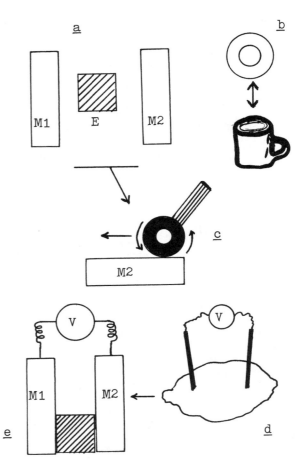

Fig. 7 - The evolution of the ETG physicochemical sensor system

Since the beginning of western civilization, few subjects have held such sway over the mind, as geometry. There is reason to believe this was even true long before recorded history.

In his famous Erlanger Programm in 1872, Felix Klein defined a geometry as the invariant theory of a transformation group. This means that a geometry is defined by a set of any objects and a group of transformations to which the set of objects may be subjected. Since from this point of view (after H. Eves)geometry came to be farther removed from its former intimate tie-up with physical space, and it became a relatively simple matter to invent new and perhaps bizarre geometries.

In 1906 Maurice Fréchet developed a theory of abstract spaces. A space became merely a set of objects together with a set of relations in which these objects are involved.

Returning to Fig. 7 we begin with 3 objects (M1,M2, and E) as shown in (a), viz., two electrodes and an electrolyte, in this case, a rubber like polymer. In the language of the topologist R. Thom [8] (Catastrophe Theory), Fig. 7a refers to a static form while Fig. 7e refers to a metabolic form , the system is now capable of producing an electrical current.

When the term topology is mentioned, students often think of either a Möbius Strip or the famous topological transformation where a donut is changed into a coffee cup, Fig. 7b. In Electrotopography, Figures 7a and 7c, we see the quasi-topological transformation of a physicochemical system, a phenomena that probably is quite common at the level of molecular biophysics and chemistry.

In Fig. 7c, one of the electrodes (M1) is now in the form of wheel, and the other electrode (M2) is represented by a fatigue or   other mechanical property test specimen. The electrolyte (E) is in the form of a tire around (M1). In the static condition so-to-speak, the sensor (M1/E) would be resting on the surface of test sample (M2), and metabolically the situation would be analogous to Fig. 7e.

However, once the sensor goes into motion, and which might be referred to as a second order metabolic process, an ETG signal or waveform is generated as the sensor travels from one end of the sample to the other. An electrotopographic signal may therefore be described as a....... rotational contact, nonequilibrium, volumetric,electrochemical potential, that is chemically specific.

The ETG waveform is simply a voltage distance(or time)plot as shown in Fig. 8; scan speeds are usually between 3 and 30 cm per second, and contact pressures are moderate.

It is of considerable interest to note,that an electrotopographic waveform can look almost exactly the same before and after failure (mechanical separation) and this is important for two reasons (1) under the proper circumstances one can see so-to-speak, the actual crack travel path before the fact, and (2) fragmented

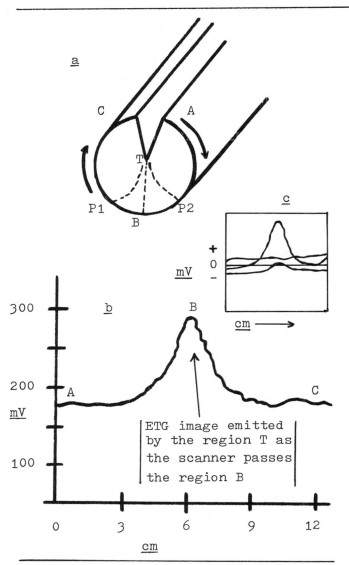

Fig. 8 - The ETG waveform. Chromium crack-tip energy profile for stainless steel extrusion

pieces can be pressure-fitted back together again, followed by scanning in order to recreate the waveform just prior to mechanical failure.

In Fig. 8 the sample shown refers to a stainless steel extrusion (1 inch dia. and 20 ft in length) that failed within 24 hours of manufacture and during shipment to the customer.

The waveform shown is for chromium, however, similar scans can be made for any of the other elemental constituents (major, minor, or trace) of the alloy.

In Fig. 8a, the two additional dotted lines show other possible paths that the crack could take, if it were to be propagated. It is also possible from waveforms of this type to get insight on the relative activity of the respective elements.

This is indicated by Fig. 8c and although all ETG waveforms are unique, there is inherently a certain degree of _redundancy_ always present. However, as in the case of the written word, it can be a factor that enriches a language.

A brief summary of ETG facts is given in Table 3.

Table 3 - ETG Summary

1 Any number of sensors - one or more types,may operate on a test specimen at any given time

2 Hard rust, stray electromagnetic radiation (magnets can be mapped), moderate dust levels, mechanical vibration, do not interfere with these measurements

3 Analog waveforms can be directly transformed into 3-D fishnet mappings

4 With reference to ETG information vectors and matrices, the working statistic is usually the simple waveform average(arithmetic mean of the digitized waveform)

5 Some sensors are very sensitive to changes in chemistry while others are sensitive to changes in stress (internal stored energy), others are in between

6 All members of product population (a collection of identical mechanical property test specimens) must be treated equally or in a similar manner; includes references

7 Training or learning sets in and of themselves are no substitute for a Knowledge Base

8 A _systems analysis_ is in some ways unrelated or independent of concentration or percentage composition by weight and so on, for any element

9 Depending on the problem and the number and type of references that are available, it is often not necessary to know anything about the chemical composition of either the ETG sensors or the material being tested.

10 ETG waveforms have _direction_ characteristics, that is, they are similar to taking photographs of a house and what you obtain or see, depends on where you are standing (the particular surface being scanned). In Fig 3c the physicochemical anomaly shown can be characterized from 6 different planes of the alloy steel cube. These are referred to as "vantage planes"; it should be noted however, in passing, that groups of such waveforms or

information matrices,possess certain _invariance_ properties, and for certain problem categories, represent a very powerful analytical tool

STATE-SPACE REPRESENTATION OF METAL FATIGUE - A sample of stainless steel (309) approximately 0.4 x 4 x 36 cm was subjected to 90° bending fatigue till separation. The operation was monitored by _ETG sensors_ for iron, nickel, chromium, and carbon as indicated by MAT A.

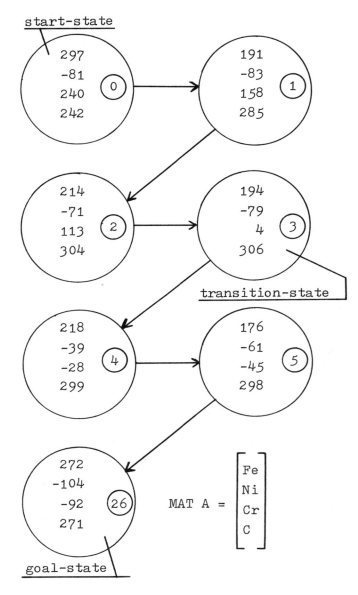

Fig. 9 - State-space mapping of metal fatigue. Inner circle indicates cycle, numbers indicate _waveform averages_

If one studies fatigue behavior by means of an ordinary optical microscope, the criteria for decision-making is obviously, _morphological_ in character.

In the present procedure, the state of fatigue is being determined by chemical bonding and/or electronic phenomena.

In this particular situation, the element <u>chromium</u> is known in ETG as a "detector" element since the waveform average, roughly follows the stress cycles in a linear manner. The <u>third</u> stress cycle is of particular interest, since the element chromium is about to undergo a <u>polarity transition</u> (change in chemical oxidation state). This type of situation can at times be interpreted as a so-called "danger" or "failure" state. How far in advance of morphological evidence this may be at times, depends on (1) the sensor, (2) the material under investigation and (3) on the nature of the testing or experimental conditions. For example, ETG data can be obtained as a function of mechanical vibrational frequencies, that is while a specimen is mounted for vibration testing.

The data contain another interesting feature, viz., a tendency for the values (as the system approaches mechanical separation or the goal-state) to return towards those in the <u>start</u> or <u>initial</u>-state.

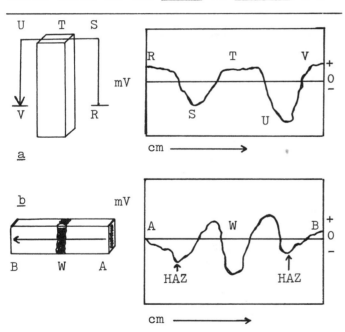

<u>a</u>

<u>b</u>

Fig. 10 - Elemental characterization of <u>edge energies</u> in a block of steel, and the heat-affected-zone in a weld

In Fig. 10a, the scan begins at R and ends at V. In the travel path shown, the block is <u>not</u> physicochemically <u>symmetrical</u> and therefore, will not dissipate energy in a manner expected. In Fig. 10b, the HAZ is clearly indicated. Any weld can provide a large number of unique <u>signatures</u> by this procedure. In work of this kind, there are no limitations

with reference to size, weight, length, geometry, and type of metal or alloy.

Primary limitations concern reasonable cleanliness, physical access, and the object must be a metallic conductor; graphite, silicon, boron, and germanium can all be mapped. Nonmetallic elements employ proprietary sensors. Returning to Figures 7a and 7c, it's obvious that one could make a sensor with a platinum wheel with an electrolyte doped with a platinum salt (you now basically have a platinum sensor), map a steel specimen and thus end up with a Pt waveform average for the steel object which obviously does not contain any platinum. At times these so-called <u>pseudoelemental</u> waveform averages are more valuable (for reasons that may not be known) than the conventional ones.

DISTANCE FUNCTIONS - In Fig. 9 the various state-spaces are in a particular sequence. When information of this kind is available, it makes another form of measurement possible viz., the n-dimensional physicochemical <u>distance</u> between two <u>events</u> or <u>states</u>, or any set of discrete entities.

A <u>Minkowski metric</u> can now be determined for phase transitions, stepwise, cold work or stress-relief etc.

DEEP KNOWLEDGE - In AI work, the term <u>Surface Knowledge</u> (also known as heuristic knowledge) is knowledge that has been gained or acquired from experience, and is used to solve practical problems, usually on the factory floor. It is this sort of rules-of-thumb knowledge that knowledge engineers in many fields are attempting to capture before all of the old-timers vanish. In the metallurgical industry these are the time-honored men who can make tough on-the-spot decisions about <u>reworking</u> a particular product, by simply visual inspection at times. This knowledge is usually restricted to a very narrow domain or task.

<u>Deep Knowledge</u> is knowledge of basic theories, first principles, axioms, analogical representations (an airfield, a telephone exchange, and the adsorption of a gas on a porous solid have much in common; land, taxi, and depart) and even rules about rules (meta knowledge). The following problems involve deep knowledge.

<u>Problem (I)</u> The operator of a fatigue machine hands an engineer 20 identical fatigue specimens, all of which appear to be similar and are unbroken. The operator knows exactly how much <u>Time</u> each coded sample spent on the machine. The engineer must <u>rank</u> the samples correct by NDT in a matter of minutes.

<u>Problem (II)</u> In this situation, all of the variables have been randomized, that

is, load and time may vary, the geometry of the specimens may vary, and the chemistry of the material may also vary. If the total population is large enough to be broken down into subpopulations, then the individual members of these populations should be ranked. This is basically a problem in nonlinear accumulative fatigue where both sample configuration as well as chemistry are allowed to vary. The problem concerns the useful-remaining-life of each sample, if subsequently subjected to a particular regimen of constraints.

In AI, satisfice refers to a process during which one seeks a solution that will satisfy a set of constraints. In contrast to optimization, which seeks the best possible solution, when one satisfices, one simply seeks a solution that will work. Successful ranking is obviously a solution, but will the same algorithm work all the time ? How does one design a knowledge base capable of solving the above two classes of problems ? These problems cannot be solved by the time-honored practice of going through one or two so-called Training Sets. The concept of a Knowledge Base to the metallurgical industry may have the appearance of just another R&D program or expenditure. On the hand, how without AI backup, can anyone expect a robotic inspection station to provide the required levels of machine intelligence and decision-making in the unattended environment ?

Before we begin to design a Knowledge Base, consideration must be given to the concept of a PROBLEM SPACE. This is a conceptual or formal area (after P. Harmon and D. King) defined by all of the possible states that could occur as a result of interactions between the elements and operators that are considered when a particular problem is being studied. Question: can a single simple training set possibly satisfy this requirement ? What is an operator ? An operator is something that can move an object from one state to another within a given problem or search space. Therefore, operationally speaking, an operator is a set of instructions (heat treat procedure is a good example).

In theory it is possible for a single training set to be large enough to fill a problem space (that was small enough); it all depends on the breadth as well as the depth of the machine decision-making that must be made. When the training set is large enough to fill the problem space, then for all practical purposes we are essentially dealing with a knowledge base (although it would still be limited and might not be able to satisfy all questions given it). If one is forced to work

with a small learning set as perhaps in some preliminary feasibility study, it may be possible to leverage the training set (extend it so-to-speak in order to probe the potential search or problem space a bit more fully) by means of search operators. A crossover operator[4] is a means of generating new structures in search space.

SUMMARY AND CONCLUSIONS

Steel, is the greatgrandfather of AI since it was a study of this material (corrosion) that laid the groundwork for the invention of the transistor. In some ways history has made a full circle, and AI can now become a powerful tool for metallurgical investigations, and fatigue behavior in particular.

However, the power of an expert system is no greater than that of its sources of information (knowledge base), and the establishment of the latter involves a major investment, and there are no short-cuts, and it can be a considerable ordeal. Ironically, advances in sensor technology, such that ETG represents, make it possible for experts systems to be developed in the complete absence of human experts, provided that the required representative samples are available and/or can be manufactured; their subjective knowledge or guidance is always a major asset and should be sought.

REFERENCES

1. Ensanian, M., Heat Treating, XVI, 9, 33-41 (1984)

2. Bosworth, R.C.L., "Transport Processes in Applied Chemistry", p. 233, 361, John Wiley & Sons, New York (1956)

3. Brillouin, L., J. Appl. Phys. 22, 1152-63 (1953)

4. Ensanian, M., Proc. Intl. Conf. Ind. Applic's. of AI (Artell'85), Phila., Nov. 5-7, Access Conf. Associates, Gaithersburg (1985). In press. A similar situation arises in the burning of solid rocket propellant, viz., acoustical resonance

5. Ensanian, M., Robotics World, 2, 3, 24-27 (1984)

6. Ensanian, M., Proc. Am. Control Conf., San Diego, June 6-8, Vol. 3, 1006-1012, Invited paper (1984)

7. Ensanian, M., Proc. AI and Adv. Compt. Tech. Conf. (AI'85), Long Beach, April 30, May 1-2, Tower Conf. Manag., 109-126 (1985)

8. Thom, R., "Structural Stability and Morphogenesis", p. 102, Benjamin/ Cummings, Reading, Mass. (1975)

FAILURE ANALYSES OF STEEL BREECH CHAMBERS USED WITH AIRCRAFT CARTRIDGE IGNITION STARTERS

Philip C. Perkins, Raymond D. Daniels
University of Oklahoma
Norman, Oklahoma, USA

A. Bruce Gillies
Oklahoma City Air Logistics Center
Tinker AFB, Oklahoma, USA

Abstract

Cartridge-pneumatic starter systems are used
on military aircraft. In the cartridge mode
used for alert starts, the starter turbine is
driven by hot gases produced through the controlled burning of a solid propellant cartridge within a closed chamber (the breech ch-
amber/cartridge chamber assembly). Premature
failures of steel breech chambers have been
prevalent enough to cause serious concern.
The failures have taken several forms, including fracture and unzipping of the chamber
dome, burn-through of the dome, and shearing
of bayonet locking lugs. Factors identified
as significant in the failures are the pressure developed in the chamber and internal
corrosion of the chamber in an environment
that can produce stress corrosion cracking.
The interior configuration of the chamber and
the stress distribution also have a bearing
upon the failure modes. Several failures are
reviewed to illustrate the problems.

CARTRIDGE-PNEUMATIC STARTERS are mechanical
systems used to accelerate jet engine turbines
to a rotary speed sufficient to start the engine. The starters are used in two modes:
the cartridge mode, in which a rapidly burning solid propellant charge is ignited within
a closed chamber (the breech chamber/cartridge
chamber assembly) and the rapidly expanding
gases provide the energy to start the engine,
and the pneumatic mode, in which air from
another engine or a ground support compressor
provides the necessary energy. The cartridge mode has two distinct advantages for
military aircraft. First, it provides a self-
sufficient starting capability not dependent
on ground support. Second, it provides a
quick start capability, in which all engines
can be started simultaneously, an obvious
advantage under alert conditions.

In recent years, premature failures of steel
breech chambers have been prevalent enough to
cause serious concern. Twelve failures of
steel breech chambers on military aircraft
have been investigated at the University of
Oklahoma since 1978. Early-on, corrosion was
identified as a significant factor in most of
the failures, and stress-corrosion cracking
was suggested as a possible failure mechanism
(1). A special study of breech chambers used
on the B-52 aircraft was undertaken to determine the conditions within the chamber
during operation (2). The work included
measurement of temperatures, pressures and
stresses that occur in the chamber during
actual firings, a finite element stress analysis based on these actual operating conditions, and studies of the propensity of the
steel to initiate and propagate cracks under
the combined influence of stress and the residues left in the chamber after firing of a
charge. These stress corrosion cracking
studies are reported in another paper at this
conference (3).

The breech chamber/cartridge chamber
assembly is shown in Figure 1. A cartridge
can be seen protruding from the cartridge
chamber. The breech chamber is fabricated
from a 4340 steel forging heat treated to a
hardness in the range Rockwell C 40 to 45.
The breech chamber and cartridge chamber are
joined by means of a bayonet closure. The
cartridge is fired electrically and the hot
gases generated exit through the exhaust port
on the top of the breech chamber. To prevent
overheating of the chamber dome there are two
Inconel heat shields welded inside the chamber.
The configuration can be seen in the cut-away
section of a chamber, Figure 2. The heat
shields are not sealed to the exhaust port and
are not pressure containing components. The
volume between the heat shields and the chamber dome is initially pressurized by the
cartridge ignition. There is no continuous

flow within this volume and after pressuriz-ation it acts as a dead air space or thermal barrier to prevent excessive temperatures from occurring on the external surface of the dome. Cartridge combustion products can de-posit between the heat shields and the chamber dome making cleaning and removal of residues virtually impossible. The external surfaces of the chamber are plated with electroless nickel. Since the process involves dipping the chamber into the plating bath, the sur-faces between the heat shields and the dome in the area around the exhaust port are also plated.

The hemispherical portion of the steel dome has a nominal wall thickness of 2.3 mm (0.90 in). The wall thickens near the dome shoulder in the area of the transition (called the Knuckle region) from the hemispherical dome to the cylindrical side walls. The maximum tensile stress in the chamber occurs on the inside surface of the dome at a radius of app-roximately 7.5 cm from the apex, near the Knuckle region. The maximum internal pressure developed in normal firing is about 8.3 MPa (1200 psi). This pressure remains relatively constant for the burn time of about 18-20 seconds. The wall stress developed under these conditions is about 1.14×10^3 MPa (165 ksi). The surface of the dome reaches a temperature in the range of 260°C to 425°C (2).

OVERVIEW OF FAILURE MODES

Splitting open of the breech chamber dome is the most frequently observed failure mode. The failure typically initiates at a location just above the dome shoulder, in the Knuckle region, on the side of the dome diametrically opposite the exhaust port. Subsequent to initiation, the fracture propagates around the circumference to the exhaust port. If the failure occurs early in the cartridge burning cycle, the internal pressures are sufficient to bend the fractured dome away from the ch-amber to form a flap. Figure 3 shows a B-52D chamber with such a fracture. In this case, the outer heat shield has been exposed by the failure. A top view of an F-111 chamber, Figure 4, illustrates the extent of the crack propagation. The failure origin is located at the seven o'clock position.

Typically, the fracture is normal to the dome surfaces at the origin. The fracture surface is usually broken up into steps, suggesting that several small cracks are com-bining to form the fracture. As the dome un-zips, the configuration changes to shear. This is illustrated in Figures 3 and 5. In each instance, the fracture surfaces have been heavily oxidized by the escaping hot combustion gases.

A less catastrophic failure occurs when the dome burns through at one or more locations. Multiple burn-throughs are seen on B-52G and FB-111 chambers in Figures 6 and 7, respectively.

The burn-throughs are frequently located above the Knuckle region, near the exhaust port. They have also been observed on the dome at various locations away from the exhaust port. The surrounding, exterior electroless nickel plating is discolored and blistered, a result of the extreme temperatures reached when the combustion gases are released through these restricted openings.

Burn-throughs located near the exhaust port are similar in configuration to dome fractures, which indicates that they form from cracks. The exhaust port acts to stiffen the dome so that these openings do not grow to more than 5 cm. in length. There is also no sig-nificant thinning of the dome around the failures.

Burn-throughs elsewhere on the dome occur as holes, up to 2.5 cm. in diameter. The dome wall immediately surrounding a hole is thinned to nothing. These failures initiate from areas of severe pitting on the interior surface of the dome.

One failure investigated involved shearing of the locking lugs on the cartridge chamber. The breech chamber/cartridge chamber assembly is held together by three bayonet locking lugs as can be seen in Figures 1 and 2. These lugs can be sheared off if the assembly is severely overpressured by a nonuniform burning cart-ridge. Figure 8 shows a failed lug from a B-52G cartridge chamber. The left hand portion has been deformed downward in shear while the right hand portion has been broken away.

FACTORS CONTRIBUTING TO FAILURE

Subsequent to an external visual examin-ation of a failed breech chamber, the dome is cut away below the Knuckle region. The con-dition of the chamber interior, including the underside of the dome, can only be evaluated by separation of the dome from the chamber. The points of interest are: (a) The presence of combustion and corrosion products. (b) The distribution and severity of pitting on the dome interior surface. These vary in degree from chamber to chamber and, therefore, give some indication of a chamber's service history.

Combustion and corrosion products ty-pically collect in the gutter between the heat shields and the chamber sidewall. An example is shown in Figure 9. The product is dark grey or black in color. It occurs either as loose granules or as a hard packed solid.

The hard packed solid, when it exists, is always found in an area immediately sur-rounding the exhaust port. The combustion products which are forced through the gap around the exhaust port tend to settle out here. Corrosion studies using the products removed from several chambers showed the pro-duct to be corrosive to 4340 steel and capable of producing stress corrosion cracking (3).

In all breech chambers, the interior

surface of the dome corrodes since the bare steel has no protection against the corrosive environment. The corrosion varies from a light rust to a thick, flaky, rust-colored scale. Rust is found in areas where little surface pitting has taken place. Pitting beneath the corrosion scale is much more severe. Examples of scale and associated pitting appear in Figures 10 and 11. These figures are from the burn-through failure shown in Figure 6. This is an extreme case, in which the pitting corrosion has locally penetrated the dome. Normally, the pits vary in size from pinholes to 6 mm in diameter. Metallographic cross sections through the dome have revealed pits up to 1 mm in depth. The pits cover a majority of the dome interior, being more concentrated along the shoulder and around the exhaust port. Figure 12 shows typical pitting along the edge of a dome fracture.

Polished and etched cross sections through the dome at various locations illustrate corrosion damage of the steel. Figure 13 shows pitting corrosion beneath a layer of electroless nickel plating on the inside surface of the dome. The plating, which should be uniform in thickness and free of porosity, has been degraded by the high temperature combustion gases. This is an area close to the exhaust port. More extensive pitting takes place in these areas because of the galvanic effect between the steel and plating, once the protective layer has been penetrated. Electroless nickel plating is intended only to protect the chamber's external surface against atmospheric corrosion. On the inner dome surface it provides no protection and even aggravates the pitting corrosion problem.

Figure 14 shows small cracks growing from the oxidized inner surface. These are thought to be stress corrosion cracks which are the cause of the dome splitting failures. It is interesting to note that these cracks can initiate from a surface which is essentially free of pits.

Oxidation and hot gas erosion have made examination of the fracture surfaces impossible. Recently, an unfailed chamber, which had been in service, was hydraulically pressurized to failure. The fracture surface at the origin of the failure was found to contain three pre-existing semi-elliptical cracks, Figure 15. The cracks had penetrated about 60% through the dome wall. Since the cracks had not suffered erosion and heavy oxidation, it was possible to remove enough of the oxide layer to examine portions of the fracture surface. Figure 17 shows the pre-existing cracks to be intergranular. This is very similar to fractures produced in stress corrosion tests of 4340 steel (3), providing evidence that this is the failure mode for dome rupture.

One failure incident involved a through-wall crack along the weld joining the exhaust port to the dome. This was a new chamber that had undergone only five test firings when the

failure occurred. In Figure 17, the failure is viewed from over the dome shoulder, behind the exhaust port. Immediately surrounding the crack, the electroless nickel plating has been melted and blistered. This suggests that the temperature generated by the escaping gas was at least $890^{\circ}C$ ($1630^{\circ}C$), the melting temperature of electroless nickel (4).

A cross section through the failure, Figure 18, shows the weld at the upper left corner. The crack runs to the edge of the weld bead. The right side of the failure has been displaced upward as a result of the internal pressure. Two significant observations are that the electroless nickel lines the walls of the crack and that the walls exhibit no oxidation. The coating covers the entire fracture surface and completely fills the secondary cracks. It was concluded that the failure occurred at a pre-existing crack at the weld interface. It had been filled with electroless nickel when the chamber was plated. When the crack reopened, the nickel melted and coated the remaining surfaces before they had an opportunity to oxidize.

When a breech chamber fails, the adjacent dome area is overheated by the escaping combustion gases. As seen in Figure 19, the steel to the right of the fracture is heavily tempered and etches more readily than the unaffected steel at the extreme right. The temperatures involved were close to the A_1 (eutectoid) transformation temperature of the steel. The A_1 is shifted locally by variations of alloy composition resulting from segregation. The lighter bands in the overheated area have been transformed to austenite upon heating and then to fresh martensite on cooling. The darker bands are heavily tempered martensite. The observed alloy segregation must have been present in the manufactured part. Microhardness measurements vary from as low as 350 KHN (35 HRC) in the dark areas to as high as 690 KHN (58 HRC) in the light areas.

DISCUSSION

Corrosion is the primary cause of breech chamber failures. The problem is twofold. First, the solid products of the combustion process are corrosive to the steel. Second, the design of the chamber prohibits cleaning of the area between the heat shields and the dome where the products collect. Inability to clean behind the heat shields also means that moisture condensing behind the heat shields cannot be removed.

The main grain of the cartridge consists of NH_3NO_3 and gum rubber. The igniter and booster chemistries are proprietary, but are known to contain perchlorates, nitrates, and sulfur. In addition, there is a good possibility of an accumulation of chlorides in the chamber dome from moisture condensation in certain operational environments.

The moistened products of combustion provide an aggressive medium for both pitting corrosion and stress corrosion cracking. For qualification, the starter assembly is required to survive 400 cartridge starts, but no records are kept of the number of cartridge starts in service over a span of time. The evidence suggests that the corrosion process is a continuing one and that much of the damage is produced by the environment present between firings.

The electroless nickel plating is designed to provide ambient temperature corrosion resistance to the steel breech chamber. The corrosion resistance derives from the presence of a continuous, pore-free nickel-phosphorus coating. This coating, which as-deposited is amorphous, crystallizes at a temperature of about 320°C (4). This process produces a two phase structure of nickel and Ni_3P which is brittle and readily cracks. On surfaces that are heated above 350°C during firing, the coating crystallizes, cracks, and promotes the galvanic dissolution of the steel substrate. This is not a great problem on the external surface of the dome, but it is a problem in those areas of the internal surface on which the plating is deposited.

ACKNOWLEDGEMENT

This work was supported under U.S. Air Force Contract No. F34601-85-C-0791 and several earlier contracts.

REFERENCES

1. W. R. Coleman, R. J. Block, R. D. Daniels, "Corrosion Proglems in Aircraft Components--Case Studies of Failures", Proceedings 1980 Tri-Service Corrosion Conference, AFWAL-TR-81-4019, Vol. II, 241-270, (1981).
2. D. M. Egle, A. S. Khan, A. G. Striz, R. D. Daniels, "A Comprehensive Stress and Life-Cycle Analysis of Jet Engine Starter Breech Chambers", AMNE Report 85-5, University of Oklahoma, USAF Contract No. F34601-83-C-3448, (1985).
3. K.J. Kennelley, R. D. Daniels, "Stress Corrosion Cracking of 4340 Steel in Aircraft Ignition Starter Residues", Int. Conf. on Fatigue, Corrosion Cracking, Fracture Mechanics, and Failure Analysis, Salt Lake City, Utah, (1985).
4. W. D. Fields, R. N. Duncan, J. R. Zickgraf, "Electroless Nickel Plating", Metals Handbook, 9th Ed., Vol. 5, American Society for Metals, Metals Park, Ohio, (1982).

Fig. 1. Cartridge Chamber (left) and Breech Chamber (right). Solid Fuel Cartridge is Protruding From the Cartridge Chamber.

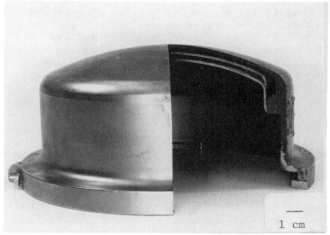

Fig. 2. Cut-away View of a Breech Chamber.

Fig. 3. Dome Rupture Failure (B-52 Chamber).

Fig. 4. Dome Rupture Failure (F-111 Chamber).

Fig. 7. Burn-through Failures (F-111 Chamber).

Fig. 5. Fracture Surface of the B-52 Chamber Failure.

Fig. 8. Failure of Locking Lug on a Cartridge Chamber.

Fig. 6. Burn-through Failures (B-52 Chamber).

Fig. 9. Chamber Dome Cut Away to Reveal Combustion Product Buildup on the Gutter (Bottom).

Fig. 10. Heavy Corrosion Scale on a Dome Interior Surface.

Fig. 11. Severe Pitting of the Dome Revealed Upon Removal of the Corrosion Scale.

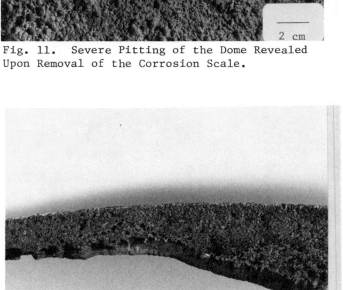

Fig. 12. Typical Pitting Along the Edge of a Dome Fracture.

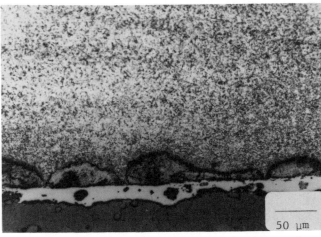

Fig. 13. General Pitting Corrosion Beneath Electroless Nickel Plating Which Has Been Degraded by High Temperature Exposure.

Fig. 14. Cracks Growing From the Dome Interior Surface.

Fig. 15. Partial Through-Wall cracks Found in a Chamber Which Failed in a Hydraulic Test.

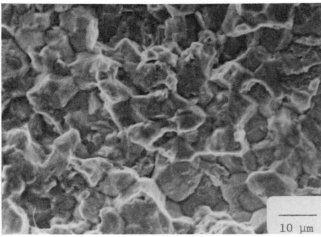

Fig. 16. Intergranular Fracture Surface of the
Cracks Shown in Figure 15.

Fig. 17. Small Fracture Along an Exhaust
Port Weld.

Fig. 18. Cross Section Through the Failure
Shown in Figure 17. The Weld is at the
Upper Left.

Fig. 19. Overheating of the Steel Caused by
Escaping Gases.

EVALUATION OF THE VENT HEADER CRACK
AT EDWIN I. HATCH UNIT #2
NUCLEAR POWER STATION

Carl J. Czajkowski
Brookhaven National Laboratory
Upton, New York, USA

ABSTRACT

A metallurgical failure analysis was performed on pieces of the cracked vent header pipe from the Edwin I. Hatch Unit 2 Nuclear Power Plant. The analysis consisted of optical microscopy, chemical analysis, mechanical Charpy impact testing and fractography. The general conclusions drawn from this analysis were: 1) the material of the vent header met the mechanical and chemical properties of ASTM A516 Gr. 70 material and that the microstructures were consistent with this material; 2) the fracture faces of the cracked pipe were predominantly brittle in appearance with no evidence of fatigue contribution; 2) the NDTT (Nil Ductility Transition Temperature) for this material is approximately −60°F (−51°C); and 4) the fact that the material's NDTT is significantly out of the normal operating range of the pipe suggests that an impingement of low temperature nitrogen (caused by a faulty torus inerting system) induced a thermal shock in the pipe which, when cooled below its NDTT, cracked in a brittle manner.

ON FEBRUARY 3, 1984, GEORGIA Power Company informed the United States Nuclear Regulatory Commission (U.S. NRC) Region II that a crack had been discovered in the vent header of the Edwin I. Hatch Nuclear Plant Unit 2. This notification prompted the U.S. NRC to issue IE Bulletin No. 84-01, "Cracks in Boiling Water Reactor Mark I Containment Vent Headers," on February 3, 1984. This bulletin informed Mark I containment owners of the problem and requested visual inspections on any plants which were in cold shutdown at the time.

On March 5, 1984, another information notice was issued -- No. 84-17, entitled "Problems with Liquid Nitrogen Cooling Components Below the Nil Ductility Temperature." This notice documented the results of the investigations required by Bulletin No. 84-01 (there were no additional cracks discovered), and gave a more detailed summary of the observations to date.

Preliminary inspections revealed that the crack was a brittle fracture which resulted from cooling the vent header material to a temperature below the Nil Ductility Transition Temperature (NDTT) by the uncontrolled release of liquid nitrogen. The nitrogen discharge system is needed to evolve an inert environment in the torus. This system was designed to maintain the nitrogen temperature (leaving the system) at approximately 100°F (38°C). The licensee reported problems maintaining the operating parameters of this system.

As a result of the original notification by the utility, U.S. NRC Region II initiated an independent failure analysis of some sections of the vent header at Brookhaven National Laboratory (BNL).

The analysis was to encompass an evaluation of the failure mechanism and a confirmation of the vent header's mechanical/chemical and impact properties. The test methods used in this analysis were:

a) Visual/Photography
b) Optical Microscopy/Metallography
c) Chemical Analysis/Tensile/Hardness Testing
d) Scanning Electron Microscopy (SEM)/ Impact Testing

VISUAL/PHOTOGRAPHY

One of the three pieces of the Hatch #2 vent header received at BNL was marked NRC-1 and was approximately 12" x 12" x 0.250" in dimension. This specimen had a crack in it approximately 15" long (Figure 1). The area from the vent header where this section was removed is shown in Figure 2. Additionally,

BNL received two opposing pieces separated by a "through wall" crack. The location of these specimens on the vent header are shown in Figure 3. The fracture face of Specimen NRC-1 was observed visually; it was brittle (no observable ductility evident) and had no gross indications of fatigue evident (beach marks, etc.). There were some marks (rachet-type) on the fracture surface, but these would be typical of a fast fracture.

OPTICAL MICROSCOPY/METALLOGRAPHY

A metallurgical cross section of the vent header pipe and seam weld was cut, mounted and polished. The specimen was etched in a 10% Nital solution and then examined. The weld appeared to have been made using a single weld pass from each side (Figure 4a). The base metal, weld metal and heat-affected zone (HAZ) were all consistent with that of a 0.2% carbon steel welded in this manner (Figures 4b, c, d).

CHEMICAL ANALYSIS/TENSILE TESTING/HARDNESS TESTING

A chemical analysis was performed on the base metal from the vent header. Table 1 is a comparison of the chemical analysis performed versus that which is required of an ASTM A516 Gr. 70 steel. All of the chemical requirements met the standard, with the exception of the carbon which was 0.01% greater than the maximum. This very slight deviation could easily have been high due to the relative position of the cut specimen, and is considered to be acceptable.

Four subsize tensile specimens were pulled for this analysis. These were made up as flat specimens with two being cut longitudinal to the rolling direction and two cut transverse to the rolling direction. The average value of the four tests was 72.9 ksi, which meets the ASTM specification requirements (Table 1).

Microhardness readings were also performed on the seam weld, HAZ and base material (Table 2). The base material had an average hardness of 180 KN, while the weld metal attained a value of 197 KN (average), with the HAZ reaching an average value of 181 KN. All of the readings were consistent for this type and grade of steel.

SEM/IMPACT TESTING

A total of ten fracture face specimens were examined by SEM in order to determine if the failure mode could be ascertained by fractography. The specimens were cut from both NRC-2 (Figure 5) and NRC-1 (Figure 6). In all cases the fracture faces exhibited a quasi-cleavage type fracture, which is indicative of a brittle fracture. This type

fracture would be expected if the steel was at or below its NDTT. There was no evidence of fatigue marks on any of the fracture faces.

Since the fracture morphology suggested a rapid, brittle failure of the material (indicating low or cryogenic temperature), a series of 18 Charpy "V" notch impact tests was performed on the vent header material. These tests were conducted on subsize specimens at varying temperatures; −270°F to +100°F. This range includes the temperature of liquid nitrogen (−270°F) and the operating conditions supposedly seen by the vent header 0°F to 100°F. Nine specimens were machined longitudinal and nine transverse to the rolling direction of the plate. The results are given in Table 3. The material's impact resistance is relatively good in the range of 100°F down to 0°F (the vent header's operating range) for all of the longitudinally-cut specimens (specimens 1-6). The longitudinal specimens also had reasonable impact resistance at −35°F and −60°F, with the lowest energy appearing naturally at −270°F. The fracture faces for these specimens showed some ductility in all but the −270°F test. The transition temperature of a material is normally defined as the temperature at which the impact specimen's fracture surface shows a 50% ductile and 50% cleavage structure. By this definition, the transition temperature for this material appears to be approximately −60°F (Figure 7). All of the nine transverse specimens tested (specimens 10-18) had values of absorbed energy significantly lower by a factor of approximately 3 than the longitudinally-cut specimens. This difference is not totally unexpected, as the 8th Edition of the Metals Handbook cites similar energy differences in 0.12% C steel plate. The fracture faces for these specimens showed the tremendous amount of directionality associated with the rolling direction of this plate. However, as with the longitudinal specimens, all the fracture surfaces with the exception of the −270°F test showed some ductility. The amount of ductility observed on the −60°F longitudinal test (specimen #8) was very similar to that observed on many of the fracture faces from the failure. A plot of all the impact data is shown in Figure 8 for convenience of viewing.

CONCLUSIONS

1. The Hatch #2 vent header material meets the chemical and mechanical properties of ASTM A516 Gr. 70 material.
2. The hardness of the weld metal, base metal and HAZ of the vent header material in the area of the seam weld is consistent with the microstructures expected in A516 Gr. 70 material.

3. All of the fracture surfaces examined on the cracks were quasi-cleavage (brittle) in nature and did not exhibit any evidence of a fatigue contribution. The cracking was characteristic of very low temperature cracking.

4. The NDTT for this material is considered to be approximately -60°F (-51°C) based upon the results of the impact tests performed and fracture faces examined.

5. Since the transition temperature for this material is significantly outside the normal operating conditions for the vent header 0°F to +100°F (-18°C to 38°C), it is considered likely that the failure occurred at or below the transition temperature. This inordinate cooling probably caused a thermal shock to the pipe, resulting in a fast brittle fracture of the vent header.

ACKNOWLEDGMENTS

The author wishes to thank U.S. NRC Region II (FIN A3500) for funding this work. Additional thanks to L. Gerlach, A. Cendrowski, D. Horne and O. Betancourt for their help, and Dr. J. R. Weeks for his constant support.

Table 1 – Chemical and Mechanical Requirements of A-516 Gr. 70 Steel

Chemical Requirements

Element	Specification %	Actual Analysis %
Carbon (max.)	0.27	0.28
Manganese	(Product) 0.79 - 1.30	1.21
Phosphorous (max.)	0.035	0.0086
Sulfur (max.)	0.040	0.021
Silicon (max.)	(Product) 0.13 - 0.45	0.24

Mechanical Requirements	Specification (ksi)	Actual (psi)
Tensile Strength	70-90 (483-620 MN/m^2)	72,478 (500 MN/m^2) (L)
		76,581 (528 MN/m^2) (T)
		70,427 (486 MN/m^2) (L)
		72,478 (500 MN/m^2) (T)
	Average:	72,991 (503 MN/m^2)

(L) = Specimen cut longitudinal to plate rolling direction
(T) = Specimen cut transverse to plate rolling direction

Table 2 - Knoop Hardness Readings on Hatch #2 Vent Header

Base Metal

| 179 | 180 | 174 | Avg. = 180 KN/50 gms |
| 193 | 175 | | |

Weld Metal

| 188 | 193 | 214 | Avg. = 197 KN/500 gms |
| 194 | 195 | | |

Heat Affected Zone

| 174 | 174 | 175 | Avg. = 181 KN/500 gms |
| 193 | 193 | | |

For reference purposes, 196 KN/500 gms is equivalent to R_B 87.1

185 KN/500 gms is equivalent to R_B 85.0

175 KN/500 gms is equivalent to R_B 81.7

Table 3 - Results of Charpy "V" Notch Testing

Specimen #	Test Temperature °F (°C)	Energy Absorbed ft-lbs. (Joules)	
1. (L)	+100 (38°C)	21.0	(28.5)
2. (L)	+73 room temp. (23°C)	20.0	(27.1)
3. (L)	+73 room temp. (23°C)	13.0	(17.6)
4. (L)	+32 (0°C)	20.0	(27.1)
5. (L)	0 (-18°C)	10.5	(14.2)
6. (L)	0 (-18°C)	13.5 Plate 2	(18.3)
7. (L)	-35 (-37°C)	20.0	(27.1)
8. (L)	-60 (-51°C)	18.75 Plate 2	(25.4)
9. (L)	-270 Liq. Nitrogen (-168°C)	0.5	(.68)
10. (T)	+100 (38°C)	7.0	(9.5)
11. (T)	+73 room temp. (23°C)	5.0	(6.8)
12. (T)	+73 room temp. (23°C)	5.75	(7.8)
13. (T)	+32 (0°C)	5.0	(6.8)
14. (T)	0 (-18°C)	4.5	(6.1)
15. (T)	0 (-18°C)	5.5 Plate 2	(7.5)
16. (T)	-35 (-37°C)	4.5	(6.1)
17. (T)	-60 (-51°C)	5.75	(7.8)
18. (T)	-270 Liq. Nitrogen (-168°C)	0.5	(.68)

L = Specimens cut longitudinal to the plate's rolling direction
T = Specimens cut transverse to the plate's rolling direction

Fig. 1 – Optical photograph of "as received" section NRC-1 from the Hatch #2 Vent Header

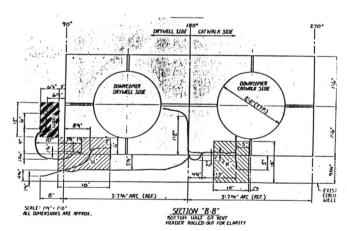

Fig. 2 – Layout of bottom half of the vest header where section NRC-1 (shaded area) was removed

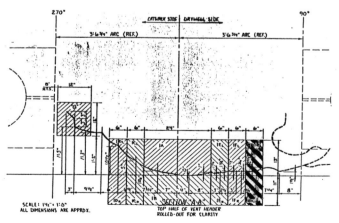

Fig. 3 – Layout of the top half of the vent header where sections NRC-2R and NRC-2L were removed (shaded areas)

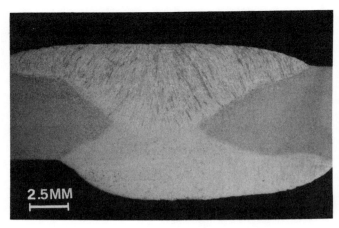

Fig. 4a – Low magnification photomicrograph showing the weld, base metal and HAZ

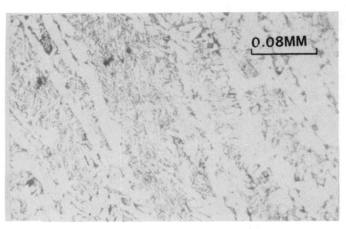

Fig. 4c – Photomicrograph of the weld metal having a typical dendritic structure

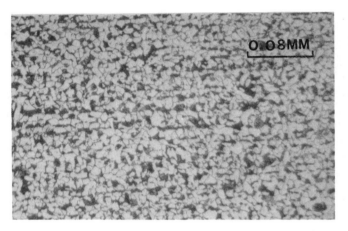

Fig. 4b – Higher magnification photomicrograph depicting a typical ferrite + pearlite base material

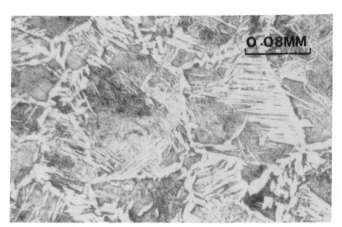

Fig. 4d – The HAZ for the weld composed of fine pearlite, typical of low temperature transformation

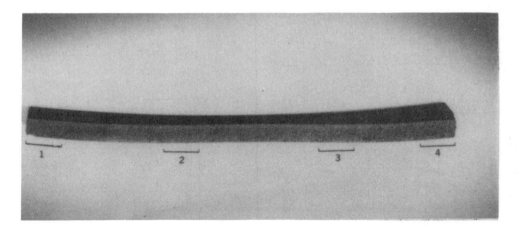

Fig. 5 – Photograph of fracture face of NRC-2. The approximate locations of the samples removed for SEM evaluation are numbered 1-4

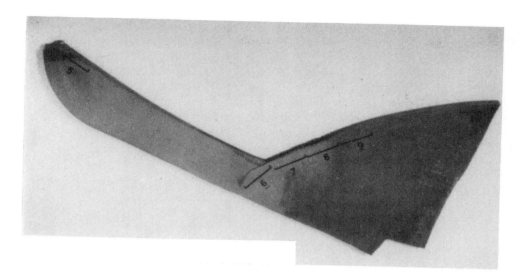

Fig. 6 – Photograph of specimen cut from NRC-1. The locations of SEM samples 5-10 are marked

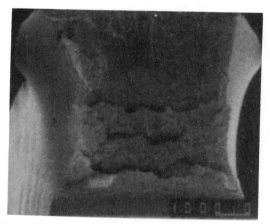

Fig. 7 – Fractograph of impact specimen #8 (-60°F; 50% cleavage, 50% ductile)

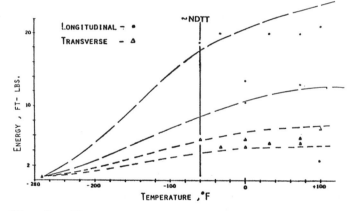

Fig. 8 – Plot of the results of the charpy impact tests performed on the Hatch #2 Vent Header Pipe

APPLICATIONS OF DAMAGE TOLERANCE ANALYSIS TO IN-SERVICE AIRCRAFT STRUCTURES

D.R. Showers, R.L. Jansen, T.F. Christian, Jr., J.A. Wagner
Warner Robbins Air Logistics Center
Robbins AFB Georgia, USA

ABSTRACT

This paper presents an improved method of aircraft structural analysis using finite element analysis and fracture mechanics techniques at the Warner Robins Air Logistics Center, Robins AFB GA. Various analytical aspects of durability and damage tolerance analysis (DADTA) are discussed for U.S. Air Force transport and fighter aircraft. The improvement over previous force management approaches is developed and the flight-by-flight crack growth methodology is examined. Examples of operational aircraft structural analyses are given to illustrate analytical efforts to insure timely answers to in-service aircraft structural considerations of fatigue safety limits and maintenance inspection intervals.

INTRODUCTION

The operational readiness of aircraft weapon systems has assumed greater importance recently. In the determination of operational readiness, a key factor is the status of the airframe itself. Accurate, rapid analysis of structural problems in time for early planning of repairs and inspections, as well as for crisis management, is mandatory. Such quick, accurate investigations are essential to the effectiveness of a modern military air force, since uncertainties concerning airframe status invariably lead to overinspection which unnecessarily strains both operational readiness and manpower.

The U.S. Air Force has met this challenge by applying durability and damage tolerance analysis as an integral part of its Aircraft Structural Integrity Program (ASIP). The functional tasks within ASIP which ensure adequate airframe quality are design criteria, design analysis and development tests, full scale testing, force management data package and force management as described in

MIL-STD-1530A [1]. These activities are intended to substantiate the structural integrity of the airframe design, assess the in-service status of individual aircraft, determine logistics needs and improve future design methods. Since the inception of damage tolerance and durability concepts, significant progress has been made by the U.S.A.F. Systems Command and the aircraft manufacturers in the employment of these concepts to new and existing airplanes. Recently, durability and damage tolerance analysis has been utilized organically at the U.S.A.F. Logistics Command's Air Logistic Centers (ALCs) to enhance their force management of operational aircraft. This paper will address the concepts of damage tolerance and durability and how they can be applied to in-service aircraft to determine safety limits and inspection intervals. These endeavors will be shown with examples from operational aircraft, which illustrate the benefits for advanced structural analysis to provide crisis solutions, as well as long range planning.

THE EVOLUTION OF USAF DURABILITY AND DAMAGE TOLERANCE ANALYSIS CONCEPTS

The U.S. Air Force defines damage tolerance as the ability of the airframe to resist failure due to the presence of flaws, cracks, or other damage for a specified period of unrepaired usage [1]. The damage tolerance of an aircraft structure is assessed by a damage tolerance analysis (DTA) which investigates the growth of cracks and the residual strength possessed by the damaged structure. This damage tolerance analysis focuses upon primary or "safety of flight" aircraft structure which is assumed to have undetected flaws either missed during initial manufacture or induced during operational service. The safety limit at a given location on the airframe is then the number of flight hours required for a crack to grow from the assumed flaw size to a critical length causing failure. Durability, on the other hand, is an economic life, not a safety life concept. It is concerned with the cost of in-service repair such that flaws, which are not safety of flight, will not grow large enough to require extensive repair before one design lifetime [2].

These concepts of durability and damage tolerance represent a significant change from the U.S. Air Force philosophy used until the 1970's. The previous "safe life" approach was predicated on the assumption of a flaw free initial structure and a scatter factor was used to account for variable factors such as manufacturing quality, material variability and operational environments.

The deficiencies of this methodology were brought into sharp focus during the early 1970's with several aircraft suffering unanticipated difficulties. Of serious concern was the fact that operational problems could happen due to initial manufacturing flaws in spite of a four design lifetime fatigue test with satisfactory results. The possible presence of flaws had to be included in the structural integrity methodology in order to realistically predict maintenance actions. This was done with the introduction of the damage tolerance design philosophy in MIL-STD-1530A, MIL-A-83444, MIL-A-8866B, MIL-A-8867B and AF Regulation 80-13 [1,3,4,5,6]. Airplane damage tolerance requirements for flight safety structure are addressed by (a) initial and in-service flaw assumptions accounting for initial quality data and nondestructive inspection (NDI) procedures; (b) residual strength requirements as a function of the degree of inspectability and load factor exceedance data; (c) inspection intervals; (d) damage growth limits; and (e) minimum periods of unrepaired service usage. Airplane durability requirements include the economic life of the airframe exceeding the design service life when experiencing the design load spectra and design procedures to minimize features leading to cracking problems.

Experimentally derived information is crucial to properly apply the DADTA philosophy. The first necessary experimental data comes from a complete teardown inspection following the two design lifetime full-scale fatigue test. The teardown inspection accomplishes several key DADTA tasks by (a) identifying critical areas or "hot spots" not known previously; (b) confirming computer crack growth predictions; (c) verifying NDI procedures; and (d) providing an estimate of initial quality. The second source of necessary major experimental data is service usage recording. Known as the three year Loads/Environmental Spectra Survey (L/ESS), initial in-service usage is measured by multichannel flight load recorders installed on 10% to 20% of the aircraft in the operational force. The flight loads monitored and recorded in this survey form the baseline operational

loads spectra which is used by the airframe contractor to update the Force Structural Maintenance Plan originally developed from design loads. This plan projects inspections and modifications, provides information to estimate the future cost of maintenance actions and identifies potential operational readiness impacts. Since future maintenance resource budgeting and scheduling is predicted on this plan, any significant changes in actual operational usage must be quickly known and factored into the plan. The importance of this in-service usage data has resulted in the present procedure of continual monitoring well beyond the initial three year survey period.

The variation in individual aircraft usage from the baseline operational usage must also be investigated experimentally. Since the baseline usage averages the total aircraft force, some aircraft must be experiencing flight loads which are more or less severe than the baseline loads. Obviously, those aircraft flown less severely than anticipated may have their inspection intervals extended and modifications delayed while the more severely flown ones will compress the time of both. An individual aircraft tracking program using flight logs, counting accelerometers and strain recorders can determine the specific usage of each aircraft. The methodology of durability and damage tolerance analysis can then be utilized to determine inspection intervals and modification times tailored specifically to each aircraft.

DAMAGE TOLERANCE ANALYSIS METHODOLOGY

The damage tolerance analysis methodology will now be discussed with the focus on the initial or baseline entire airframe DADTA but this will illustrate the process to be followed for a structural component or a repair as well. As shown schematically in Figure 1, a DADTA is a sequence of procedures which lead to a prediction of operational service time before failure of a structural component.

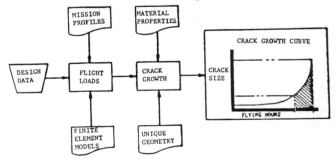

Figure 1. Damage Tolerance Analysis Sequence

The process employs the sequence of flight and ground loads encountered by the airframe during operation service. Specialized computer programs then relate these flight loads to mathematical models of the critical areas of the airframe. Crack growth rates at these critical locations are then calculated and a curve of crack length as a function of flight hours is finally obtained. This curve usually describes a period of slow crackgrowth during which time it is possible to inspect and find the crack. As the crack approaches its critical length, however, its behavior changes from slow growth to an unstable or rapid uncontrolled propagation which results in complete fracture. Thus, the safety limit in flying hours for a given structural location is determined by the critical crack length at that location as illustrated in Figure 2.

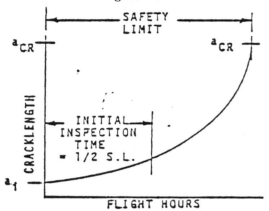

Figure 2. Safety Limit and Initial Inspection Interval

A residual strength analysis is required to establish the critical crack length and the crack growth analysis traces out the path to it. As is also seen in Figure 2, cracks generally grow stably such that inspections can be conducted well before the safety limit is reached. The baseline or entire airframe DADTAs that have been performed during the past decade have usually required a score of contractor and USAF engineers from one to three years of effort depending on the airframe being studied [7]. Whatever the specific aircraft's complexity, three basic assessment tasks must be completed before any crackgrowth curves can be calculated. Critical areas must first be identified on the airframe, next the most probable initial manufactured quality of the airframe must be estimated for those critical areas and, finally, stress spectra must be developed for the locations.

Several factors must be considered in the identification of the critical areas to be analyzed such as the working stress level, the material fracture properties, redundant load paths, failure consequences and the degree of inspectability. Of the most value, however, are the experimental results from the full scale fatigue test article or actual in-service failures from operational aircraft. As a rule, the larger the aircraft the greater the number of critical areas, with a bomber/cargo aircraft generally having over 100 critical areas while a fighter/attack aircraft may only have 50 critical areas.

It is usually difficult to estimate the initial quality of the as-manufactured airframe. The distribution of initial flaws is, in general, an unknown probability function which would require a significant sample size to quantify. Hence, it is mandated in safety limit calculations to assume an initial flaw size based on the NDI detectability limit. Therefore, the largest flaw that could be missed by the NDI technique being applied must be assumed to be present and used as the origin of the crack growth curve. The experimentally derived values resulting from the fractography inspection of cracks either from the full scale fatigue test article or in-service failures are also used, particularily to yield the distribution of equivalent initial flaw sizes for durability or economic life predictions. In both safety limit and economic life analyses the most severe orientation and flaw shape for the location being analyzed must be used.

The flight loads spectrum definition requires dividing the various aircraft missions such as air to air combat, ground attack, training, etc. into distinct phases or segments like taxi, take-off, climb out, cruise, etc. During each segment the airframe will experience loads due to flight maneuvers, turbulence and cabin pressurization, to name but a few. The sequence of loads experienced by the airframe results in a specific stress distribution whose sequence and magnitude governs crack growth. By convention, the sequence of mission segments within a given mission is referred to as the mission profile and the percentage of total operational time spent in each mission profile is called the mission mix. This information must be experimentally derived, frequently using a counting accelerometer at the aircraft center of gravity to measure load factor exceedances. This data is vital to realistic crack growth predictions since variations in mission mix can result in significant differences in crack growth estimates. The culmination of this experimental study is a flight-by-flight load spectrum, typically 1000 flight hours, of a randomly distributed sequence of missions weighted by mission mix. The stress spectrum at the critical location under consideration is determined from external/internal loads analysis of the relationship between the loading history and the stress response of the airframe. Unit loads are often applied to finite element models which simulate the structure in these critical areas. These then lead to stress-to-load ratios which can be used to convert loads acting on the aircraft to localized stresses.

After identification of the critical

locations, estimation of initial quality and establishment of the stress spectra is completed, a flight-by-flight crack growth analysis can be undertaken to generate a crack growth curve at each critical location. From this curve, the crack length for any given number of flight hours during the component's service life can be determined. This allows NDI techniques to be selected and inspection intervals to be set. Conducting this flight-by-flight crack growth analysis is, of course, the heart of the DADTA process as shown in Figure 3 [8]. This process incorporates the following elements.

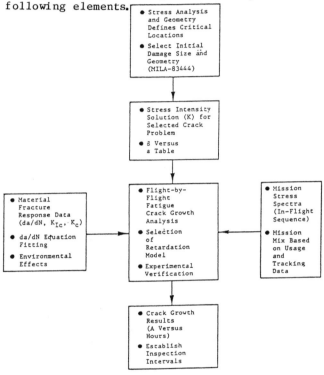

Figure 3. Features of Crack Growth Analysis

    a. the initial flaw distribution

    b. the airframe stress spectra

    c. the crack tip stress intensity factors as the crack extends

    d. the material fracture toughness and constant amplitude loading crack growth rate data

    e. the damage integration and load interaction models

    f. the fracture criterion

The first two elements have already been discussed and the remaining four elements can all be expressed in terms of the crack tip stress intensity factor, K. Independently

developed in 1957 by Irwin and Williams, this parameter interrelates the local stresses in the region of the crack with crack geometry, structural geometry and structural load level [9]. It is defined as

$$K = \beta \, \sigma \sqrt{\pi \, a}$$

where β = a geometric correction term to account for structural configuration variations as a function of crack length

    σ = applied stress

    a = crack length

Generally speaking, determination of the stress intensity factor means just calculation of since β may be considered as a correction factor relating a crack in an actual structural configuration to the case of a central crack in an infinite sheet. Considerable experimental and analytical work has been done to obtain the stress intensity factors for typical airframe structural configurations. Sometimes the stress intensity factor can be found directly using a crack tip element in a finite element model which gives K directly as a function of load and crack length.

Experimental studies are used exclusively to develop crack growth rate data for the various aircraft structural materials. To do this, coupon specimens are subjected to constant amplitude cyclic loads and the results plotted on the basis of growth rate, da/dN (the increment in crack extension for one load cycle), as a function of stress intensity range, ΔK (where $\Delta K = K_{MAX} - K_{MIN}$). The fracture toughness data for airframe material (K_{IC} for plane strain and K_C for plane stress conditions) is generated experimentally as well.

With the crack growth data available, the damage integration model then performs the integration required to grow a crack from its assumed initial flaw size to its critical crack length. This is accomplished by calculating the crack growth increment for each cycle in the stress spectrum. Each cyclic growth increment, da, is equal to the constant amplitude crack growth rate, da/dN, that corresponds to the stress intensity factor range, ΔK, for that stress cycle. The ΔK range for a given cycle is expressed as the difference in stress intensity factors for the maximum and minimum stresses in that cycle. Known as linear cumulative crackgrowth, the results from this procedure may not agree with experimentally observed data. Often, the analytical results will be overly conservative and predict much shorter lives than experimentally obtained. This observed phenomenon, called retardation, occurs when a high stress is followed by lower stresses in the spectrum [10]. The fact that applying a high stress slows crackgrowth is believed due to the plastic zone size at the crack tip. Retardation is evaluated using load interaction or retardation models to modify the stress intensity range, K, to account for the history of previous stresses in the spectrum but the analytical predictions should always be verified by experiment [9].

Finally, the fracture criterion establishes the final or critical crack length as a function of the fracture toughness of the material being considered. The crack length at the safety limit is determined by setting the stress intensity factor equal to the fracture toughness and solving for the crack length resulting from the greater of either the design limit stress or the maximum stress in the spectrum.

DAMAGE TOLERANCE STUDIES AT AN AIR LOGISTICS CENTER

The advantages of damage tolerance analysis to airframe maintenance should be readily apparent. The crack growth curve of crack length versus flight hours is used to establish intervals for the analyzed area. This curve also helps determine the best NDI method to be employed since it predicts the length of crack which is expected to be found during a given inspection. The DTA methodology is also very useful in determining how different operational utilizations such as higher gross weights, more assault landings, etc. will affect service life. By repeating the analysis at a given location for several different mission mixes, a parametric investigation can be performed to forecast the trends in inspection requirements and economic life. This is essential to develop the information needed for trade studies and decisions of operational benefits versus maintenance burdens. By doing these sensitivity studies, future maintenance needs can be projected accurately and unnecessary cost avoided. This also gives the mission planners at the operating commands such as the Military Airlift Command (MAC) and the Tactical Air Command (TAC) visibility as to the impact of mission variations on service life. By showing the possibly significant reductions in economic life and consequent greater maintenance actions that might result from slight changes in missions, DTA is a preventive measure to avoid maintenance difficulties resulting from lack of operator knowledge of the consequences of mission variations.

The primary emphasis of damage tolerance studies at an Air Logistics Center must be aircraft readiness through force management enhancement. This means having a capability to ensure structural integrity, maximize airframe availability and minimize life cycle maintenance costs. For instance, damage tolerance analysis is a significant aid in determining if a structural problem can be isolated to only a manageable number of aircraft within the total

force. This could occur if only a given group were used for a particularly severe mission or if only a certain number had a structural configuration which was more susceptible to cracking. The ability to release aircraft for unrestricted service or to develop a plan that has minimum interruption to normal operations after discovery of an airframe problem is the great advantage afforded by damage tolerance analysis.

The wide range of situations in which damage tolerance analysis is applied to operational aircraft is best illustrated by the following examples. These specific examples were chosen because they also demonstrate the strong contribution of finite element analysis and experimental techniques to the entire damage tolerance analysis process.

Example 1. Crack growth analysis of an engine aft truss mount tang.

A crack growth analysis was performed at WR-ALC on an Air Force medium transport engine aft truss mount tang which was discovered severed at 19,000 flight hours. A metallurgical analysis verified that this part failed from fatigue induced by a corner edge flaw. This initial flaw then propagated through the tang's thickness of 0.7 inches in the transverse grain direction. The material was verified as 7075-T6 aluminum as required by the design drawings.

A crack growth analysis was accomplished using the medium transport aircraft specific DTA runstream currently residing on WR-ALC VAX 11/780 minicomputer system. This runstream includes aircraft loads, mission profiles/mix, flight/ground loads criteria, material data, etc., in order to develop a flight-by-flight stress spectrum and a crack growth curve. The initial flaw used was .050 inch corner flaw, and the crack growth phase was a quarter circular crack propagating through the tang's thickness. Stress intensity solutions were developed for this crack model. Forman's crack growth rate equation with a crack retardation model were used in this crack growth analysis. A crack growth curve was developed for the specific aircraft that had the failure by using its specific past mission utilization. This curve was then adjusted utilizing the part's failure time and crack length. Then, a crack growth curve for the medium transport force using average mission utilization was developed as shown in Figure 4. Hence, the safety limit at the critical crack length of 0.7 inches is 24,435 hours. As defined in MIL-A-83444 (USAF) specification, the recurring inspection shall occur at one half the time for a crack to grow from a missed maximum flaw size to critical crack length. Using this definition and the curve in Figure 4, Figure 5 was developed, which presents the inspection interval as a function of NDI detectable crack length.

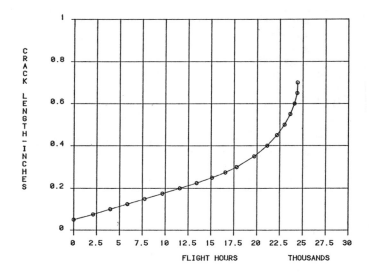

Figure 4. Crack Growth Curve for Truss Mount Tang

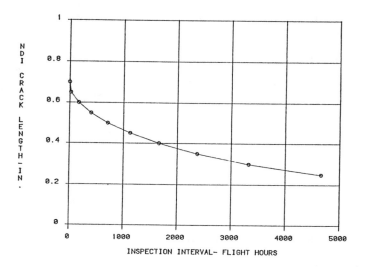

Figure 5. Inspection Interval Curve for Truss Mount Tang

A typical NDI inspection procedure is to perform an ultrasonic surface scan around the truss mount tang's base to pick up a detectable flaw length of 0.25 inches. By using this inspection technique, the inspection interval is approximately 4660 flight hours. Also as defined in MIL-A-83444 (USAF) specification, the recommended initial inspection time shall occur at one half the safety limit, which equates to 12,217 flight hours, and Figure 4 shows that the crack would grow to a length of approximately 0.20 inches. However, the above inspection technique could miss this 0.2 inch crack. Hence, the initial inspection should occur when the NDI procedure will detect a crack length

greater than 0.25 inches, and referring to Figure 4, the initial inspection time is now 15,115 flight hours. From a force management standpoint, the new inspection requirement for the aft engine truss mount tang will be to inspect all aircraft initially at 15,000 flight hours with a reinspection interval of 4600 flight hours.

Example 2. Stress concentration factor calculation by finite element method.

Recently the need has arisen for means of calculating stress concentrations for nonclassical configurations such as gouges, machining grooves, dents, scratches, and other types of surface damage to airframe components. It appeared clear that a highly detailed finite element model of the surface damage should yield accurate stresses from which the appropriate stress concentration factor could be obtained. Such stress concentration factors can then be used in the crack growth calculation programs to determine the time for a crack to grow from such a surface flaw. In a recent instance, where surface grooves were under magnification, accurate measurements of the grooves were made to derive dimensions to be used in the finite element model. A finite element model of a thin slice of this cross-section was then analyzed. CQUAD8 elements were created using the 2-D mesh generator available in our preprocessor, SUPERTAB. These were then restrained at one end and a uniform load was applied at the other end. The resulting stress contour plots showed that these defects had caused a 30% increase in stress at the bottom of each groove. The component was originally designed to withstand large tensile stress loads and increasing these stresses by 30% in some areas would lead to unacceptable early cracking.
To test out this method of analysis, we performed similar analysis on a typical standard size notch for which a known solution was available. Using data given by Peterson for a semi-circular notch, a stress concentration factor of 2.138 was predicted for a notch of certain dimensions. A finite element model, made using these same dimensions, predicted a stress concentration factor of 2.1378. This correlation verified the method we had used to obtain the stress concentration factor of 1.3 resulting from the machining grooves we had found. As a result of this analysis, this method of calculating stress concentration factors has been validated.

Example 3. Medium transport aircraft strake fitting.

One particular example of our efforts aimed at solving existing problems occurred on a strake fitting. The Strake Modification Program was aimed at reducing drag on the aircraft, thereby leading to fuel savings. It involved the addition of two small airfoils to the underside of the aft fuselage to streamline the vortices produced there. Recently, a failure occurred in one of the two fittings used to attach this strake to the fuselage. Metallurgical examination of the fracture surface determined that a semi-circular fatigue crack had initiated at opposite upper corners of the attachment lug hole in the tang. This examination also determined that the crack had become a thru-crack and covered approximately 95% of the cross-sectional area of the tang before instantaneous failure occurred. The number of flight hours leading to failure were rather small. Most puzzling was the fact that the original far field stress had to be large enough to initiate the crack in a short period of time, yet instantaneous failure did not occur until only 5% of the cross-section remained. This phenomena was understood once a "Shake Test" had been performed on a strake fitting which was attached to the fuselage. Accelerometers and a shaker were attached to the strake and it was tested for its natural frequencies. It was found that the strake had its first natural frequency at 68 HZ. Unfortunately, the engine's operating RPM multiplied by four propellor blades was also approximately 68 HZ. The strake fitting was in tune with the prop wash. As the crack grew, the natural frequency of the strake changed and it was no longer in resonance with the prop wash. This reduced the load acting on the strake and allowed the crack to grow slowly until almost no metal remained.

In addition to this resonance problem, finite element analysis also uncovered a high stress concentration at the corner of the hole where failure had initiated. An MSC/NASTRAN finite element model was created using QUAD4 shell elements and RBAR rigid bar elements. A unit bending load was applied to the strake, which simulated the vibrational loads. A stress contour plot of the resulting stresses, which shows an area of high stress adjacent to the hole on the forward fitting, was then obtained. The combination of these two problems: resonance and high stress concentration led to the premature failure of the strake attachment fitting.

Several steps were taken to change the natural frequency of the strake and eliminate the resonance problem. These changes will not be discussed here except to point out that the natural frequency was moved above the resonance frequency. The attachment fitting was modified to eliminate the high stress concentration by using a tension type bolt attachment (with a locking barrel nut) in lieu of the shear type bolt. A finite element model was created of this redesigned fitting using CHEXA solid brick elements which showed that the high stress had in fact been eliminated. At the present time we are modifying aircraft with this redesigned attachment fitting to increase range and fuel efficiency.

Example 4. Long range transport aircraft aft pressure bulkhead.

The aft pressure bulkhead in a long range transport is a closure bulkhead in the aft fuselage which resists large pressure loads due to cabin pressurization, yet allows numerous cables and hydraulic lines to pass through it for control of rudder and stabilizers in the tail. Reports from field units describing defects found that the aft pressure bulkhead recently began to show a high trend of failure in this area. At this time, a sample inspection performed on all aircraft which were available at depot confirmed this high failure trend. An immediate fleetwide inspection was performed on all aircraft which did indeed uncover defects. The defects found could be placed into one of three categories: in the top flange, center web, or bottom flange. Repair and redesign to prevent failures in the top flange was accomplished rather easily, without the use of advanced structural analysis techniques and will not be discussed here. Modifications to the other two areas, however, was quite complex and did require advanced analysis techniques.

The fatigue cracks found in the center web of the aft pressure bulkhead, though large in size, were not considered to be a safety of flight problem. As these cracks reached a length of approximately 16 inches, full pressurization of the aircraft was unattainable and the resulting maximum loads became limited. An equilibrium point would eventually be reached where the pressure loads attainable would not be large enough to drive the crack any further. However, it is imperative that our aircraft be able to meet their pressurization requirements. Maintenance costs to continually repair this area were high enough to justify a fleetwide structural modification.

A finite element model of the bulkhead was created using shell elements (CQUAD4), bar elements (CBAR), and rigid elements (RBAR). A symmetric half model was built which included the area 20 inches fore and aft of the bulkhead itself. Cabin pressure loads were applied to the inner surface of the fuselage, appropriate boundary conditions were applied to the edges of the modeled area, and this information was then input to MSC/NASTRAN for solution. The resulting stress contour plots showed a high level of stress in the web which aligned itself properly with the failures that had been found. By shrinking the stress contour information and overlaying the bulkhead frame on top of it, it was obvious what had caused these defects. There were two stiff areas in the bulkhead which were independent of each other. They were connected by a very weak area of web which had no stiffening frame to support it. A redesigned model was then created which added a stiffener to cross this weak section, thus tying these two stiff areas together. The resultant stress pattern showed a 50% decrease in the level of stress reached in the web. As the aircraft had been in service for many years before the fatigue had become a problem, this decrease would be sufficient in eliminating this particular problem. However, this modification to the center web would have a detrimental effect on other areas of the bulkhead by redistributing the existing load in the web to new areas. This redistribution had to be analyzed to insure that it had not created any new problems. By following the load redistribution throughout the frame, it was found that the lower areas of the bulkhead had load increases of between 6% and 30%. To counteract increases of between 6% and 15% in one stiffener, a simple C-channel repair will be added. The bottom flange, which already had been experiencing failures, saw a load increase of 29% as a result of the repair of the center web. This required a modification to the bottom flange of the bulkhead which not only eliminated the existing fatigue problem, but also withstood the increased loads.

SUMMARY

Durability and damage tolerance analysis is being used successfully by the U.S. Air Force Logistics Command's Air Logistic Centers to enhance the structural maintenance and, hence, readiness of operational aircraft. The judicious application of these techniques during the service lives of operational aircraft will yield sound force management decisions to provide for flight safety, operational readiness and economic repairs.

REFERENCES

1.   Military Standard, "Airplane Structural Integrity Program, Airplane Requirements, "MIL-STD-1530A (USAF), ASD/ENFS, Wright-Patterson AFB, Ohio, December 1975.

2.   Coffin, M. D., and Tiffany, C. F., "New Air Force Requirements for Structural Safety, Durability, and Life Management", Journal of Aircraft, pp. 93-98, February 1976.

3.   Military Specification, "Airplane Damage Tolerance Requirements", MIL-A-83444 (USAF), July 1974.

4.   Military Specification, "Airplane Strength and Rigidity Reliability Requirements, Repeated Loads and Fatigue", MIL-A-8866 (USAF), August 1975.

5.   Military Specification, "Airplane Strength and Rigidity Ground Tests", MIL-A-8867B (USAF), August 1975.

6. .   Air Force Regulation, "Aircraft Structural Integrity Program", AFR 80-13, July 1976.

7.   Tiffany, C. F., "Analysis of USAF Aircraft Structural Durability and Damage Tolerance Notebook", Proceedings of ASIP Conference, Wright-Patterson AFB, Ohio, 1978.

8.   Smith, S. H., et. al., "Development of a Plan for Organic Damage Tolerance Analysis Capability", Battelle Columbus Laboratories, Columbus, Ohio April 1981.

9.   Wood, Howard A., and Engle, Robert M. Jr., "USAF Damage Tolerant Design Handbook: Guidelines for the Analysis and Design of Damage Tolerant Aircraft", AFFDL-TR-3021, March 1979.

10.   Grandt, R. F., "Lecture Notes on Linear Elastic Fracture Mechanics", George Washington University Continuing Engineering Education Course No. 888 (Damage Tolerance Analysis of Aircraft Structures), April 1983.

11.   Peterson, R. E., "Stress Concentration Factors, John Wiley & Sons Inc., New York, New York, 1974.

HARVARD BRIDGE EYEBAR FAILURES

Ronald F. Brodrick
Teledyne Engineering Services
Waltham, Massachusetts, USA

ABSTRACT

This century-old bridge spans the Charles River between Boston and Cambridge. About half of the 23 spans are suspended by wrought iron eyebars. Recent failures of some of these eyebars are examined. Failure is attributed primarily to corrosion and seizure in regions which are inaccessible for inspection or lubrication.

THIS PAPER DESCRIBES failures of supporting eyebars on the Harvard bridge which spans the Charles River between Boston and Cambridge Massachusetts. The bridge carries Massachusetts Avenue traffic across the river to a point at the location of MIT. The bridge also carries communication equipment, including several thousand telephone circuits. This bridge is almost one hundred years old, having been opened in the early 1890s when the major traffic was that of street railway vehicles. It consists of four lanes, two in either direction, plus sidewalks along either side. The original design was such as to support the heavy street railway car traffic along the center two lanes and then lighter traffic which, in those days consisted mostly of horse and carriage traffic, at the outer two lanes. The foot traffic on the sidewalk was carried beyond the outer two lanes and was pretty much incidental as far as loads are concerned. As will be mentioned later, the design of the supporting structure reflects this load distribution.

Several changes to the structure have been made over its century of life. Perhaps the major change was the elimination of a draw section near the middle and its replacement by a fixed span similar to those other sections of the bridge which we shall discuss shortly. Several episodes of repair and refurbishment of the bridge have occurred. The major supporting structure, however, is pretty much as it was originally built.

In the early days, the Charles River was a tidal basin at the point where the Harvard bridge spans it such that, although the Charles itself is fresh, salt water came in with the tide. Later, a dam was built downstream of the bridge preventing salt water from reaching this portion of the river. The water passing under the bridge is now fresh although anyone entertaining thoughts of swimming in it might take exception to this description. A considerable pollution problem has existed in the past but is now pretty well corrected. In any case, the bridge in its early days was exposed to the presence of salt water in its vicinity. Currently, although the water is fresh, the distance to the Boston harbor is rather small so that the salt air and its corrosive capabilities are present. In addition, modern useage calls for the application of road salt in the wintertime. This also contributes to the considerable corrosion that is now observed on the bridge. The corrosion is worse toward the sides of the bridge since these are exposed more to rain water than is the underneath section. Also, the road salt tends to wash off and run down the sides. We mention the corrosion since it played an important part in the eventual failure of the eyebars which are being discussed here.

This report covers the first two eyebar failures to have been discovered on the bridge and the study of the fracture of one of the eyebars as well as the determination of the underlying causes of these failures.

GENERAL DESIGN FEATURES

The bridge is comprised of some 23 spans which are supported on stone piers which are, in turn, founded on a solid surface well below the river bottom. The piers alternate in spacing between about 23 meters and 32 meters (76 feet and 104 feet), this alternation being carried on all the way across the river. Cantilever girders extend out both sides of each pier. In the cases of the piers which are closer together, these girders join to form a continuous structure between the piers. In the cases of the wider spacing, the gap between the ends of the paired cantilevers is filled by a suspended span. The bars which attach the suspended span to the cantilevered ends are the eyebars of interest in this study. This arrangement is shown in Figure 1.

Fig. 1 - Hanger Configuration (Schematic)

A cross section of the structure, shown schematically in Figure 2, shows that there are four main longitudinal girders. This girder structure extends throughout the length of the bridge. It is noted that the center girders are considerably heavier than the outer girders, this being consistent with the original intent for the bridge to carry the heavy street railway cars in the center and the lighter horse, carriage and pedestrian traffic at the sides. The deck and sidewalk structure are placed on top of the longitudinal girders, the walk extending out beyond the outer girders. The deck and walk structures have been subjected to several stages of evolution throughout the life of this bridge. These changes are not relevant to the present topic, however.

At the points where the suspended spans join the cantilevered spans, each girder is suspended by two eyebars. Thus, with the four main longitudinal girders, there are eight eyebars at each of these locations for a total of 16 eyebars for each of the suspended spans. With 10 completely suspended spans plus a half-span at the Cambridge end, there are 168 eyebars in the bridge. (The region of the original draw section does not have the bars.) The center eyebars at each location, i.e., those which attach to the heavier center girders are larger than the outer ones. These center bars are 35 mm (1.5 in) thick by 203 mm (8 in) in width and 760 mm (3 ft) on pin centers. The outer bars are shaped like eyebars, i.e., they have a central section of 25 mm (1 in) thickness by 101 mm (4 in) width, the width increasing in the region of the pins to approximately 250 mm (10 in). This shape can be seen in Figure 3. It is these outer eyebars which have exhibited failures. No failures of the heavier central eyebars have been discovered.

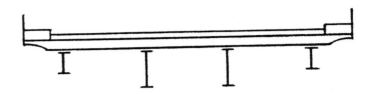

Fig. 2 - Cross Section of Bridge (Schematic)

The method of attachment of the eyebars is shown in Figure 4. It is seen that the eyebars articulate with a pin at either end and that this pin passes through two eyebars as well as the upper or lower girder attachment, as the case may be. Note that the recessed nuts which retain the pins create a cavity which encompasses the outer portion of the pin and part of the outer surface of the eyebar. Otherwise, the different portions of the structure are clamped together by the nut forces. As will be discussed later, this arrangement creates a crevice between each eyebar and the mating girder as well as a crevice at the point where the recessed nut contacts the eyebar. These crevices then become regions in which corrosive material can be collected, such as water or salt solution. They are completely inaccessible for purposes of lubrication or inspection.

The design obviously was intended to have the major load on the eyebar be that of tension arising from the dead and live loads imposed on the suspended spans and carried through the eyebars to the cantilevered spans. The eyebars may

Fig. 3 - Downstream Eyebar

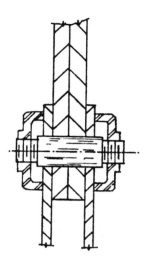

Fig. 4 - Pin and Nut Assembly

carry some side load resulting from fitup errors or from lateral components of any forces created by the live loads. They also carry bending loads in the plane of the eyebars as the bridge expands and contracts. These bending loads are restricted to the values which can be carried through pin structures, i.e., normally only the friction forces of rotation at the joints.

FAILURES

The bridge is subjected to periodic inspections. The responsible authorities are fully aware of the age and general condition of the bridge. They are alert to any hazardous condition which might arise from the corrosion or other wear and aging aspects. The deterioration of the bridge over time is apparent in many areas. Thus, the eyebars represent only one of numerous structural features which must be examined and on which safety judgements must be made on a periodic basis.

During an examination shortly after the failure of the Mianus River bridge in Connecticut, particular attention was paid to the suspending eyebars. At this time, two failed eyebars were discovered. These eyebars were both on the same end of the same span. One of the bars was on the outside face of the downstream longitudinal girder. The other failed eyebar was on the inside face of the upstream longitudinal girder. Figure 3 shows the downstream failure which is near the bottom end of the straight midsection of the eyebar. Figure 5 shows the other eyebar failure which is of similar appearance except that it is near the upper end of the straight section of the eyebar. In both cases the eyebar was completely fractured

Fig. 5 - Upstream Eyebar

so that it could no longer carry any load. It should be recalled that there are a total of eight eyebars transversely across the span at each location. Thus there is sufficient redundancy in the system so that a catastrophic failure is not likely unless two eyebars fail as a pair at a single joint. Nevertheless, the responsible authorities wisely chose to restrict traffic on the bridge until the failures could be discussed and, possibly, corrected. Since these bars were near the outer edges of the span and, as you will recall, the center two lanes are supported by heavier structure, the traffic on the bridge was restricted to the center two lanes. Furthermore, commercial traffic has been prohibited until further corrective action can be taken.

The long end of the upstream eyebar was removed for examination. This removal was accomplished with great difficulty since the entire joint assembly was badly rusted such that the joint had seized up and could not be moved. The generous application of heavy sledgehammer blows over a considerable period of time finally led to success in loosening the joint sufficiently such that one end of the eyebar could be removed.

FRACTURE ANALYSIS

The failure analysis was limited to visual examination, including low power magnification, some metallurgical examination and some simple calculations. Although fracture tests and some crack growth studies could have been performed, it was felt that the fundamental causes of the failures were determined without these investigations. Budgetary limitations also influenced this decision. Detailed examination of the fracture surface by electron microscope or other sophisticated means was not reasonable since most of the fracture surface area was very badly rusted and many of the features such as fatigue striations had been obliterated. Visual examination of the surfaces, such as are seen in Figure 6, indicates that two basic failure initiation sites were present. Although these sites cannot be located exactly, it would appear that they are near the edges of the bar and that the cracks progressed in from either edge until the two cracks approached each other leaving insufficient material in between them to carry the load on the eyebar. One of these cracks progressed much further than the other so that the final fracture zone, as seen in Figure 6, is much closer to one edge than to the other. The nature of the crack surfaces is typical of fatigue failure or of crack growth under conditions of stress corrosion cracking. The rough area at the final fracture is typical of a tensile failure which would occur suddenly at the time of the final break of the bar. Figure 7 shows where these two modes joined. It can be seen that the longer of the cracks has progressed a little bit beyond the sheared surface which joins the fatigue-type crack to the final tensile failure surface.

Fig. 7 - Fracture Surface Showing Extending Crack

Fig. 6 - Fracture Surface

Blocks of material were removed from the vicinity of the fracture surface. The microstructure of this material is shown in Figure 8. The records for the bridge had indicated that this material was wrought iron. Figure 8 confirms this information. It is further confirmed by chemical analysis which shows a very low content of carbon and any alloying elements. The major features to be seen in the micro of wrought iron are the numerous stringy slag inclusions. This is inherent in the process of making wrought iron as it was produced back in the days of the late nineteenth century. Figure 9 is a view of the fracture surface, i.e., a

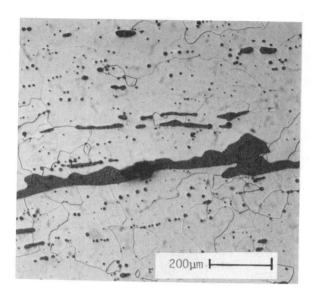

Fig. 8 - Microstructure

Fig. 9 - Curved Slag Inclusion

cross section of the fracture surface. The view is from the narrow edge of the beam with the fracture surface being at the top of the micro section. This view is taken away from the final fracture and shows that there is no discernible distortion to the material structure. This further indicates that the crack was of a progressive nature such as would result from fatigue loading. It also shows an interesting feature wherein the slag formation in one area has a curved configuration such that it has a component which is transverse to the bar rather than being longitudinal as would be expected. It is seen that a tensile load along the axis of the eyebar would load across the weak direction of this slag stringer. This would have the effect of reducing the strength of the eyebar below the nominal value. It would also serve as a likely spot for a fatigue or stress corrosion crack to initiate.

The normal tensile strength of wrought iron is about 345 MPa (50 ksi) with a yield strength of 207 MPa (30 ksi). Typical elongation is 25% and reduction of area is typically 40%. It is this good ductility which gave wrought iron a good advantage over most other available structural materials back in the late 19th century.

The loading of the eyebar was examined in order to determine a source of stresses which would be sufficient to cause the progressive type of fracture which was observed. Three types of load were identified. One, of course, was the axial load for which the bar was originally designed. In addition to this, two bending loads were identified. One of these is a bending load arising from a lateral offset of the eyebar, this being of the nature of a parallel offset of one end of the bar relative to the other. The second bending load was of the nature of edgewise bending of the bar as a result of seizure of the joints. First, consider the normal tensile load on the bar. Although we do not have good weight information of the span, we can be pretty sure that the original designers used a safety factor on yield of approximately five. With a nominal yield strength of 207 MPa (30 ksi) then it is safe to assume that the nominal longitudinal stress in the bar is about 41 MPa (6 ksi) from dead and live loading. It is estimated from the available portion of the bar that the two eye ends were offset by 19 mm (3/4 in). We can only speculate as to the reasons for this offset and, at this time, cannot state whether or not such an offset is typical of the eyebars on the bridge. Nevertheless, with the two eye ends of the bar constrained to remain parallel by virtue of the pins and the recessed nuts being clamped against the girders it is seen that a bending stress is created in the bars as axial load is applied. With a 25 mm (1 in) bar thickness and a 19 mm (3/4 in) offset, this flatwise bending stress turns out to be 4-1/2 times that of the nominal axial stress. Thus, if the design stress was 41 MPa (6 ksi), this offset adds another 184 MPa (27 ksi) for a total of 225 MPa (33 ksi) at the points of the junction of the straight section and the eye. This total stress is slightly above the yield strength and is well above any estimate of fatigue strength of the wrought iron material. The third load which is the edgewise bending, turns out to be even more critical. The expansion displacement at each eyebar location is estimated to be about 12 mm (1/2 in), assuming a temperature range of 44C for the Charles River area. If any of you are ex-MIT students who have walked across this bridge in January you probably would insist that this estimate of temperature range is far too conservative. In any case, if the eyebars are siezed up so that they cannot function to absorb this expansion, then most of that expansion will have to be borne directly by the eyebars, they being the most compliant link in the structure. Taking this deflection and applying it as an edgewise offset of the eyebar, we calculate a total stress range of 1100 MPa (160 ksi) or a plus or minus stress of 550 MPa (80 ksi) if the loading is symmetric. This stress either adds or subtracts from the previously mentioned stresses, depending on which corner of the beam is being considered. The location will alternate as the bridge either expands or contracts. When expanded, one edge of the beam will be subjected to a summation of tensile stresses from the three loads while, when the bridge is contracted, the opposite edge will be subjected to the maximum summation of the three loads. Thus, under this extreme case, a combined load of the order of 760 MPa (110 ksi) is conceivable. Lower temperature excursions will create correspondingly lower combined stresses but, in any case, it is easy to see that these can greatly exceed those required to initiate and propagate a crack in the material. If we use the old rule of thumb of fatigue strength being on half of ultimate tensile strength, then we would estimate fatigue strength of about 172 MPa (25 ksi) and any combined stress higher than this would initiate and then propagate cracks.

SUMMARY AND CONCLUSIONS

It is concluded that there are three interacting causes of failure. The primary cause of failure is the seizure of the joints at the eyebar pin locations as a result of the instrusion of water and salt and the consequent heavy corrosion of the joint. The seizure of these joints led to high edgewise bending stress in the bars as the bridge underwent thermal movement. The cracking was enhanced by the presence of the corrosive medium so that one could only say that the cracks were initiated and caused to grow by some combination of corrosion fatigue and stress corrosion cracking, the former probably being predominant. A secondary cause of the failure was the out-of-plane offset of the bar. This led to an even greater stress in the bar and contributed to fatigue as well as to stress corrosion cracking. The third contributor to the failure was the weakening effect of the slag inclusion which lay partially across the direction of the primary load. At this time it is not known whether these latter two contributors are unique to this particular bar. If not, it is still probable that seizure of joints in other bars would result in cyclic stresses sufficiently large to cause additional failures. In fact, since the investigation of these first failures, very thorough studies of the numerous other bars have been made involving detailed inspection, including dye penetrant inspection. A few other failures have occurred. Early cracks have been detected by these inspections and have been observed to grow with time until the point where failure was imminent. All of these failures have the same characteristics in that cracks grow in from the edges.

FOLLOW-ON WORK AND CORRECTIVE ACTION

The periodic inspection is, of course, continuing. With the observation of the formation of new cracks which could lead to incipient failure it appears that the stage in the life of the bridge has been reached wherein corrosion has achieved its goal of causing the joints to seize up. The resulting fatigue cracks from thermal movements are on the increase. Methods have been developed to allow replacement of broken or weakened eyebars. This replacement is difficult because of the possibility of overloading a remaining eyebar and causing further damage. Precautions are taken to avoid this situation.

The restriction of traffic is still in effect so that the loads are confined to the region of highest strength of the bridge and so that the high loads from commercial traffic are forbidden.

The ultimate goal is to replace the entire structure except for the stone piers which appear to be in excellent condition. Since great esthetic value is placed on this bridge, the proposed replacement will result in a bridge of nearly identical external appearance to the current structure although the fundamental design is quite different.

In closing let us say that, with all due respects to our ancestors who had the courage to design such structures, they should have anticipated the possibility of seizure of the joints and should have provided some means for lubricating these joints. The design did not permit either lubrication or inspection of the joints and thus was doomed to ultimate failure. It is hoped that we will learn a lesson from this and recognize that, if such a structure is to last for such an extended period we must ensure that all components be accessible for such routine maintenance as would be required to enhance its life.

ACKNOWLEDGEMENTS

The author wishes to acknowledge the contributions by Donald Messinger and James Rivard of Teledyne Engineering Services. The interest of David Lenhardt of the Metropolitan District Commission, which authorized this work, is also acknowledged.

APPLICATION OF FRACTURE MECHANICS
TO PIPELINE FAILURE ANALYSIS

Walter L. Bradley
Mechanical Engineering
Texas A&M University
College Station, Texas, USA

ABSTRACT

A pipeline explosion and subsequent fire significantly altered the pipeline steel microstructure, obscuring in part the primary cause of failure; namely, coating breakdown at a local hard spot in the steel. Fracture mechanics analysis used in conjunction with fractographic results confirmed the existence of a very hard spot in the steel prior to the explosion, which was significantly softened in the insuing fire. This finding allowed the micromechanism leading to fracture to be identified as hydrogen embrittlement resulting from cathodic charging.

IN THE FALL OF 1978, a natural gas pipeline ruptured, and subsequently, the escaping gas was ignited, apparently by a hot water heater pilot light in a nearby trailer park. The resultant fire burned for over one hour before the line was finally shut off. The effect of the fire was to alter the metallurgical evidence to such a degree as to obscure in part the primary cause of failure. The purpose of this paper is to explain how fracture mechanics analysis used in conjunction with the more typical information supplied by failure analysis procedures was able to identify the cause of failure.

The standard failure analysis procedures used in this investigation along with the results they produced will be presented next. Then the fracture mechanics analysis used in this failure analysis will be summarized. Finally, the synthesis of the conventional failure analysis results with the fracture mechanics will be given as a basis for identification of the primary fracture mechanism, along with the necessary service manufacturing and maintenance conditions that allowed this mechanism to be operative.

FAILURE ANALYSIS PROCEDURES AND RESULTS

The failure analysis began with the reassembling of the fragmented pieces from the ruptured section of the pipeline followed by a macrofractographic examination to determine the probable crack origin site. Metallography, hardness testing, chemical analysis and microfractographic examination followed. Tensile testing and a flattening test completed the experimental portion of the investigation.

INITIAL SITE INSPECTION - The failure analysis began with an initial inspection of the failed section of pipeline on location. This included collecting the several fragments of pipeline steel produced by the explosion. The ruptured portion of pipeline was about 18'(5.5m) in length, terminating on one end at a girth weld and on the other end near the center of the 40'(12.2m) pipeline section were moved to a laboratory for cleaning and macrofractographic examination.

Additional information collected during the initial site investigation included the following: pipeline outside diameter was 30"(0.12m); wall thickness was 0.375"(0.95cm); operating pressure at time of explosion was 560 psi (3.9MPa), which was normal operating pressure for this pipeline; and pipeline was put into service in 1953 at an average depth of 40" (1.0m) in the ground. It was subsequently learned that corrosion protection provisions included an organic coating of Lion Oil Company E-120 followed by wrapping with a fiberglass/asbestos wrapping material. This barrier protection was supplemented with cathodic protection using rectifiers. Additional rectifiers were added in 1978.

IDENTIFICATION OF CRACK ORIGIN LOCATION- Using the ruptured pipeline section and the several fragments as patterns, paper replicas of these pieces were cut and then reassembled, as shown in Figure 2. All fractured surfaces were then examined to determine whether chevron marks and/or shear lips were present. Where chevrons were noted on the various fractured surfaces, they were indicated on the paper model. The clear chevron pattern found on many of the fractured surfaces allowed the fracture origin site to be unambiguously identified, as shown in Figure 2 and indicated schematically in Figures 1 and 3. Smaller pieces (4-1 and 5-1, Figure 3) containing the crack origin location were subsequently cut from the larger pieces to allow metallograph, hardness testing and microfractographic examination to be made. Note in Figure 3 that the shaded area represents that portion of the ruptured pipeline that was covered by dirt, and thus, partially protected from the heat of the fire. Further note that crack origin location on piece 5-1 was under the dirt while the complimentary crack surface location on piece 4-1 was directly exposed to the fire.

METALLOGRAPHY - Specimens were cut from pieces 4-1 and 5-1 at the fracture origin location and from piece 5-1 at 1" circumferentially behind the fracture initiation site and 1" circumferentially/6" axially removed from the fracture origin site. These were prepared using standard metallographic technique and etched using a nital etch. The results are presented in Figure 5.

It is clear that the fire reaustenitized the steel in piece 4-1, giving it a completely different microstructure from that piece 5-1 at the fracture origin site. The microstructure for piece 5-1 is highly irregular for a pipeline steel, appearing at the crack origin location to be quenched and tempered martensitic, or possibly a bainitic microstructure. At the two locations removed from the fracture origin site, the quench was less severe, but still gave an excessive amount of fine pearlite compared to what one would expect for a pipeline steel.

HARDNESS TESTING - A combination of Rockwell and Tukon hardness testing was used to map the hardness around the fracture origin site contained in piece 5-1, with the results presented in Figure 4. The fracture origin site is exceedingly hard for a pipeline steel, namely, as high as 350 KHN. The hardness values for piece 4-1 were uniformly 170 KHN, which is about what one would expect for this pipeline steel with a proper normalizing heat treatment. The hardness in piece 5-1 approaches this value at the furthest extent from the fracture origin site.

CHEMICAL ANALYSIS - Chemical analysis was made on pieces cut from the portion of the pipe that did not fracture during the explosion and from piece 5-1 which contained the fracture origin site. Both pieces were found to have 0.30% carbon and 1.2%Mn with sulfur and phosphorus impurities acceptably low.

MICROFRACTOGRAPHIC ANALYSIS - Specimens were cut from piece 5-1, which contained the fracture initiation and initial unstable crack extension, and prepared for examination in the scanning electron microscope. This preparation consisted of cleaning with a mild detergent and water followed by ultrasonic cleaning in acetone. This left the fracture surface free of dirt and clay, but still coated with an oxide film formed during the fire.

Selected photographs from the SEM examination are presented in Figure 6. Included are SEM fractographs of the fracture initiation site and of some adjacent secondary cracks which were opened and examined to obtain information on a cracked surface that was less heavily scaled. These fractographs indicated some intergranular fracture; but generally, transgranular fracture was observed. The SEM examination in combination with the earlier macrofractographic examination using a stereo-microscope defined the crack size at the fracture origin site at the time that the pipeline ruptured to be 0.175" (0.44cm) deep (maximum) by 0.165" (0.419cm) long.

TENSION AND FLATTENING RING TESTS - Three tension test coupons with 1.5" (3.8cm) wide by 4" (10.2cm) long reduced sections were machined from the portion of the pipeline that did not fail in service. Two of these specimens were in the transverse direction with one containing the longitudinal weld. A 4" (10.2cm) wide flattening ring was also cut from this same section of pipe. The measured yield strength for all three specimens were approximately 60 Ksi (414MPa), the tensile strength was approximately 85 Ksi (587MPa), and the elongation was 33%. The tension coupon containing the longitudinal weld did not fracture in the weld. These results are all well above the specified minimum values.

It should be noted that the mechanical properties in the region where the fracture originated are quite different that those determined in these tests. The hardness results indicate an ultimate tensile strength of at least 150 Ksi (1035MPa) with a much lower ductility.

The flattening of the ring occurred without any observed cracking in the longitudinal weld, confirming the earlier observation from the tension test that the weld was of excellent quality and was not a factor in the pipeline failure.

APPLICATION OF FRACTURE MECHANICS ANALYSIS

The application of fracture mechanic in this failure analysis was as follows: (1) use calculated principal normal stress and measured critical flaw size to determine the critical stress intensity of steel at fracture origin site; (2) use calculated critical stress intensity to infer apparent hardness, yield strength and tensile strength of steel prior to fire; and (3) note whether this strength (or hardness) was sufficient to make steel susceptible to hydrogen embrittlement.

CALCULATION OF CRITICAL STRESS INTENSITY- For thin walled cylinders, the principal normal stress is the tangential stress, or hoop stress, which may be calculated using the relationship

$$\sigma_1 = \frac{Pr}{t} \qquad (1)$$

where P is the operating pressure of the pipeline (approximately 560 psi or 3.9MPa for this pipeline at the time of rupture), t is the wall thickness (0.375" or 0.95cm), and r is the pipeline radius (15" or 38.1cm). Substituting these values into Eq. 1 gives a value of 22Ksi (152MPa) for the principal normal stress.

The critical flaw size at the time of fracture as determined by macrofractography and confirmed by microfractography has previously been noted to be 0.175" deep by 0.165" wide (0.44cm x 0.419cm). The critical stress intensity of the steel at the fracture origin site may then be calculated using the standard relationship

$$K_1 = \sigma_1 C \sqrt{\pi a} \qquad (2)$$

where σ_1 is the principal normal stress (22Ksi or 152MPa), "a" is the crack depth (.175" or .44cm) and C is a factor which depends on flaw shape and the size of the flaw relative to the wall thickness. Using the results of Rice and Levy (1) for a part through crack in a plate, C for the pipeline flaw in question is approximately 1.8. Substitution of these values into Eq. 2 gives a value of 29.4Ksi in for the critical stress intensity of the pipeline steel at the fracture origin site. The observed critical stress intensity value of 29.4Ksi in (32MPa m) implies a very high tensile strength of 270Ksi (1863MPa), which corresponds to a knoop hardness of 600KHN(2). Thus, it may be concluded that significant tempering of a much harder origin material in piece 5-1 occurred during the fire. It has already been noted that piece 4-1 was heated to a much higher temperature, which reaustentized it, totally removing any evidence of a hard spot.

The results of the stress and fracture mechanics analysis may be summarized as follows: (1) The static stress in the pipeline was at 42% of the specified minimum yield strength of the steel; (2) The pipeline steel at the fracture origin site had a hardness in excess of 600KHN and an ultimate tensile strength in excess of 270Ksi (1863MPa), which was subsequently tempered to the observed hardness values and microstructure by the fire.

DISCUSSION OF RESULTS

The important results of the conventional failure analysis and the subsequent fracture mechanics analysis are as follows:
(1) the fracture occurred at a 0.175" (0.44cm) by 0.165" (0.42cm) crack;
(2) the steel at the fracture origin site had a hardness of approximately 600 KHN;
(3) the cracking that preceded final rupture was in part intergranular;
(4) additional secondary cracking was noted on the surface of the steel pipe in the vicinity of the hard spot.

These findings suggest that the critical flaw initiated and grew to a critical size as a result of either stress corrosion cracking or hydrogen embrittlement. The basis for this contention will be developed in what follows.

STRESS CORROSION CRACKING OR HYDROGEN CRACKING - At present there are no known unique features associated with hydrogen cracking which would distinguish it from other types of intergranular fracture such as stress corrosion cracking (3-5). In general, stress corrosion cracking (SCC) may be identified by the following distinctive features:
1. An intergranular pathway for carbon steel (usually);
2. Abundant branching (usually);
3. Multiple surface initiation sites (usually);
4. Pronounced deep crevices or secondary cracking;
5. Corrosion product on fracture surface;
6. Crack origin at corrosion pit or other stress riser.

The grain size of the tempered martensite at the rupture origin site was too fine to allow one to determine from metallography whether the fracture was transgranular of intergranular. Intergranular cracking was however observed in several places using SEM. Very little branching was observed in either the primary or the secondary cracks. There were multiple surface initiation sites; however, the secondary surface cracking was to a depth of only .01" - .02", which would hardly qualify as "pronounced deep crevices or secondary cracking". No corrosion product was noted at the rupture origin. However, one cannot exclude the possibility that the corrosion product was removed in cleaning; removal of clay baked onto the fractured surface required some harsh cleaning procedures. A dark film was seen on some but not all of secondary cracks that were opened in pieces cut from piece 5-1. This may be a corrosion product but it could also be oxidation product formed during the fire on surfaces of the cracks of sufficient width to allow oxygen availability to the crevice walls. Finally, the primary crack did originate at a corrosion pit. In summary, the rupture origin site shows some but not all of the distinctive features usually associated with SCC.

Hydrogen cracking (HC) is distinguished by the following general features:

1. Single initiation site which may be subsurface or surface;
2. Limited or no branching;
3. An absence of corrosion product on fracture surface (usually);
4. Generally intergranular fracture for carbon steels.

The observed fracture clearly had only one significant initiation site, with the secondary cracks being less than 1/8 as large. As was previously mentioned there was essentially no branching. Corrosion product was not observed at the origin. While it is possible that this was removed during the cleaning of the clay from the fractured surface, this is somewhat unlikely, particularly if the corrosion product was Fe_3O_4, which is quite adherent to steel. There was some evidence of intergranular fracture in the SEM, though one could not say it was generally intergranular fracture. However, it is not uncommon in SCC or HC failures to have intergranular, quasi-cleavage and ductile rupture, all in the same sample (6). This would better describe our results. Thus, the general, distinctive features for hydrogen cracking would seem to be satisfied.

While suggesting that HC is better supported than SCC by our results, it should be noted that this tentative conclusion is based on two speculative interpretations of the evidence: first, that the dark film seen on the surface of some of the opened secondary cracks was Fe_3O_4 formed during the fire rather than as a part of the SCC process; and second, that the small surface cracks near the origin do not constitute the significant degree of secondary cracking which usually accompanies SCC. If one takes the view that the observed secondary cracking was significant and the Fe_3O_4 formed as a corrosion by-product, then a strong case for SCC can be made. Thus, identification of HC or SCC as the primary micromechanism of rupture initiation seems certain. Since the conditions principally responsible for HC and SCC in high strength steels (recall that this pipeline steel at the rupture initiation site was locally a high strength steel) are quite similar, distinguishing between the two may not be necessary anyway.

PRINCIPAL CONDITIONS REQUIRED FOR HC OR SCC IN HIGH YIELD STRENGTH STEEL - For medium and high strength alloy steels, there is increasing agreement bordering on unanimity, that a hydrogen embrittlement mechanism is involved in the stress corrosion cracking of these alloys (7). It is thought that hydrogen

may be produced as a by-product of the corrosion reactions occurring at the tip of a stress corrosion crack (8,9); however, in hydrogen cracking of pipeline steel, the hydrogen is introduced at the surface via cathodic charging. In either case, hydrogen embrittlement would be the primary mechanism of fracture.

It is well known that SCC and HC of carbon steels are favored by high yield strengths where the hydrogen embrittlement mechanism is operative (10,11). In fact, hydrogen induced HC or SCC in carbon or lower alloy steels is seldom encountered for yield strengths less than 100,000 psi. Generally a yield strength greater than 150,000 psi is required for hydrogen embrittlement (12,13).

Thus, one may state the three conditions for HC or SCC of high strength pipeline steel via hydrogen embrittlement as follows:

1. Sufficiently high tensile stress;
2. Hydrogen Absorption at defect in pipeline coating via
   a. Cathodic charging or
   b. Corrosion pits or crevices precipitating stress corrosion cracks with hydrogen evolution at the stress corrosion crack tip;
3. Steel with a yield strength in excess of 150,000 psi.

Evidence that these conditions were satisfied in the section of pipeline that failed will be summarized next.

SATISFACTION OF CONDITIONS FOR HC OR SCC AT FRACTURE ORIGIN SITE - Earlier in this paper the static service principal normal stress was calculated to be 22Ksi (152MPa) or 42% of the specified minimum yield strength. These results implicitly assumed a round section of pipe containing no residual stresses. Recent studies (14), however, have shown that hard spots may have tensile stresses from 40% to 100% greater than the nominal tensile stress due to internal pressure alone. Residual stresses may develop in the region of a hard spot during the fabrication of flat plate into round pipe as a result of the heterogeneous deformation pattern the hard spot produces. Furthermore, the out of round condition of the hard spot would cause the local stresses due to internal pressure to be much higher than the 22Ksi calculated for an assumed round pipe. Therefore, the 22Ksi nominal tensile stress calculated is very likely a quite conservative estimate of the actual tensile stress at the hard spot, which may be as high as 30,000 to 45,000 psi or 60 to 80% of the specified minimum yield strength. Such a stress level would increase the previously calculated K_{1c} for the steel in the hard spot to 50Ksi in (54.5MPa m), which still implies a tensile strength in excess of 220Ksi (2).

A necessary requirement for hydrogen absorption at the surface of a pipeline is a condition of coating rupture. The rupture may take the form of larger holidays (or holes) in the coating or very fine pinhole porosity. Local regions of disbonding may also occur. Cracks or pinholes in the coating may allow ground water to reach the surface of the steel in a region where the cathodic protection system would not have enough "throwing power" to supply complete protection, allowing general corrosion, pitting, or SCC to occur.

Where the cathodic potential is sufficient to maintain cathodic protection at the local coating breakdown, hydrogen charging will clearly be a problem. The coating breakdown decreases the cathodic potential to a level where the steel is susceptible to general corrosion, pitting or SCC, depending on the environmental conditions and the local cathodic potential. Hydrogen may be produced at the surface as a by-product of the corrosion reactions. Thus, hydrogen in sufficient quantity to cause hydrogen embrittlement may occur as a result of coating breakdown through cathodic charging or as a by-product of corrosion.

The coating in the section of pipeline which ruptured in service was experiencing some deterioration as was clearly seen during inspection of the pipe section that did not rupture. This piece was clearly experiencing coating disbonding and/or rupture, as evidence by numerous locally corroded regions found on the pipe surface. The recent addition of new rectifiers to maintain the proper cathodic potential (1978) is also a good indication of coating breakdown.

The fracture toughness (K_{1c}) at the time unstable crack propagation produced the service rupture was used to infer a local tensile strength of 220,000-270,000 psi in the steel contained in the hard spot. Such a high strength steel would certainly have a yield strength in excess of 150,000 psi. Thus, all three conditions for hydrogen embrittlement via HC or SCC as stated earlier were met in the fracture initiation site of the pipeline.

SUMMARY AND CONCLUSIONS

The total metallographic and fractographic evidence has been interpreted to indicate hydrogen embrittlement via hydrogen cracking or stress corrosion cracking to be the primary cause of failure. This failure mechanism is only operative in high strength steels at a sufficiently high service stress in the presence of absorbed hydrogen. A local hard spot can both increase the strength of the steel to the required level locally as well as increase the effective service stress level through the residual stresses that always accompany hard spots. Some breakdown of the coating is required to allow hydrogen absorption at the surface. Thus, it may be concluded that the macroscopic cause of failure was the combined effect of a coating breakdown at or very near to a local hard spot in the pipeline steel.

The use of fracture mechanics allowed the lower bound of hardness and strength in the hard spot prior to rupture to be estimated. A direct measure of this value was not available due to tempering back during the fire.

REFERENCES

1. J. R. Rice and N. Levy, "The Part Through Surface Crack in Elastic Plate," J. of Applied Mech., 1972, Vol. 39, Ser R, pp. 185-194.

2. Richard W. Hertzberg, "Deformation and Fracture Mechanics of Engineering Materials," 1976, Wiley, New York, pp. 339.

3. A. Phillips, V. Kerlins, R. A. Rowe and B. V. Whiteson, Electron Fractography Handbook, Battelle Laboratories, Columbus, Ohio, 1976, pp. 3-35.

4. Metals Handbook, Vol. 10, American Society for Metals, Metals Park, Ohio, 1975, pp. 216.

5. M. Russo, "Analysis of Fractures Utilizing the SEM, "Metallography in Failure Analysis, Plenum Press, New York, 1977, pp. 76-82.

6. W. W. Gerberich, "Effects of Hydrogen on High Strength and Martensitic Steels," Hydrogen in Metals, Metals Park, Ohio, pp. 12.

7. A. R. Troiano, "General Keynote Lecture," Hydrogen in Metals, American Society for Metals, Metals Park, Ohio, pp. 12.

8. I. LeMay, "Failure Mechanisms and Metallography," Metallography in Failure Analysis, Plenum Press, New York, 1977, pp. 27.

9. A Phillips, V. Kerlins, R. A. Rowe and B. V. Whiteson, Electron Fractography Handbook, Battelle Laboratories, Columbus, Ohio, 1976, pp. 3-35.

10. Metals Handbook, Vol. 10, American Society for Metals, Metals Park, Ohio, 1975, pp. 217.

11. B. F. Brown, Stress Corrosion Cracking in High Strength Steels and in Titanium and Aluminum Alloys, Naval Research Laboratory, Washington, D.C., 1972, pp. 85-108.

12. Metals Handbook, Vol. 10, American Society for Metals, Metals Park, Ohio, 1975, pp. 233.

13. T. P. Groeneveld and A. R. Elsea, "Hydrogen-Stress Cracking in Natural-Gas-Transmission Pipelines," Hydrogen in Metals, American Society for Metals, 1974, pp. 727.

14. T. P. Groeneveld and R. R. Fessler, "Hydrogen Stress Cracking," 5th Symposium on Pipe Research, 1974, Houston, Texas.

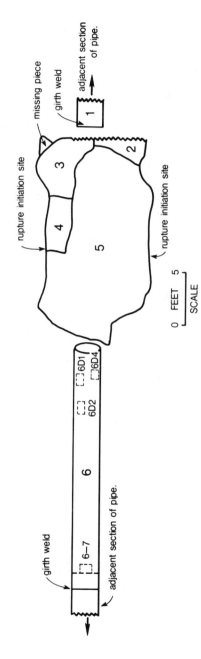

Fig. 1: Schematic of section of pipeline which ruptured in service.

Fig. 2: Paper replica of failed section of pipeline indicating crack path and direction as determined from Chevron markings.

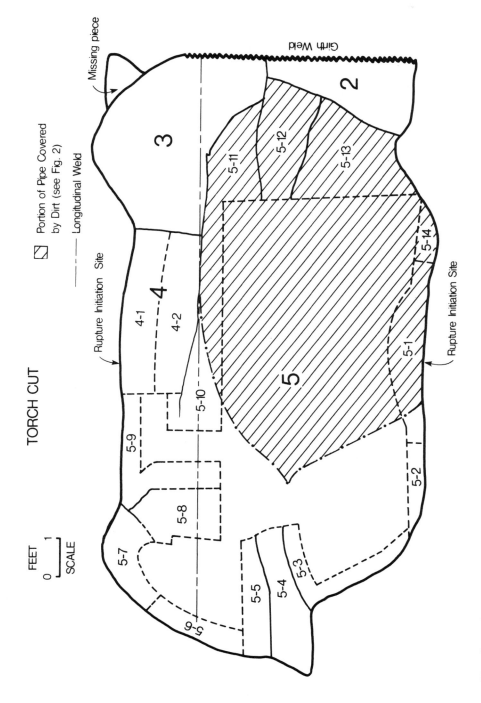

Fig. 3: Schematic of ruptured portion of section of pipeline that failed in service. Dotted lines indicate subsequent torch cuts. Shaded region was partially protected from heat of fire by a dirt cover.

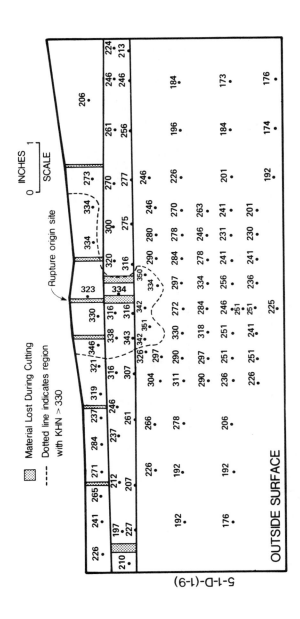

Fig. 4: Map of hardness of outside surface of piece 5-1-D, compiled from hardness measurements using Rockwell hardness tests (subsequently, converted to Knoop hardness) and Knoop hardness measurements.

Fig. 5: Microstructures of pipeline steel at various locations relative to the fracture origin site: 10' from fracture origin, 400X (upper, left); 8", 400X (upper, center); 1.5", 400X (upper, right); at fracture origin, 1000X (lower, left); and at fracture origin, 3600X (lower, right).

Fig. 6: Fractograph showing (a) secondary cracking on the pipe surface near the fracture origin location (upper left and left-center at 225X and 60X); (b) a secondary crack after opening it with higher magnification of crack extension region (upper, center-right and right at 60X and 2000X) and of the initial crack showing intergranular cracking (lower left, 2000X); and (c) fracture surface at crack origin (lower, center and lower right at 1500X and 15X).

QUANTIFYING RELIABILITY
FROM FAILURE ANALYSIS, WARRANTY,
AND SALES DATA

Roy G. Baggerly
Kenworth Truck Co.
Kirkland, Washington, USA

ABSTRACT

Information provided with field failures returned from a large active population may not provide sufficient background to determine a corrective action plan. Additional factors which can help assess total part reliability are warranty, manufacturing build quantities, and after-market sales data. Examples will be discussed showing use of this additional information.

INTRODUCTION

THE GROUND TRANSPORTATION INDUSTRY has experienced significant downweighting over the last decade. Greater fuel efficiency has been the driving force for this trend in automobiles whereas weight reduction for the highway tractor/trailer has been motivated primarily by the ability to increase payload since the total vehicle weight is monitored closely by each state. Utilizing the fullest potential of structural materials however, does result in a certain element of risk. There are over 130,000 Kenworth trucks operating in the U.S. today and since there is such a wide variability in service loads, various components will undoubtedly be subject to premature failures. When one component from a large field population fails, the pressing question that is asked is "Will it happen again?". The immediate task then is to analyze the failed component to determine if:

1. the failure was due to abusive use,

2. a manufacturing defect was present,

3. service degradation such as corrosion, wear, fatigue, etc. was responsible,

4. a design flaw is present,

5. the material was defective.

It may be unclear as to which cause is primarily responsible but ultimately a conclusion has to be reached as to whether a field problem exists and if it does, what is the requisite action that needs to be taken.

In order to understand the overall liability of a failure, some form of analysis is required to place the problem in perspective. The concept of reliability provides the basis for such an analysis and a comprehensive definition has been provided by Kececioglu (1). Reliability involves specifying a conditional probability at a given confidence level where the equipment will perform without failure for a certain length of time at some designated stress level. The relationship between failure and reliability is complementary such that their sum is unity; therefore, knowing one of these functions, the other may be derived. Various mathematical distributions can be applied to analyze failure but the Weibull function seems to be a generally useful tool to describe a wide variety of failure rate conditions (2,3).

The failure distribution can only be completely defined after the total population has suffered failure. Since this is impractical if timely corrective action is required, information obtained from early failures can often be utilized to predict with reasonable certainty the failure course of the remaining population. Hence, the failures as well as the successes need to be incorporated in the analysis. This can be accomplished by using the unfailed population as suspended data in a Weibull analysis.

Information contained in truck warranty claims generally includes the age or date in service, the mileage and the date of repair. Since many structural components are covered by warranty for 200,000 or 500,000 miles, the truck may be several years old when a claim comes in to the service department. The truck population will have been reduced due to the retirements and the mileage of the remaining truck population will be distributed over some mileage probability

distribution. The retirement and mileage probability distributions need to be quantitatively
defined in order to rank the suspended data
relative to the failure data.

TRUCK RETIREMENTS

Sources of information for the truck population can be obtained from the transportation
census conducted every few years by the U.S.
Dept. of Transportation (DOT) and an independent
research firm of R.L. Polk. Annual U.S. truck
registrations are recorded by R.L. Polk according
to vehicle manufacturer and age. Knowing the
difference in truck registrations between two or
more years for a specific age and comparing this
to the original number of trucks manufactured for
those years allows the retirement function to be
quantified. This is shown in Figure 1 for
Kenworth trucks manufactured from 1969 to 1984.
The data follows a three parameter Weibull
function and the resulting relationship between
reliability/survivability and failure/retirement
can be projected to longer time periods as shown
in Figure 2.

MILEAGE PROBABILITY DISTRIBUTION

Determining a probability distribution for
mileage as a function of age is not quite as
straightforward since the DOT census data only
reports the average mileage for a particular
truck age and the shape of the distribution is
not defined. However, knowing that there are a
few occurrences of trucks achieving 3,000,000
miles in 10 years of service allows a second
point to be estimated on a Weibull plot,
Figure 3. Since the slope seems reasonable at 2,
this value was assumed for the mileage
probability distribution. The resulting function
is close to a normal distribution with a slight
skew to higher mileages as shown in Figure 4.

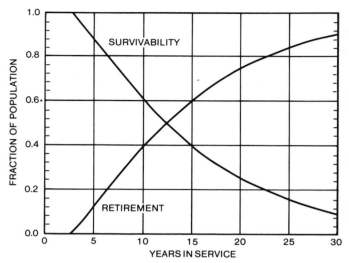

Fig. 2 - Truck retirement and survivability.

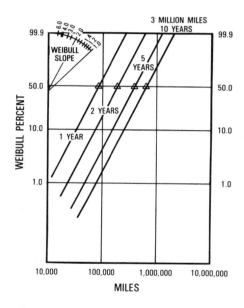

Fig. 3 - Weibull analysis of truck mileage.

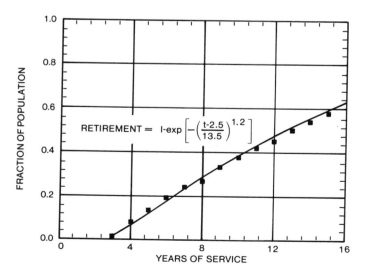

Fig. 1 - Truck retirement probability.

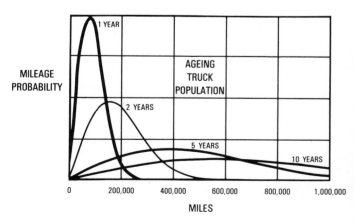

Fig. 4 - Mileage probability distribution for trucks of various ages.

RELIABILITY OF TWO PROPRIETARY STEELS

The use of this retirement and mileage probability information can be better appreciated by analyzing a specific case. Two proprietary steels from different suppliers were considered equivalent because of their stated minimum yield strengths. However, published data from each supplier showed that one steel had lower actual values for hardness, yield strength, ultimate strength, and endurance limit. The difference in the reversed bending endurance limit was approximately 28 MPa (4 KSI). Fatigue failures were being recorded in Warranty claims and the question of durability between the two steels arose. The engineering S-N fatigue curve can be related in a qualitative way to the duty cycle and miles of operation for a vehicle. Since the S-N curve can be linearized by plotting log S vs. log N, a similar qualitative treatment was used with duty cycle and miles as shown in Figure 5. More failures would be expected from steel B over a longer portion of the service life prior to retirement as compared to steel A.

The miles to failure for components fabricated with steel A are shown as a complete distribution without suspended items in Figure 6. A three component Weibull function is fitted but since the duty cycle is variable, the curve represents a composite of many distributions as discussed in reference 2. This enables a comparison to be made between the miles to failure distribution and the mileage probability distribution for the truck population. Figure 7a indicates that most of the failures for steel A occur at early mileages suggesting that as the median of the population mileage distribution advances to higher mileages with time, including subsequent retirements, the failure rate will decrease. However, when the failures for steel B are analyzed in a similar way, the median failure miles are closer to the median population

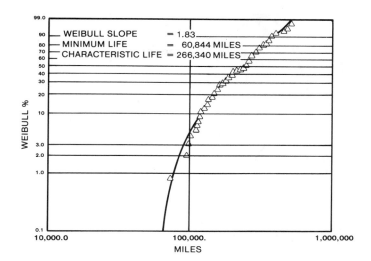

Fig. 6 - Weibull analysis of steel A failures.

mileage, Figure 7b. Since the service life and population mileage distribution are defined quantitatively, the number of unfailed components can be evaluated as a function of miles and used as suspended data in the Weibull analysis. This procedure has been well docmuented in reference 4. The complement of the failure function which is the reliability is shown plotted in Figure 8. The differences between the two failure distributions are dramatically portrayed in these reliability curves which have been extrapolated considerably past the median population mileages. This suggests that the failure rate for steel B will increase relative to steel A as the truck population ages. It is interestinbg to note that within the warranty period, the reliability is nearly equivalent for the two steels. However, the decision to use steel A over steel B will depend on several factors but significantly on whether the component is safety related as well as the relative cost and availability between the two. Oftentimes failures are related to severe service conditions where the increased productivity gained by exceeding the design limitations justifies the occasional failure. A cost penalty passed on to the customer may not be justified in this instance since the truck observing a normal duty cycle would never experience a similar failure.

REPLACEMENT PARTS SALES

Replacement parts not covered by warranty may be more difficult to analyze since the recording of mileages at time of replacement is seldom practiced in the industry. The field experience or reliability for these parts will remain unknown or vague unless a failure is brought to the attention of a warranty analyst. In an attempt to overcome this problem a method was developed to analyze the replacement part sales history since this would be an indication

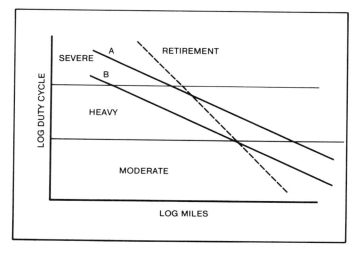

Fig. 5 - Qualitative log S - log N or log duty cycle - log miles plot for two proprietary steels A and B.

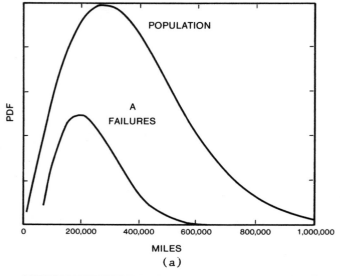

(a)

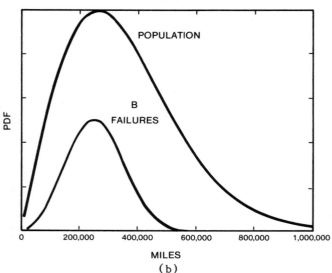

(b)

Fig. 7 - Mileage probability distributions of truck population and part failures: (a) steel A, (b) steel B.

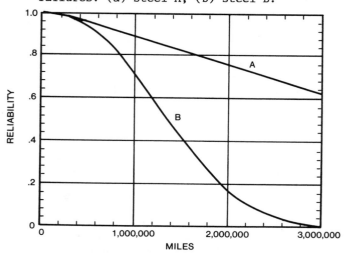

Fig. 8 - Reliability of trucks with steel A components and those with steel B.

of reliability with time instead of mileage. If the quantity being manufactured or put into service over successive time intervals were known along with the composite number of replacement parts being sold for each time interval, then presumably a replacement or failure function could be estimated that would satisfy the sales data. This method requires that the sales history be available from the beginning of manufacture for the part. This information at Kenworth has been available since the late 1960's. Figure 9 gives a brief outline for the method where each monthly quantity of parts produced are acted upon by a constant failure rate. A sale of a replacement part occurs whenever there is any failure from quantities placed in service. Since the time in service prior to failure is not recorded, the sales record represents only a composite of the total replacement parts which is available on a monthly basis. A computer program performs an iterative calculation for each monthly quantity manufactured. The Weibull slope and characteristic life for a trial sample failure rate are held constant during each iteration and then incrementally changed. The RMS error is calculated for each iteration and the Weibull parameters producing the least RMS error represent the best fit to the cumulative replacement sales data.

This method can be better explained by describing an actual case of a failed aluminum die casting. The part happened to fail during the warranty period and so the appropriate record was available for this one part. Analysis of the fracture surface revealed a high percentage of shrinkage porosity as shown in Figure 10a. The production of die castings represent a well

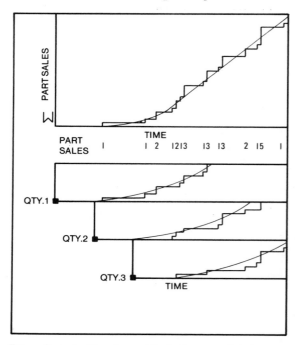

Fig. 9 - Analysis method for replacement part sales.

regulated and repetitive process. If one part is discovered with shrinkage porosity, then there is a high probability that other parts will also exhibit this effect. Suspecting there may have been other unreported failures, the replacement part sales records were searched and analyzed. Figure 11 shows the resulting Weibull fit to the cumulative sales data. The slope or shape parameter of 2.64 indicates a near normal distribution and the resulting reliability for the part is shown in Figure 12.

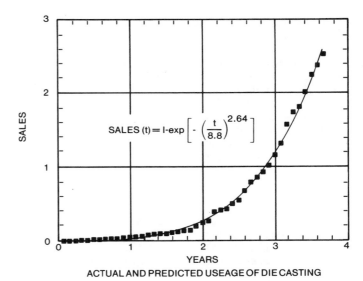

$$SALES (t) = I\text{-}exp\left[-\left(\frac{t}{8.8}\right)^{2.64}\right]$$

ACTUAL AND PREDICTED USEAGE OF DIE CASTING

Fig. 11 - Actual and predicted replacement part sales for an aluminum die casting.

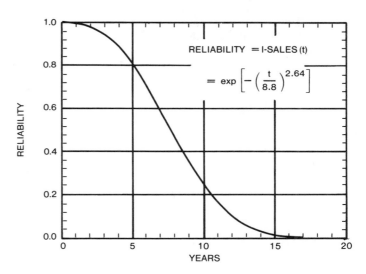

$$RELIABILITY = I\text{-}SALES (t)$$

$$= exp\left[-\left(\frac{t}{8.8}\right)^{2.64}\right]$$

Fig. 12 - Reliability of an aluminum die casting.

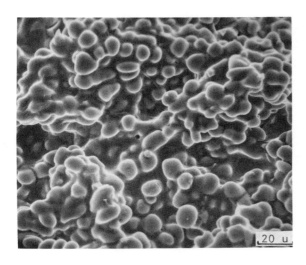

(a)

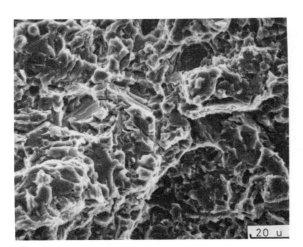

(b)

Fig. 10 - Fracture surfaces of aluminum die castings: (a) with shrinkage porosity, (b) without shrinkage porosity.

SUMMARY

Two case histories have been presented in an attempt to describe the procedures used in determining the service reliability of truck components. Statistical treatments of truck retirement and population mileage probability were required so that complete failure and reliability distributions could be estimated. This information was evaluated to determine if parts have acceptable durability. In the case of the two steels, an appropriate corrective action was taken to improve the fatigue endurance level of steel B. The shrinkage porosity in the aluminum die casting was related to gating and die design which was subsequently revised in order to eliminate this condition.

Determining the cause of a specific failure is essential but may be too limiting in scope for a valid corrective action plan if a large service population exists. The total reliability should be evaluated to obtain a complete understanding of the consequences of a failure and this can usually be accomplished if all available information is analyzed.

ACKNOWLEDGEMENTS

The author wishes to acknowledge J.R. Strong and R.A. Drollinger for many helpful discussions and for writing the source codes for the various computer programs used in these analyses.

REFERENCES

1. D. Kececioglu, Mechanical Reliability And Probabilistic Design for Reliability, UCLA Short Course Lecture Notes, 1984
2. J.H. Bompass-Smith, Mechanical Survival, p.59, McGraw-Hill Book Co., London, 1973
3. C. Lipson and N.J. Sheth, Statistical Design And Analysis of Engineering Experiments, p.36, McGraw-Hill Book Co., New York, 1973
4. M.A. Vasan, Reliability Estimation And Failure Prediction of Highway Tractor Components, SAE Paper 820979, 1982

MICROSTRUCTURAL EXAMINATION
OF FUEL RODS SUBJECTED TO
A SIMULATED LARGE-BREAK LOSS OF
COOLANT ACCIDENT IN REACTOR

A. Garlick
United Kingdom Atomic Energy Authority
United Kingdom

ABSTRACT

A series of tests has been conducted in the
National Research Universal (NRU) reactor,
Chalk River, Canada, to investigate the
behaviour of full-length 32-rod PWR fuel
bundles during a simulated large-break loss
of coolant accident (LOCA). In one of these
tests (MT-3), 12 central rods were pre-
pressurized in order to evaluate the ballooning
and rupture of cladding in the Zircaloy
high-$\alpha/\alpha+\beta$ temperature region. All 12 rods
ruptured after experiencing < 90% diametral
strain but there was no suggestion of coplanar
blockage.

Post-irradiation examination was carried out
on cross-sections of cladding from selected rods
to determine the azimuthal distribution of wall
thinning along the ballooned regions. These
data are assessed here to check whether they
are consistent with a mechanism in which fuel
stack eccentricity generates temperature
gradients around the ballooning cladding and
leads to premature rupture during a LOCA.

After anodizing, the cladding microstructures were examined for the presence of prior-
beta phase that would indicate the $\alpha/\alpha+\beta$
transformation temperature (1078K) had been
exceeded. These results were compared with
isothermal annealing test data on unirradiated
cladding from the same manufacturing batch.

IN THE UK, considerable attention has been
devoted to a study of the so-called cladding
ballooning phenomenon since it is a potentially
adverse influence on coolability of the fuel in
the unlikely event of a major loss of coolant
accident (LOCA) in a PWR. During such a
postulated accident, reactor depressurisation
occurs and the helium pressure in the fuel rods
can lead to deformation (ballooning) of
Zircaloy cladding during the subsequent
temperature transient. If the magnitude and
axial extent of the deformation were to become
excessive it could result in a restriction of
coolant flow between the rods and thus reduce
the effectiveness of the emergency core
cooling system.

As part of this study, the UKAEA became
a participant in a series of large-break LOCA
simulation experiments sponsored by the USNRC
that were conducted on full-length PWR fuel
bundles in the National Research Universal
(NRU) reactor, Chalk River, Canada, by
Battelle Pacific Northwest Laboratory. In
one of the experiments (MT-3) for which the
UK specified the test conditions, the
thermal-hydraulics and ballooning
behaviour of the rods were investigated during
heat-up, reflood and quench phases in order
to provide information relevant to the design
of the Sizewell B PWR, construction of which is
presently under consideration in the UK. The
results of the MT-3 test,[1][2] taken together
with data from many studies worldwide, have led
to the conclusion that ballooning, and the
consequent degree of coolant channel blockage,
will be limited by non-uniformities in cladding
temperature. This is basically because the
creep rate of the Zircaloy cladding is strongly
dependent on temperature. Furthermore, the
magnitude of the temperature non-uniformities is
believed to be governed by heat transfer from
fuel to cladding, thus emphasising the importance
of the eccentricity of the fuel stack relative
to the ballooning cladding.

Post-irradiation examination has been
carried out on selected fuel rods from the
MT-3 test assembly to determine whether
the azimuthal distribution of wall thickness
strain in the ballooned cladding is consistent
with the proposed fuel stack eccentricity model.

The results of that work and their assessment
are described here along with observations

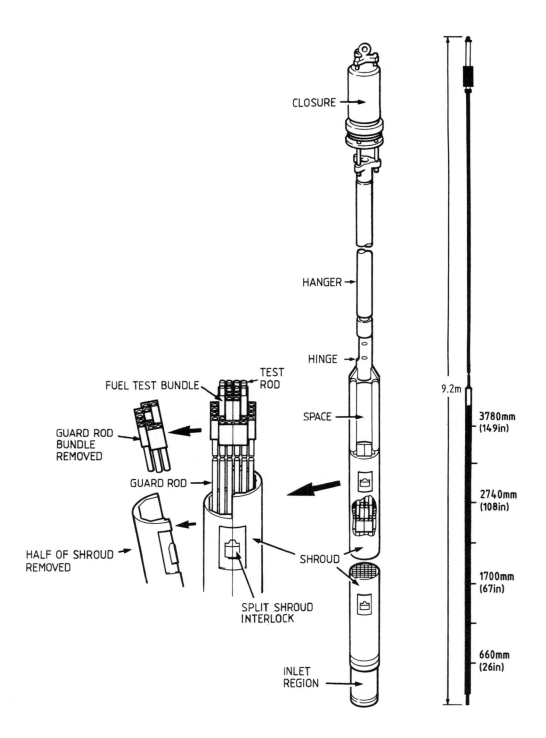

Fig. 1 - Schematic diagram of MT-3 test train (grid locations indicated):   reproduced from Ref. 1.

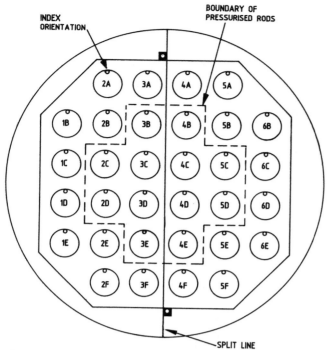

Fig. 2 - Section through MT-3 test assembly

TABLE 1
Fuel rod design parameters

Cladding material	Zircaloy -4
Cladding outer diameter	9.63mm (0.379in)
Cladding inner diameter	8.41mm (0.331in)
Fuel pellet diameter	8.26mm (0.325in)
Fuel pellet length	9.53mm (0.375in)
Fuel stack length	3.66mm (144in)
Rod-to-rod pitch	12.75mm (0.502in)
Helium pressure in rods	3.9MPa (565 psia)

TABLE 2
Typical composition of Zircaloy-4 cladding

Element	Concentration (wt%)
Sn	1.2 - 1.7
Cr	0.07 - 0.13
Fe	0.19 - 0.24
Fe+Cr	0.28 - 0.37
O	0.10 - 0.15

of cladding microstructural changes that occurred during the transient. These changes are compared with data from out-of-reactor isothermal annealing of unirradiated cladding from the same manufacturing batch in order to allow estimation of the cladding temperature during ballooning.

THE MT-3 LOCA SIMULATION TEST

The MT-3 test was one of a series performed in the NRU reactor on instrumented full-length fuel arrays consisting of 6x6 segments (Tables 1 and 2), of a 17x17 PWR fuel assembly. Because the fuel was irradiated for only a brief period before the test, nuclear heating was applied during the quench and reflood periods in order to simulate the decay heating that would have been generated in more highly irradiated fuel rods.

The LOCA simulation test train, consisting essentially of the fuel assembly, split stainless steel shroud, hanger bar and top closure, is illustrated schematically in Figure 1: the locations of the grids are also indicated. A detailed description of the MT-3 test train has been reported previously[3].

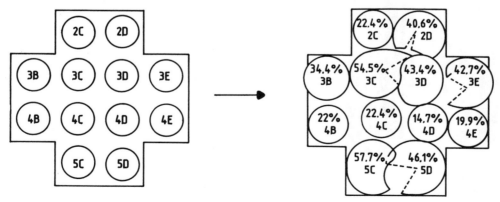

Fig. 3 - Cladding diametral strains resulting from the MT-3 test (axial location 2.69m from the bottom of the test train): reproduced from Ref 1.

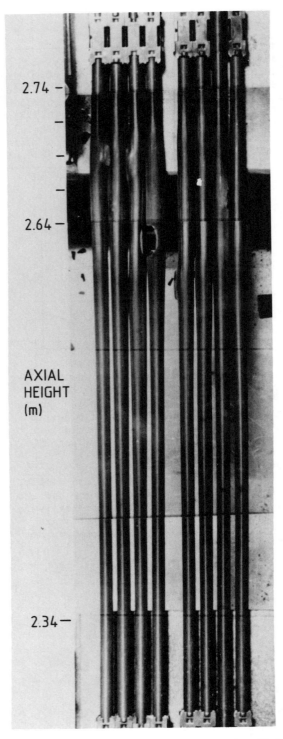

Fig. 4 - Cladding deformation in the MT-3 assembly (from left to right, Rods 3E, 3D, 3C, 3B, 4B, 4C, 4D, and 4E): reproduced from Ref 1.

The test fuel rods were arranged in a 6x6 array (Figure 2) but the four corner rods were omitted. Also shown in Figure 2 is the boundary of the 12 internally pressurized test rods; the other 20 rods acted as a guard ring to reduce heat losses.

The aim of the MT-3 experiment was to produce ballooning of the 12 test rods and to investigate the extent of mechanical interaction and flow channel restriction under conditions relevant to a postulated large-break LOCA in the Sizewell B PWR. For this purpose a transient was chosen in which the fuel cladding would be most susceptible to ballooning, namely a so-called 'flat-top' transient where the cladding temperatures are maintained in the Zircaloy high-alpha phase region (~ 1073K) for an extended period (~250s). Based on computer code runs, the test rod internal pressures were set at 3.9MPa (565 psia) to ensure that cladding hoop stresses were within the ballooning 'window'. The test resulted in rupturing of all 12 rods and considerable deformation of the cladding, as shown in Figures 3 and 4, but there was no evidence of coplanar blockage and the reflooding process proved to be relatively unaffected by ballooning.

EXAMINATION OF SELECTED RODS

A post-irradiation examination programme was undertaken on Rods 2C, 3C, 3D, 4C, 4D and 5C to provide information on azimuthal distributions of cladding wall thickness strains. The cladding sections from Rods 3C, 3D and 4D were also subjected to detailed microstructural examination after anodizing in order to allow estimation of local temperatures that had been reached during the transient. The sections were taken at 25mm intervals at the following positions relative to the bottom of the test train (Figure 1): Rod 2C, 2.44-2.54m; Rod 3C, 2.49-2.72m; Rod 3D, 2.36-2.62m; Rod 4C, 2.53-2.73m; Rod 4D, 2.49-2.72m; Rod 5C, 2.44-2.54m.

The positions of fuel assembly grids, two of which are visible in Figure 4, are shown in Figure 1. The relative azimuthal orientations of adjacent sections were preserved throughout the cutting: locations of the sections with respect to the rod diametral strain profiles can be determined by reference to Table 3.

After metallographic preparation of each section, the cladding wall thickness was measured at ~1mm intervals around the whole circumference.

TABLE 3
Summary of cladding wall thickness and diametral strain results

Rod identifi- cation	Section number	Height of section from bottom of test train (m)	Cladding wall thickness (mm)		Cladding diametral strain (%)
			Range	Mean	
2C	1	2.44	0.043	0.586	4.0
	2	2.46	0.051	0.578	4.5
	3	2.49	0.048	0.572	5.4
	4	2.51	0.054	0.567	6.5
	5	2.54	0.074	0.555	8.8
3C	1	2.49	0.063	0.552	9.7
	2	2.51	0.124	0.523	14.5
	3	2.54	0.165	0.483	22.1
	4	2.57	0.132	0.460	29.4
	5	2.59	0.198	0.455	31.1
	6	2.62	0.109	0.523	16.4
	7	2.64	0.107	0.497	20.4
	8(a)	2.67	0.322	0.434	37.9
	9	2.69	0.249	0.446	58.2
	10	2.72	0.036	0.568	8.0
3D	1	2.36	0.030	0.598	1.8
	2	2.39	0.036	0.589	2.6
	3	2.41	0.036	0.585	3.3
	4	2.44	0.038	0.581	4.1
	5	2.46	0.036	0.575	5.0
	6	2.49	0.043	0.566	6.4
	7	2.51	0.041	0.554	8.7
	8(a)	2.54	0.038	0.537	11.4
	9	2.57	0.026	0.517	15.0
	10	2.59	0.083	0.495	20.1
	11	2.62	0.139	0.472	25.7
4C	1	2.53	0.150	0.546	10.8
	2	2.55	0.236	0.515	17.0
	3(a)	2.58	0.310	0.475	26.9
	4	2.60	0.478	0.458	48.3
	5	2.63	0.252	0.478	26.0
	6	2.65	0.200	0.469	27.7
	7	2.68	0.083	0.489	23.8
	8	2.71	0.050	0.516	16.4
	9	2.73	0.071	0.575	5.4
4D	1	2.49	0.035	0.559	7.7
	2	2.51	0.018	0.542	10.8
	3	2.54	0.033	0.521	14.7
	4	2.57	0.076	0.508	17.6
	5	2.59	0.147	0.474	26.3
	6	2.62	0.272	0.449	33.4
	7(a)	2.64	0.335	0.434	37.6
	8	2.67	0.317	0.469	26.6
	9	2.69	0.153	0.527	14.7
	10	2.72	0.071	0.565	6.6
5C	1	2.44	0.051	0.578	4.6
	2	2.46	0.066	0.570	5.9
	3	2.49	0.059	0.560	7.5
	4	2.51	0.056	0.547	9.5
	5	2.54	0.088	0.535	12.8

Note: (a) Site of cladding rupture

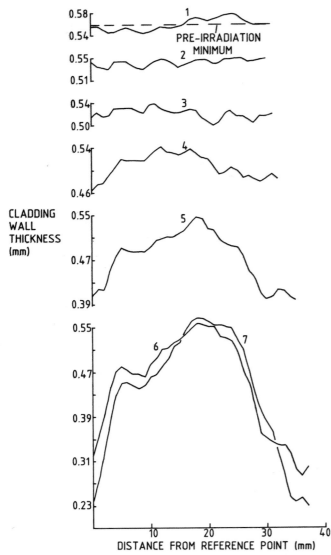

CLADDING WALL THICKNESS (mm)

PRE-IRRADIATION MINIMUM

DISTANCE FROM REFERENCE POINT (mm)

Fig. 5- Cladding wall thickness distributions around cross-sections from Rod 4D (identification numbers refer to sectioning positions in Table 3).

The sections from Rods 3C, 4C and 4D were subsequently anodised in Picklesimer's solution[4] to reveal any prior-beta phase, and then photographed at high magnification.

ASSESSMENT OF RESULTS

Cladding wall thicknesses measured on sections from Rod 3D, as a typical example, are plotted in Figure 5: the minimum wall thickness specification for as-manufactured cladding tubes is included, for comparison purposes. Each of the wall thickness plots is identified by a number corresponding to

the metallographic section concerned; the section locations and the wall thickness data are presented in Table 3 along with the corresponding diametral strains.

It will be noted that in Figure 5 the azimuthal variation in wall thickness follows an approximately cosine distribution on those sections that had experienced low diametral strains. This is typical of as-manufactured tubing and is introduced during the fabrication process. In spite of this initial non-uniformity of the wall, the cladding deformation proceeded apparently uniformly around each circumference until diametral strains exceeded 10% on most of the rods.

The onset of non-uniform deformation of the cladding is indicated more clearly in Figure 6 where the range of the wall thickness measurements for each section from Rod 3D is plotted against diametral strain. An interesting feature of these results is that the range of wall thickness values for Section 9 (Table 3) is rather lower than for the preceding sections. This is taken to mean that localised wall thinning was initiated in the thicker wall region on Rod 3D. The wall thickness data from the other rods were analysed in the same way and similar behaviour to that in Rod 3D was observed on sections from the top end of the balloon in Rod 4C and the bottom end of the balloon in Rod 4D. Values of diametral strain at onset of localised wall thinning are summarised in Table 4. It is clear that local deformation occurs at much lower diametral stains than the 32.5% at which the cladding of adjacent rods would come into contact. These results support the view that the temperature gradients, that are directly responsible for the localised wall thinning, result from azimuthal differences in heat transfer from fuel to cladding rather than between adjacent rods.

Since localised wall thinning was initiated in the thicker wall region on some of the rods it is possible to calculate the minimum azimuthal temperature gradient around the cladding from a knowledge of the Zircaloy secondary creep equation and the range of wall thicknesses around each section. An equation [5,6] describing secondary creep of alpha phase Zircaloy in steam in the temperature range 973-1073K takes the form:

$$\dot{\varepsilon} = A\sigma^n \exp (-Q/RT) \text{ ----------------- (1)}$$

where $\dot{\varepsilon}$ is the secondary creep rate

A is a constant

σ is the initial hoop stress

n is the stress index for secondary creep (=5)

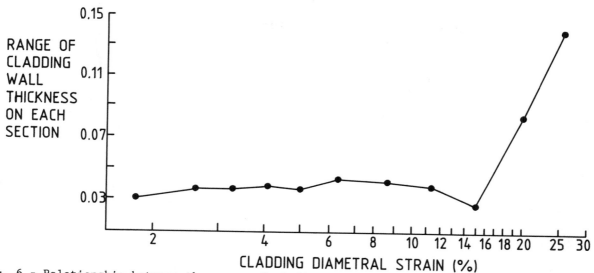

Fig. 6 - Relationship between the range of cladding wall thickness, measured on each section, and diametral strain for Rod 3D.

TABLE 4
Cladding diametral strain at onset
of localised wall thinning

Rod identity	Location on ballooned region	Cladding diametral strain at onset of localised wall thinning (%)
2C	bottom	8
3C	top	10
3C	bottom	<10
3D	bottom	15
4C	top	20
4D	top	6
4D	bottom	16
5C	bottom	12

Q is the activation energy for secondary creep (=294000 J/mole)

R is the gas constant

T is the absolute temperature

Taking the wall thickness ranges observed on Section 1 of Rod 3D and Section 9 of Rod 4C as extreme values, the minimum differences in cladding temperatures, based on Equation 1, are 6 deg K and 20 deg K, respectively. Earlier work[7,8] has demonstrated that the rupture behaviour of Zircaloy in the high alpha range is very sensitive to temperature differences of this magnitude.

The azimuthal orientations of the maximum and minimum wall thickness regions were determined on the sections and plotted against axial location. The distributions generally varied smoothly along each rod and there was a tendency towards a spiral distribution, which is possibly a reflection of the form

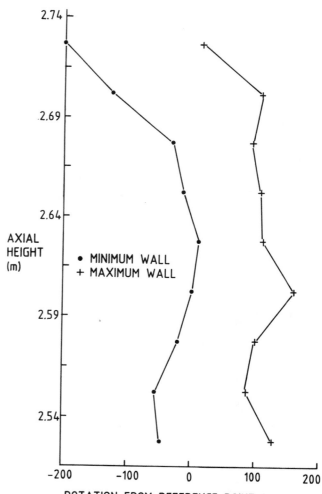

Fig. 7 - Azimuthal orientations of regions of maximum and minimum cladding wall thickness on sections from Rod 4C.

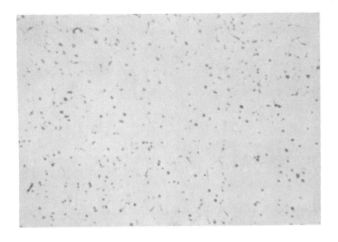

a

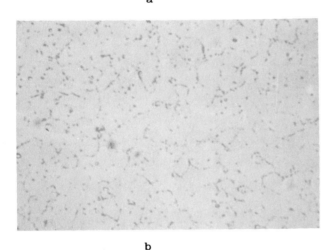

b

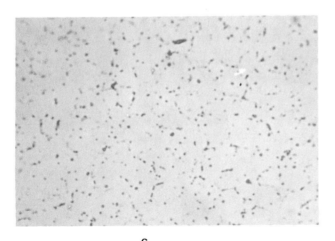

c

Fig. 8 - Microstructure of cladding sections from Rod 4D after anodizing: (a) Section 7 (2.64m), (b) Section 5 (2.59m) and (c) Section 3 (2.54m). All fields of view are in line with the cladding rupture (at 2.65m). (Bright field, x850).

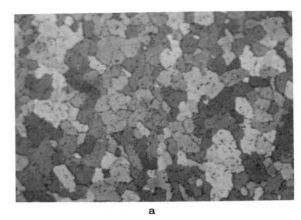

a

b

Fig. 9 - Comparison of cladding microstructure from Rod 4D with that produced by isothermal annealing of archive tubing: (a) section through Rod 4D at 2.59m (b) section through archive tubing after 200s at 1138K. (Polarised light, x 425).

adopted by the underlying fuel pellet stack during ballooning. A typical plot is illustrated in Figure 7. The average rotation of the thinned region along the balloon was in the range 4-9°/cm for Rods 4C and 4D. Results from Rod 3C were comparable after allowing for the fact that the balloon was in the form of a double bulge (Table 3). It was noteworthy that there was a much larger change (30°/cm) in the region between the two bulges on this rod.

The sections from Rods 3C, 4C and 4D[4] were anodized in Picklesimer's solution to reveal evidence of any prior-beta phase in the Zircaloy (note that beta phase is not retained on cooling below the α/α+β transition temperature, but can be observed as areas of prior-beta phase that are distinguishable from the alpha matrix). Clear indications of an intergranular network (believed to be prior-beta phase, but not positively identified) were present on 2 sections from

Rod 4D but not on any other section from the 3 rods. Photomicrographs of these two sections (from 2.54m and 2.59m) are presented in Figure 8 along with one from Section 7 (2.64m) for comparison purposes. These were compared with structures produced by isothermal annealing archive MT-3 cladding tubes for 200s over a range of temperature, from which it was concluded that intergranular networks began to form at temperatures between 1133K and 1153K.

The cladding structure from the fuel rod at the 2.59m location may be compared with annealed unirradiated cladding in Figure 9. It was further concluded from examination of the rod sections that the intergranular networks were present in regions that were in line with the cladding rupture. This is further evidence for the existence of an azimuthal temperature gradient in the cladding.

It is concluded, therefore, that Rod 4D experienced locally higher cladding temperatures than Rods 3C and 4C and this is consistent with the in-reactor measurements of cladding temperatures at the 3.02m level, the closest thermocoupled points to the ballooning regions on these rods.

CONCLUSIONS

Post-irradiation examination has been carried out on cross-sections of Zircaloy cladding from some of the fuel rods that were subjected to a simulated large-break LOCA in the MT-3 test. The following conclusions have been drawn.
(i) Even in sections that had experienced only low diametral strains there was an azimuthal variation in wall thickness, following an approximately cosine pattern, that had been introduced during tube manufacture.
(ii) Diametral strains at which onset of localised wall thinning commenced were in the range 6-20% for the 8 regions examined. These values are much lower than the 32.5% at which adjacent ballooning rods would contact and they lend support to the view that temperature gradients, that are directly responsible for localised wall thinning, result from azimuthal differences in heat transfer from fuel to cladding rather than between adjacent rods.
(iii) There were 3 instances in which localised wall thinning commenced in the initially thicker wall region. Calculations based on the known temperature dependence of Zircaloy secondary creep and the measured range of wall thickness on each section show that, for this to occur, the minimum temperature difference between the thick

and thin regions on each section would have had to be 6 deg K (Section 1, Rod 3D) or 20 deg K (Section 9, Rod 4C).
(iv) The azimuthal orientation of wall thinning generally varied smoothly along the ballooned regions. There was a tendency towards a spiral configuration, possibly indicative of the form adopted by the fuel pellet stack during ballooning.
(v) Microstructural examination of the cladding after anodizing revealed an intergranular network, possibly prior-beta phase, in 2 sections, in line with the cladding rupture, on Rod 4D but not on the other rods examined (Rods 3C and 4C). Comparison of this structure with those produced by isothermal annealing of archive cladding led to the conclusion that local cladding temperature on Rod 4D was higher than in the other rods and probably reached 1133-1153K.

ACKNOWLEDGEMENT

The work reported in this paper has been partly funded by the Central Electricity Generating Board.

REFERENCES

1. Gibson, I. H., P. Coddington, T. Healey and C. A. Mann, UKAEA Report AEEW-R1506 (1982).
2. Mohr, C.L., G.M. Hesson, L.L. King, R.K. Marshall, L.J. Parchen, J.P. Pilger, G.E. Russcher, B.J. Webb, N.J. Wildung, C.L. Wilson and M.C. Wismer, NUREG/CR 2528 (1982).
3. Coddington, P., UKAEA Report AEEW-R1538 (1982).
4. Picklesimer, M.L., ORNL-2296 (1957).
5. Hindle, E.D., paper presented at Water Reactor Information Meeting, Gaithersburg, USA, October 1975.
6. Donaldson, A.T., T. Healey and R.A.L. Horwood, Central Electricity Generating Board Report TPRD/B/0007/N82; presented at the IAEA Specialist Meeting on Water Reactor Fuel Element Performance and Computer Modelling, Preston, England, March 1982.
7. Erbacher, F., H.J. Neitzel and K. Wiehr. In: Zirconium in the Nuclear Industry (Fourth Conference), ASTM STP 681, American Society for Testing and Materials, 429-446 (1979).
8. Healey, T.E., A.F. Brown, A.T. Donaldson, R.A.L. Horwood, R.M. Cornell, G.F. Hines, T.J. Haste, E.D. Hindle, P.E.Pearson, A.E.Reynolds, C.Vitanza, T.Johnsen and R.Smallcalder. In: Proc. BNES Conference on Nuclear Fuel Performance, Stratford on Avon, England, March 1985.

THE ANALYSIS OF
AIRCRAFT COMPONENT FAILURES

A. Jones and C.J. Peel
Royal Aircraft Establishment
Farnborough, Hants, United Kingdom

Abstract

The Royal Aircraft Establishment has analysed
aircraft failures for at least four decades and
this paper outlines the developments in both
understanding the failures and the practical
techniques that have been applied in their
interpretation. Previous failures are reviewed
and the relative importance of the various
failure modes encountered indicated, emphasising
the dominant role of fatigue of metal structures,
and the improvements that have occurred
particularly with corrosion related problems.
Improved techniques for the quantitative analysis
of fatigue failures are explained and illustrated, including combined fractographic and
fracture mechanics analyses. It is shown how
quantitative results have led to a greater
understanding of the damage effects of defects
on fatigue performance and how the metallurgist
now frequently advises the engineer on the
integrity of his structures.

THE ROYAL AIRCRAFT ESTABLISHMENT has been
involved in the investigation of aircraft
component failures for over 40 years. During
the last 16 years, over which more complete
records have been kept, 50-70 failures in
various systems have been examined each year.
This paper reviews the nature of the failures
already examined, the techniques employed in
their investigation and highlights persistent
trends in the incidence of failures. In the
main the work is involved with new arisings
rather than those of a repetitive nature, and
the analysis of both civil and military systems
is included.

REVIEW OF PREVIOUS FAILURES

The underlying causes of some 450 failures
have been analysed and divided into 8 categories
(Fig 1) in terms of the occurrence of the
failures, possibly during Test, under a Service
Environment or, for example as a result of a
Crash. A detailed analysis of the modes of
failure in these categories quickly reveals
that fatigue failure predominates in all
categories but the Service Environment and
Crash. This important conclusion is illustrated
by a breakdown of the failures in all the
categories in terms of the numbers occurring
per year by fatigue and corrosion over the last
16 years (Fig 2). This breakdown is repeated
(Fig 3) for failures in aluminium alloy
structures only. In this latter case of
failure of aluminium alloys it would seem that
there is some evidence of a decline in the
number of corrosion related failures, particularly stress corrosion, due to the introduction
of resistant alloys but the actual decline is
not very marked. The reason for this probably
lies in the fact that there are still many
components in service that have been manufactured in stress corrosion susceptible alloys
which will take some time to phase out. It is
understood, however, that the expected trend is
being observed by undercarriage manufacturers.

A more detailed look at the failures that
have been judged to have been caused by the
Service Environment without any influence from
the other factors such as manufacturing defects
or inadvertent damage shows (Fig 4) four
principal mechanisms; fatigue, stress corrosion

cracking (SCC), corrosion and corrosion-initiated fatigue, in which the total of corrosion related failures outweighs pure fatigue. A consideration of the failures within the Design category, that is those failures, predominantly by fatigue, for which it was considered that the design was inadequate for the purposes for which it was intended, enabled a further division into sub-categories such as for example, Poor Design, High Stress Concentration (Kt), and Material Choice (Fig 5) and shows the considerable influence of stress concentrations, in both machined and fabricated conditions, surely a reflection on the way experience already gained is passed on to new designers and draughtsmen. It seems that each new generation must make the same old mistakes.

The problems of underdesign, in which the concept is right but where dimensions are inadequate, may frequently be influenced by lack of knowledge of the actual stresses involved in service and, as such, highlight the need for early strain gauging exercises on prototype aircraft. Such exercises also provide information that can be fed into the full scale fatigue test, thus enabling realistic loads to be applied from an early stage of the test. A similar situation arises when existing structure is stressed beyond its design capability because of an aircraft role change or increase in all-up weight. Poor design, in which the concept is wrong, not a prominent factor in the statistics but which occasionally gives rise to serious problems, can sometimes be brought about by the shear complexity of some modern aircraft components but in the main has been found to occur in ground equipment, ancilliary aircraft equipment and weaponry, where aircraft standards are not always cost effective.

The use of unsuitable materials is occasionally the result of not involving the metallurgist in the design stage. However, situations have arisen in which the space available, or strength levels required, dictate the use of very high strength materials typically more prone to stress corrosion cracking and brittle failure. Similarly, perhaps, in the case of surface coatings the timely advice of a materials specialist might have avoided a failure. On the other hand, failures that occur in the short transverse direction are often due to the use of integrally machined ribs and angles in wing skins and other areas such as forgings where it may be impossible to avoid stressing that direction. To some extent the ever increasing cost of integrally machined thick structure is tending to promote a trend towards built-up structures, a reversal of the trends of the last decade. More expensive materials such as the emerging aluminium-lithium alloys will naturally encourage higher material utilisation and less integral machining. It may be noted that some of the modern aluminium alloys are especially designed to offer improved short transverse properties.

PRACTICAL TECHNIQUES FOR FAILURES ANALYSIS

The examination of a failure begins with an overall low power stereoscopic examination of the failed component at a magnification of 0 to 25, if possible with a knowledge of its function, to establish possible relationships between the focal point of the failure, a crack or a fracture, and other features such as changes of section, casting, forging and machining marks, surface protection, damage, corrosion and distortion. This is an important stage in any examination in which valuable information can be gained about the nature of the failure, its origin and direction and any influencing factors.

The next stage involves examination of the surface adjacent to the failure and of the surface of the crack or fracture at higher optical magnifications. To achieve this it may be necessary to cut the component to obtain a small enough section for the microscope and this is the time that decisions should be made concerning mechanical test requirements and the positioning of the specimens to be machined.

Detailed examination of the outer surface for, eg cracks in surface coatings, fine corrosion damage, machining marks etc, can be carried out at magnifications from 25 to 500 using an optical microscope with either coaxial or external, oblique lighting. In fact the direction of lighting can be critical, particularly at the lower magnifications, in the detection and identification of features in the outer surface adjacent to a crack. At the highest practical magnification for this work, in the region of 500, it is necessary to use an objective lens with a long working distance, at least 4mm is available, to enable the greatest possible coverage of awkward shapes and rough fractures. It is at this stage that initial impressions of the fine fracture detail will be gained, ie whether there are intergranular features or fatigue striations visible and, if fatigue, how the striations are grouped and related to inclusions or microstructural features. In fact it is sometimes the case that an optical examination will reveal all available information although, since the advent of electron optics, the optical microscope is often virtually overlooked in favour of the scanning electron microscope (SEM). This is unfortunate because, as explained later, the two systems are truly complementary. Nevertheless, the vast improvement in resolution and depth of focus provided by the SEM has greatly extended the fractographers scope and dramatically increased the range of fractographic observation, so that it is now an essential tool in failures investigation. It is frequently coupled with the ability to provide chemical analysis of very small volumes of material, eg 0.1μm in diameter.

Metallography also has an important part to play in relating any microstructural deficiencies, internal defects and surface features with crack initiation, and it is often found that this stage reveals the actual reason for initiation, whereas study of the fracture itself may only provide information on what happens after initiation. The manner of sectioning and the stage at which it is carried out, particularly if only one sound half of a fracture is available, must be carefully considered in order to obtain maximum benefit without premature loss of valuable evidence.

IMPROVED QUANTITATIVE FRACTOGRAPHIC TECHNIQUES

Fatigue failures of both the Test and Service Environment types now require quantitative analysis. Such analysis relies heavily on the measurement of fatigue striation spacings which is readily achieved in the structural materials in predominant use, ie aluminium and titanium alloys. Fatigue striations are produced by the cyclic loads and if the resolution of the surface examination technique is adequate then, in principle, every load producing crack extension should be observed as a striation. In practice there are finite limits to resolving power and in this respect the appearance of the modern electron-optical microscopes has increased the fractographers ability to measure slow crack growth rates by an order of magnitude, ie crack growth rates of less than 0.1 micron per load application may be routinely measured using the scanning electron microscope compared with of the order of 1 micron per load optically. Fig 6 shows how the optical and electron optical techniques complimented each other in the examination of an undercarriage that failed during fatigue testing. At high growth rates, where the electron microscope may provide too much detail, the optical microscope is capable of identifying major load level changes, eg aircraft ground-air-ground cycles, because of changes in crack path angle and hence in optical reflectivity.

In practice both real service experience and modern major fatigue tests involve load combinations that are complex, being a mixture of a regular pattern of load frequency excursions, ie the ground-air-ground cycle, with superimposed higher frequency loads of a more irregular nature generated by manoeuvre and gust experiences. Fatigue striation patterns must therefore be recognised before fractures resulting from complex loading can be read. A technique is now readily employed with complex loading test programmes in which every increment of increasing load in each flight is summed. The sum of these "upstrokes" is a measure of the severity of each flight. Flight-by-flight spacing on the pattern of flights is predicted by this means and the position of severe load exceedances within certain flights readily deduced. Fig 7 shows a flight-by-flight pattern with the position of occurrence of severe loads indicated. Fig 8 shows the matching fracture surface which can be read by either distinguishing the high loads or, as the pattern expands. each individual flight in a mutually consistent way.

Fractures produced by fatigue cracking under service loads are not necessarily more complicated, since they too contain a ground-air-ground cycle which may be a readily recognised striation (Fig 9) whose spacing can be measured. However, major higher frequency loads within a flight may sometimes be more readily identified. For example, Fig 10 contains a trace of major changes in engine speed for a military jet aircraft engaged in a repetitive duty and shows that a typical flight contains approximately 60 such changes which produced the illustrated striations on the surface of cracks in its compressor blades. Counting the striations between limits of 0.1mm crack length to final failure indicated some 40,000 major speed changes suggesting a crack growth period of approximately 700 flights, which is consistent with the known engine history of some 2000 flights. It would be expected that, for the stress concentration at the blade root fitting, initiation of a crack 0.1mm deep would require approximately 2/3rds of the total fatigue life.

Whilst quantitative fractography can provide detailed analyses of fatigue crack growth behaviour, there are many examples in which fatigue striations are not found or cannot be read. Predictive techniques based upon Linear Elastic Fracture Mechanics are invaluable in such situations and the combination of quantitative fractography with predictions of crack growth rates provides an extremely powerful technique mutually adding to the confidence with which either technique may be applied.

FRACTURE MECHANICS PREDICTIONS

The emergence of Linear Elastic Fracture Mechanics has had a major influence on failures analysis. It is not the purpose of this paper to review the extensive background information now available in LEFM but to indicate how it is used regularly to predict crack growth rates, critical crack lengths or residual strengths in practical cases and to compare these predictions with observed results. This comparison is invaluable in increasing confidence in both quantitative fractography and LEFM predictions.

FATIGUE CRACK GROWTH PREDICTIONS. There are many models available for the prediction of crack growth rates using LEFM. For military loading spectra that generally repeat over a short range a modification of the Forman, Kearney and Engel model is frequently used:-

$$\left(\frac{da}{dn}\right)_i = \frac{C\Delta K_i^m}{[(1-R)K_c - \Delta K_i]^{\frac{1}{2}}}$$

where C and m are constants derived from laboratory data for the metal in question, K_c is the material fracture toughness, R the stress ratio and ΔK_i the range in stress intensity factor for each segment of the fatigue loading programme. The sum of each load cycle's contribution to crack growth rate then predicts the crack extension per flight at increasing crack lengths. Stress intensity factors are calculated in the normal way, eg:-

$$\Delta K_i = \frac{K_i}{K_o} \quad \Delta\sigma_i \frac{\sqrt{\pi a}}{\phi}$$

where $\Delta\sigma_i$ is the range in stress for each segment of the loading programme, K_i/K_o is a geometric correction factor usually derived from standard solutions, "a" the crack length and "ϕ" a correction factor for crack front shape. The efficacy of this relatively simple predictive method is illustrated in Fig 11 where predicted growth rates are compared with those measured by striation analysis for a crack in an under-carriage cylinder. An example of one duty cycle applied to the undercarriage is shown (Fig 12). Calculations of the individual stress ranges of the individual loadings is readily achieved with such simplified loading sequences.

The usual result of such predictions is a plot of crack length against number of flights showing both the number of flights taken to grow the cracks and the critical crack lengths (Fig 13). This type of information is of direct relevance to the engineer trying to determine safe operating procedures in terms of periodicity of inspections or safe life. It also reveals the very damaging effects of stress concentrations already identified as a major source of unexpected failures. For example, if crack growth rates are predicted and found to conform with those measured by striation spacing it is possible to identify the effects of surface defects on crack initiation. For example, an equivalent loading cycle can be defined as a single sinusoidal cycle of stress ratio 0.1, producing the same crack growth rates as a complete flight of complex loads and standard data used to determine the predicted total life of a component under this equivalent stress cycle. This predicted life can be compared with the achieved total life. Frequently it is found that in material containing metallurgical defects, or with poorly machined surfaces, cracks initiate during the first loading cycles and that the fatigue life is totally that of the crack growth period. Predictions of the life expected in the absence of such defects may be ten times greater (Fig 14). This dramatic reduction in total fatigue life because of the premature initiation of cracks at defects is the underlying reason for the introduction of damage tolerant design procedures.

It has been found that on occasions the test loading spectrum or service load spectrum has been too complicated to predict crack growth even though more advanced prediction methods are now available, including those that deal with load interaction effects. Reading striations under these circumstances is clearly dangerous if no direct connection can be made between striation pattern and loading pattern. It has been found efficacious to produce coupon test pieces cracked under the required complex fatigue loading pattern but to deliberately mark the fracture surface with a small number of constant amplitude cycles of a completely innocuous nature, (eg 10 cycles per complete programme with a peak stress not exceeding 90% of the level achieved on average in every flight). The fracture surface of the component is then read on the basis of flight-by-flight striations and the interpretation of the striations confirmed by reading the marked coupon fracture. In the example shown (Fig 15) it was found that the wing loading spectrum contained occasional flights in which compressive loads occurred in the bottom skin, ie as if the aircraft had landed, thus producing extra striations of a flight appearance and breaking up the flights into sub-flights. It was predicted that 1100 striations would be observed for every 900 flights applied. The striation spacings once corrected by this factor of 1100/900 predicted the observed crack growth rate extremely accurately, giving confidence that flight striation spacing could be used to interpret component crack growth rates.

It is considered that both quantitative fractographic techniques and Linear Elastic Fracture Mechanics are to continue to develop adding greatly to the understanding of fatigue failures in the service environment. It would seem unlikely that matching quantitative techniques will ever emerge for other failure modes such as stress corrosion.

CONCLUSIONS

1   It has been shown that over a large population of diverse failures fatigue is the dominating factor producing more than twice as many new arisings than any other type of failure.

2   Premature fatigue failures, that were not initially expected on the basis of service conditions, are predominantly attributable to local stress concentrations.

3   In the investigation of failures the emergence of modern electron optics has greatly improved fractographic capability, and quantitative fractography coupled with Linear Elastic Fracture Mechanics now provides detailed information on fatigue crack growth in both the service and test environments.

ACKNOWLEDGEMENTS

NO. OF FAILURES IN 10YRS IN TERMS OF BASIC CAUSE

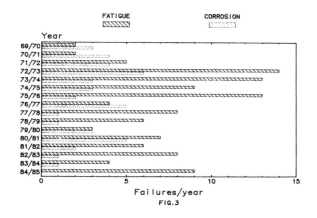

FIG.1

FATIGUE AND CORROSION FAILURES, ALL ALLOYS, 1969-85

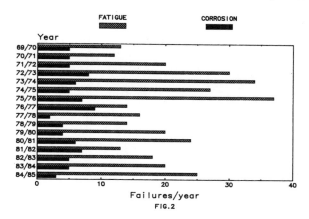

FIG.2

FATIGUE AND CORROSION FAILURES, ALUMINIUM ALLOYS, 1969-85

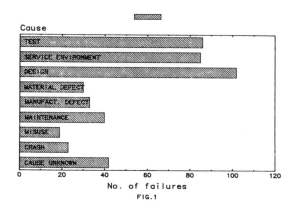

FIG.3

FAILURES IN TERMS OF SERVICE ENVIRONMENT, 1975-85
All alloys

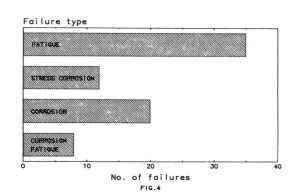

FIG.4

FAILURES OVER 10 YEARS IN TERMS OF DESIGN

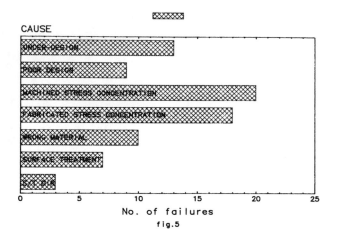

fig.5

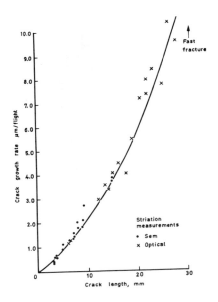

FIG.6   Crack growth rates measured by
two fractographic techniques

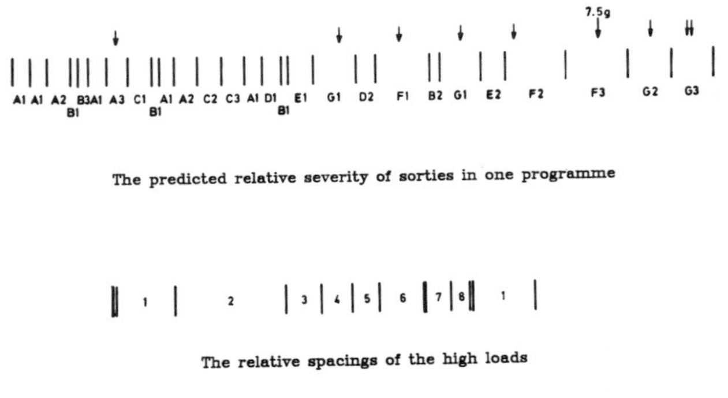

The predicted relative severity of sorties in one programme

The relative spacings of the high loads

FIG.7

FIG.8   Striation pattern produced by   X2000
programme shown in Fig.7

FIG.9   Service fracture with ground—   X360
         air—ground striations arrowed

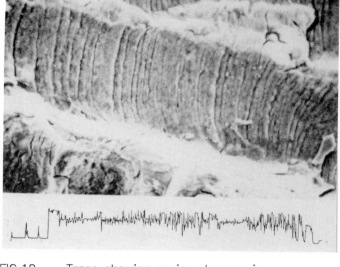

FIG.10   Trace showing major changes in
         engine speed, and associated
         fatigue striations

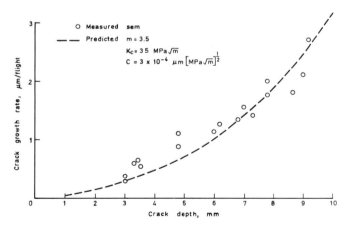

FIG.11   A comparison of crack growth rates
         measured fractographically and
         predicted by fracture mechanics

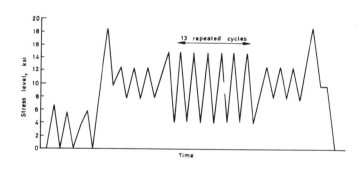

FIG.12   Typical duty cycle for undercarriage

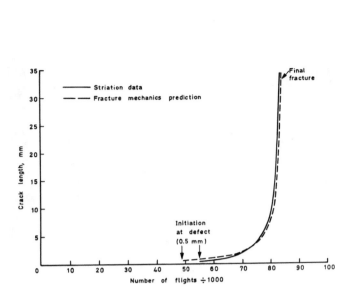

FIG.13   Measured and predicted crack
          growth behaviors

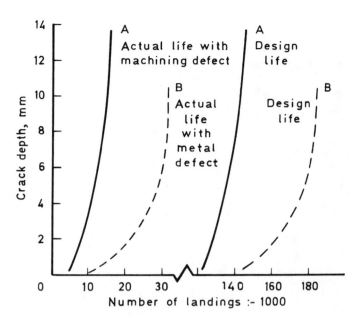

FIG.14   Effect of defect on fatigue life

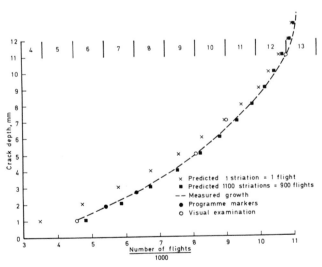

FIG.15   Crack depths in coupon test
          piece as predicted from
          striation spacings and as
          measured

PG&E DEAERATOR TANK WELD TRANSVERSE CRACK ACCEPTANCE LIMITS

J. F. Copeland
Structural Integrity
Associates, Inc.
San Jose, California
USA

A. Ferdi
Pacific Gas &
Electric Company
San Francisco, California
USA

R.D. Kerr
Pacific Gas &
Electric Company
San Ramon, California
USA

ABSTRACT

The analyses of this study show that weld transverse cracks in deaerators may be acceptable without repair under certain conditions. These are the most common cracks found during recent inspections of Pacific Gas and Electric Company (PG&E) deaerators. Such crack acceptance limits were developed in accordance with fracture mechanics methodology, and represent a viable option to costly and time-consuming repairs.

INTRODUCTION

Deaerator systems in power plants perform the function of heating and deaerating cold feed-water. A number of deaerator tanks at Pacific Gas and Electric Company (PG&E) power plants have been inspected, and cracks have been found and repaired. The majority of these cracks have been at weld toes and transverse to the weld direction, as shown in Figure 1. Most of the cracks have been at circumferential head-to-shell welds and penetration or attachment welds on the tank inside surface. These weld transverse cracks will tend to propagate into the base metal adjacent to the weld. With such a growth pattern, concern over the cracks is reduced since they are growing out of the influence of the weld residual stress field (which probably had a significant role in crack initiation) and into the generally tougher base metal of the tank.

It has been postulated that the crack propagation mechanism for these weld transverse cracks is corrosion fatigue, in the tank high oxygen water environment, driven by pressure and thermal cycling due to the operational duty cycles of the deaerators, and accelerated somewhat by the effects of weld residual stresses. As cracks of this orientation become deeper and longer, however, the influences of weld residual stresses and thermal stresses become less significant. Therefore, it follows that the weld transverse cracks may be acceptable without repair, especially for lower pressure stressed tanks, when monitored periodically by inspection to assure the crack orientation has not changed.

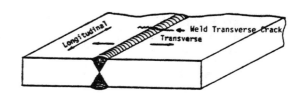

Figure 1. Illustration of Weld Transverse Crack Orientation

A crack acceptance criterion, as a function of deaerator tank stress, is thus a viable option when repairs are costly or impractical. Weld transverse crack limits are developed in this report, using established fracture mechanics methodologies, to provide such an alternative to 100% repair of cracks in the PG&E deaerator tanks. this report does not address cracks which are parallel to the weld direction, as they will be a topic for further study.

DEAERATOR TANK STRESSES

Weld residual stresses, pressure stresses, and possible thermal stresses are considered in developing flaw acceptance criteria based on deaerator tank stresses.

WELD RESIDUAL STRESSES - An assumed weld residual stress distribution [1], and a third order polynomial curve-fit to this data, is shown in Figure 2. Figure 2 represents the longitudinal residual stress on the plate surface, which would act to grow a transverse crack as shown in Figure 1. This surface residual stress distribution is assumed as constant through the thickness of the welded plate, which is conservative [1]. The slightly tensile region, in Figure 2, beyond about 15.25 cm (6") from the weld centerline, is an artifact of the curve fit and in reality would be a zero residual stress region. Although conservative, this region is beyond the distance for considering crack acceptance in this study.

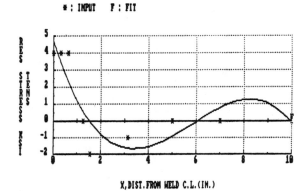

Figure 2. Weld Longitudinal Residual Stress Distribution Curve Fit (Assumed as Constant Stress Through-Wall For a Given Distance from the Weld Centerline), (6.895 MPa = 1 ksi, 2.54 cm = 1 in)

Figure 2 shows that transverse cracks longer than about 2.5 to 5 cm (1" to 2") (depending on plate thickness) from the weld centerline would run into a compressive residual stress field in the base plate. This is one of the basic reasons for reduced concern about such transverse cracks.

PRESSURE STRESSES - Applied pressure stresses in deaerator tanks are computed by the basic equations for thin-walled vessels. The tank hoop stresses, acting on weld transverse cracks at circumferential welds, are given by:

$$\sigma_{hoop} = \frac{pR}{t}$$

where p is the tank pressure, R is the tank radius, and t is the tank wall thickness. An example of hoop stresses, due to pressure, for 380 cm (12.5 ft) diameter tanks of varying wall thicknesses is given in Figure 3. In reality, for tanks designed in accordance with ASME Section VIII, Division 1, the allowable stress is limited by 1/4 of the material tensile strength (generally 12.5 to 17.5 ksi allowable stress). Thus, some of the higher stresses in Figure 3 may not be pertinent.

THERMAL STRESS DISTRIBUTIONS - Postulated thermal stress distributions due to possible high frequency temperature fluctuations were analyzed. Since it was found that such postulated stresses are a surface effect, and although they may be significant to crack initiation, it was considered reasonable to ignore them in crack growth and critical flaw size calculations, especially for the assumption of a through-wall crack which is used to establish acceptance criteria in this study.

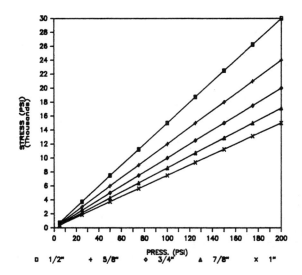

Figure 3. Calculated Hoop Pressure Stress for a 381 cm (12.5 ft) Diameter Tank of Varying Wall Thicknesses, as Shown (6.895 Pa = 1 psi, 2.54 cm = 1 in)

CRITICAL CRACK SIZE

Critical crack sizes for predicted fracture as a function of applied stress are determined by fracture mechanics methodology. Linear elastic fracture mechanics (LEFM) is used to determine critical crack sizes for fracture when applied stresses are well below the yield strength of the material. This methodology has been used for many years and is included in Appendix A of ASME Section XI for the acceptance of cracks in operating nuclear vessels and piping. However, when applied stresses are on the order of yield strength or above, and for materials thin enough that yielding can occur in the through-wall direction, elastic-plastic fracture mechanics (EPFM) is a more accurate means of predicting fracture. For deaerator tank carbon steels, with yield strengths of about 207 MPa (30 ksi) and thicknesses of 2.5 cm (1") or less, EPFM is most appropriate, especially when safety factors are applied to fracture stresses and they approach the material yield strength. Both LEFM and EPFM are employed in this analysis to provide a cross-check and comparison; however, EPFM is the basis for establishing the deaerator tank weld transverse crack size limits.

ASSUMED CRACK GEOMETRY - Through-wall cracks (TWC) are assumed for this determination of crack length limits. Although the most conservative case possible, this crack geometry is considered appropriate at this time since surface inspection by magnetic particle testing (MT) is the most practical means of inspecting deaerators, and this inspection technique does not yield any information on crack depth. Also, with the TWC assumption surface effects such as those caused by thermal stresses are not considered significant. Furthermore, although weld residual stresses are considered in the subcritical fatigue crack growth analysis, such secondary stresses would be relieved by plastic flow during ductile fracture and are not included in the EPFM determination of critical crack lengths.

ANALYSIS RESULTS - The results of linear elastic fracture mechanics (LEFM) and elastic-plastic fracture mechanics (EPFM) J-integral/tearing modulus analyses to predict critical stress for instability versus TWC length are shown in Figure 4.

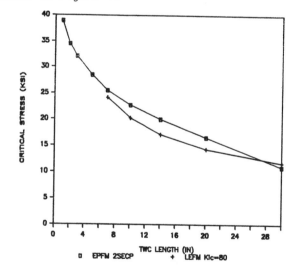

Figure 4. Predicted Fracture Stress for Through-Wall Cracks (TWC) Based on Elastic-Plastic Fracture Mechanics (EPFM) and Linear Elastic Fracture Mechanics (LEFM), (6.895 MPa = 1 ksi, 2.54 cm = 1 in.)

Figure 4 illustrates that quite large TWCs can be tolerated in low stressed tanks. The basis for the LEFM computations is the following basic equation for a TWC in an infinite plate:

$$K = \sigma (\pi a)^{1/2}$$

where K is the crack tip stress intensity

factor, is the hoop stress tending to open the crack, and "a" is half of the TWC length. It can be seen in the above equation that K increases as the stress (internal pressure) increases, for a given crack size. Also, as seen in the above equation, K increases as the crack size increases, for a given pressure stress. When K reaches a critical value, K_{Ic}, determined by testing laboratory fracture toughness specimens, fracture is predicted.

Tests run by PG&E and others on carbon steel (as-rolled A516) at 0°C (32°F) have shown that $K_{Ic} = 88$ MPa(m)$^{1/2}$ (80 ksi(in)$^{1/2}$) is a reasonable lower bound fracture toughness for use in this LEFM analysis when the crack tip is in base metal. The results of this LEFM analysis of TWCs in deaerators are shown with the EPFM results in Figure 4, as a reasonableness check.

The EPFM analysis was performed by modelling a TWC with a single edge cracked plate (SECP) and doubling the resulting critical crack size. this is a conservative model since a relatively small uncracked ligament exists in this model, as compared to the case in deaerators. Details of this approach are given in [2], and material properties for use in EPFM analyses of carbon steel are given in [3].

The expected trends between EPFM and LEFM are shown in this figure and further support the use of this EPFM solution to establish TWC limits as a function of stress. The assumed material yield stress in this analysis is 187 MPa (27.1 ksi), which along with the corresponding stress-strain curve was taken from [3] and agrees with the minimum yield strength requirement for A285 Grade B, a possible copper-bearing tank steel. These TWC size limits limits in Figure 4 now need to be adjusted to allow for possible crack growth by fatigue during subsequent service of tanks with remaining transverse cracks.

FATIGUE CRACK GROWTH

CRACK GROWTH LAW - A conservative crack growth law, represented by the following equations [4], for A36 carbon steel in 3% NaCl was solved numerically to adjust the preceding TWC limits for subsequent added service time:

$$da/dN = 1.555 \times 10^{-10} (\Delta K)^{3.419}$$

$$da/dN = 3.006 \times 10^{-12} (\Delta K)^{4.695}$$

The units in these equations for crack growth, da/dN, are in./cycle and for ΔK are ksi(in)$^{1/2}$. (1 ksi(in)$^{1/2}$ = 1.1 MPa(m)$^{1/2}$ and 1 in. = 25.4 mm). In the above equations, ΔK is the cyclic variation in K resulting from stress cycling in the presence of a crack. In this case, ΔK results from assumed pressure stress cycling, with a mean stress (stress ratio, R) effect imposed due to the constant residual stress. Again, the double SECP crack model was employed, consistent with Figure 4.

CYCLIC STRESS - the weld residual stress distribution from Figure 2 was curve-fit for a plate thickness of 16 mm (5/8"). Although not exactly corresponding to the input in Figure 2, this curve-fit is considered adequate for its purpose. Beyond the distance of 17 cm (6") from the weld centerline the residual stress would be zero, but is curve-fit to be slightly tensile, which is conservative for long cracks.

Corresponding K values for this residual stress distribution and for a pressure stress of 51.7 MPa (7.5 ksi) are shown in Figure 5. It can be seen that the compressive residual stress region in the base plate nearly counteracts the tensile pressure stress effect for a crack tip about 17 cm (6") from the weld centerline. This further supports neglecting these residual stresses for the stability analysis in the preceding section.

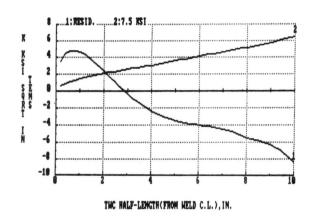

Figure 5. Stress Intensity Factors as Functions of Transverse Through-Wall Crack (TWC) Half-Length for Weld Longitudinal Residual Stresses and a Pressure Stress of 51.7 MPa (7.5 ksi)

Figure 6 shows two cases of assumed pressure cycling superimposed on the residual stress field of Figure 2 (assumed constant through the wall). Again, surface stresses due to postulated thermal stresses are ignored for this TWC assumption. Possible crack growth along the surface by thermal stresses should be monitored by periodic inspections. The two cases of pressure cycling in Figure 6 were derived from the examination of power plant control room 24-hr. charts and are considered conservative. These cases each assume 2190 major pressure cycles per year (1 start-up/shut-down per day and 5 less severe excursions per day), with Case 1 being the more conservative case.

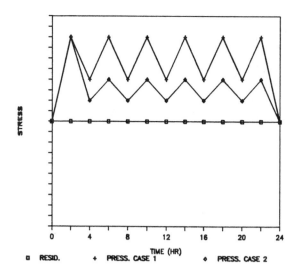

Figure 6. Schematic Representation of Two Cases of Pressure Cycling Superimposed on Weld Residual Stress for Fatigue Crack Growth Analyses

PREDICTED GROWTH RATES - Predicted crack growth rate calculations were performed for a number of different stress levels for Case 1 pressure cycling and for two Case 2 pressure loadings. From the Case 1 results in Figure 7 it is seen that the predicted crack growth is quite small (< 50 mm (2") in 92 years) for low stress tanks.

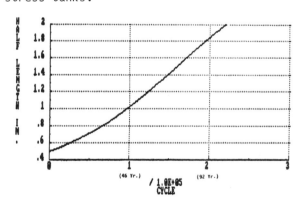

Figure 7. Predicted Fatigue Crack Growth for Case 1 Cycling With a Maximum Pressure Stress of 51.7 MPa (7.5 ksi), (2 x 10^5 cycles = 92 yr, 25.4 mm = 1 in.)

This predicted growth rate increases to a growth rate of 100 mm (4") in 23 years in the higher stressed tanks. Since the Case 2 results are enveloped by the Case 1 loadings, the Case 1 results are thus used to adjust the TWC instability limits for predicted crack growth. Note in Figure 7 that the crack growth rate is relatively independent of crack size, which is caused by the crack growing into the compressive region of the weld residual stress field as it becomes longer. This constant crack growth rate, from the curve slopes, simplifies the TWC limit adjustment for growth.

TRANSVERSE CRACK SIZE LIMITS

The instability TWC length versus critical stress data from Figure 4 are given in Table 1, with an arbitrary safety factor of 3 on stress, and adjusted for predicted crack growth for additional service periods of 5, 20 and 40 years using the crack growth rates determined in the preceding section. These results are also plotted in Figure 8. Crack growth rates (both tips of a TWC) range from 0.75 to 5 mm/yr (0.03 to 0.20 in/yr.), depending on the maximum pressure stress used in the fatigue crack growth analysis (34.5, 51.7, 69, 86.2 or 103.4 MPa (5, 7.5, 10, 12.5, or 15 ksi)). These crack growth rates have been multiplied by the number of service years indicated, and subtracted from the instability crack lengths to obtain the crack growth adjusted TWC length limit curves of Figure 8. As might be expected, the crack growth adjustment is relatively minor for 5 years additional service, but more significant for 20 years and 40 years. Note in this evaluation that the safety factor is applied to the critical flaw size determination, but not to the fatigue crack growth portion of the analysis.

TABLE 1

Deaerator Weld Transverse TWC Length Limits (Safety Factor on Stress = 3), (6.895 MPa = 1 ksi, 25.4 mm = 1 in.)

DESIGN STRESS (KSI)	S.F. TIMES DESIGN STRESS (KSI)	TWC LENGTH GROWTH (IN/YR)	0 YR ADDED SERVICE TWC (IN)	5 YR ADDED SERVICE TWC (IN)	20 YR ADDED SERVICE TWC (IN)	40 YR ADDED SERVICE TWC (IN)
5.00	15.00	0.03	22.68	22.53	22.08	21.48
7.50	22.50	0.05	10.15	9.90	9.15	8.15
10.00	30.00	0.08	4.11	3.71	2.51	0.91
12.50	37.50	0.12	1.32	0.72	0.00	0.00
15.00	45.00	0.20	0.00	0.00	0.00	0.00

These results show that tanks with relatively low pressure stresses, acting normal to weld transverse cracks, can tolerate large assumed TWCs for periods of 40 years. Tanks with pressure stresses, normal to cracks, of greater than about 8 ksi are generally tolerant of TWCs for shorter periods of service, and sometimes reduced safety factors are incurred. Of course, these conclusions should be verified and adjusted by periodic inservice inspection of the tanks.

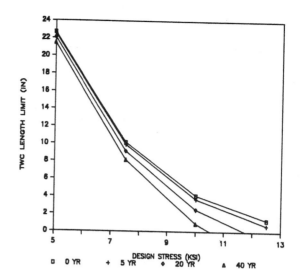

Figure 8. Weld Transverse TWC Limits for a Safety Factor of 3 on Design Stress, and Adjusted for Predicted Fatigue Crack Growth in the Additional Service Years Shown (6.895 MPa = 1 ksi, 2.54 cm = 1 in)

CONCLUSIONS

The most common cracks found during recent inspections of PG&E deaerator tanks are weld trasnverse cracks at circumferential welds and penetration and attachment welds. The analyses of this study show that such cracks at circumferential and longitudinal tank welds may be acceptable without repair for tanks with low pressure stresses. Tanks with high pressure stresses are less tolerant of such cracks but may still be acceptable under certain conditions. In either case, it is recommended that unrepaired cracks be monitored periodically (~2 yr. periods) to assure that they are not turning to grow parallel to welds, and not extending at high rates on the tank surface. This study does not address cracks parallel to welds or cracks at small penetration and attachment welds. The results of this study are presented in the form of allowable flaw length versus design pressure stress for various numbers of years of additional service life ranging from 0 to 40 years. these crack acceptance limits were developed in accordance with standard fracture mechanics methodology, and represent a viable option to costly and time-consuming repairs.

REFERENCES

1. "Review of Requirements and Guidelines for Evaluation of Component Support Materials Under Unresolved Safety Issue A-12", EPRI RP1757-2, Aptech Engineering Services, June 1983.

2. Kumar, V., et. al, "An Engineering Approach for Elastic-Plastic Fracture Analysis", EPRI NP-1931, G.E., July 1981.

3. Mehta, H.S., et. al, "Flaw Evaluation Procedure for Ferritic Piping", EPRI RP2457-2 Program Review, G.E., Presented Aug. 15, 1984.

4. Barsom, J.M. and Rolfe, S.T., "Fracture and Fatigue Control in Structures", Prentice-Hall Inc., 1977.

GROWTH OF SHORT CRACKS IN IN718

B.N. Leis, A.T. Hopper
Battelle Columbus Laboratory
Columbus, Ohio, USA

D. G. Goetz
Texas A&M University
College Station, Texas, USA

INTRODUCTION

Fatigue crack growth rate predictions based on linear elastic fracture mechanics (LEFM) have been successfully used for many years. For example, the electric utility industry has been relying increasingly on LEFM as the basis for run, retire, repair decisions on major components. But, there are sound reasons for not blindly applying damage tolerance analyses. A growing body of experimental observations support the conclusion that LEFM does not always consolidate crack-growth-rate data. Frost, Pook, and Denton [1], Ohuchida, Nishioka, and Usami [2], and Kitigawa and Takahaski [3] were among the first to present data that were not consolidated with results for standard test specimens in terms of LEFM. Similar results have since been observed in many materials; the difficulties are particularly acute for physically short cracks. Data not consolidated by LEFM often indicate higher growth rates than expected thus implying some degree of nonconservatism in certain applications.

Damage tolerance analyses based on LEFM have been used to make rational decisions concerning inspectability and continued use of cracked components, until the cracks are of near critical size. The success of LEFM in many applications has led to research to explore its ability to characterize the growth of defects in engine materials. This paper is extracted from a broader effort to explore LEFM based integrity analysis of engine components. Specifically, this study had two objectives: (1) identify the parameters that cause short cracks to behave differently than long cracks and (2) develop data for short crack growth which included the effects of the governing parameters. The first undertaking culminated in a critical literature review [4]. It was concluded there that the reported short-crack effect arose primarily because of crack tip plasticity, transients from initiation to microcracking, and incorrect or incomplete implementation of LEFM. This paper examines these aspects in the light of data developed for IN718, a material popular in engine applications. Results are presented and discussed for room temperature tests on thin plates containing a central circular notch. These results represent a portion of the data generated in a much broader based effort [5].

EXPERIMENTAL ASPECTS

MATERIAL AND SPECIMENS - The Inconel 718 material used in this investigation was 0.093 inch thick sheet with the following chemistry (in weight percent):

C	Mn	P	S	Si	Cr	Cb/Ta	Cu
0.04	0.07	0.012	0.006	0.13	18.42	5.14	0.03
Mo	Fe	Co	Ti	Al	B	Ni	
3.07	18.14	0.3	1.06	0.48	0.004	bal.	

The as-received material was cut into specimen blanks and then heat treated in batches of 10 using a duplex heat treatment including: Anneal in vacuum 954 C/Air Quench; heat 704 C/8 hours; to 621 C/8 hours; and then air cool. The heat-treated material exhibited a uniform fine grain size (ASTM 10 to 10.6), with a microstructure typical of this alloy for the indicated heat treatment. Metallographic study did not disclose any significant alloy depletion or any other undesirable surface condition as a result of any of these treatments [5].

The grain size of this material is about 10 μm. Thus, on the average, increments of crack advance measured on the surface separated by less than 5 μm will represent growth steps within surface grains. However, the crack tip observed on the surface is tied to a subsurface front that on the average crosses about 225 grains. Therefore, while on the surface the crack tip may be tied up by grain and subgrain features, this tip will be dragged along by the subsurface crack front.

The material used in this study had room temperature mechanical properties as follows: Ultimate tensile stress 1378.7 MPa; 0.2 percent offset yield 1027.3 MPa; 21.4 percent elongation; and; a modulus of 194.6 GPa. These properties were obtained from a single 5.1-cm-gage-length sheet, tensile coupon tested at 20 C at a displacement rate of 0.063 mm sec^{-1}. The average hardness was about R$_C$42.

The center-hole-notched panel (CNP) shown in Figure 1 was selected to achieve the program's plan to examine microcrack growth through notch fields. The long axis of the specimen was aligned with the rolling direction of the sheet. A center circular notch was preferred over other notch configurations because holes are commonly found in engine components. It was also preferred because a center notch facilitates tracking four surface cracks through the use of dioptric lenses. This dioptric lense optically removes the center of the camera's field of view to bring diametrically opposite cracks together; then they could be photographed at one time by the same camera.

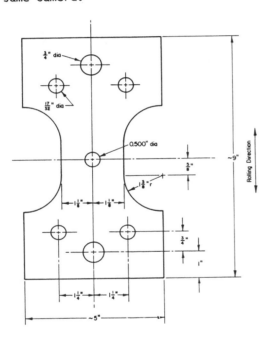

Fig. 1 - Center notched specimen

APPARATUS AND PROCEDURE - All tests were performed in a commercially available servo-controlled test system under axial load control using a sinusoidal forcing function. The load was measured via a load cell mounted in series with the specimen. The load cell was calibrated prior to the testing program to within ±1 percent accuracy.

The camera system used to track the cracks was built using commercially available optics and flash units, adapted to present focal length and magnification requirements. Shutter and flash triggering electronics were designed and built to match requirements of this and other similar studies. Cameras were mounted on X-Y slide translation mechanisms attached to the test frame to permit coarse focus and to center the camera with respect to the line of crack extension assuming symmetric growth.

Setup of each experiment involved the usual specimen installation and setting of the command function. Prior to testing, the anticipated area of crack advance was polished on each specimen with successively finer grades of paper and polishing compound, in some cases down to 0.5 μ diamond paste. Cycle interval and initial delay for the cameras were then programmed and the test initiated.

DATA REDUCTION - The raw data generated in this study consist of surface crack length, c, and the corresponding number of cycles, N, for a given crack. For each specimen containing preflaws, as many as four crack tips may be active. However, for natural initiation, multiple cracks have been found to be active at a site anticipated to generate one crack. In these cases, many more than four crack tips could be active. The CNP specimens are prefixed with CH. For the sake of crack identification, each crack is labeled as follows: specimen number, specimen face, crack location.

Several factors made editing of the raw crack growth data necessary. The automatic measurement system used did not allow decisions to be made about the admissibility of each reading as the test was in progress. Therefore, such decisions had to be made after the complete crack length versus cycles record was produced. "Admissible" in this study means meeting the criterion that the increment in crack extension between successive readings was greater than the precision of the measurement system used--enough so that unreasonable scatter would not present a problem. A criterion based on a multiple of the standard deviation in repeat measurements, $\bar{\sigma}$, was chosen to avoid scatter. Following a study of the effect of editing on the data trends, $\Delta c = 2 \bar{\sigma}$ was chosen as the standard editing increment. In all cases, editing began after the first data point.

EVALUATION OF DC/DN - It remains to decide upon analysis procedures to translate raw c-N data into a format that admits comparison of data for long and short cracks and other stress levels and specimen geometries. To this end, dc/dN must be calculated as a function of the stress intensity factor, K.

Calculating the crack growth rate dc/dN is complicated by the nature of the short crack problem. Because dc/dN may vary significantly for only small changes in N, values of dc/dN for prior or successive cycles may bias the computed "average" value found in smoothing procedures. On the other hand, simple slope calculations may lead to excessive scatter. With these considerations in mind, dc/dN has been calculated for a range of results using unedited data and data edited using $\Delta c = 1 \bar{\sigma}$, $2 \bar{\sigma}$, and $10 \bar{\sigma}$. Three procedures were compared: (1) simple slope analysis (SSA), (2) three-point-divided difference analysis (3PDDA) and (3) seven-point incremental polynomial analysis (7PIPA). The results of this study showed that the growth rate calculation procedure can accentuate or camouflage possible short crack effects. The SSA introduced the least analysis bias but was prone to high scatter. On the other hand, 7PIPA tended to reduce SSA scatter but in so doing may smooth out what are real short crack effects. The 3PDDA lies between the extremes of SSA and 7PIPA and suffers (to a lesser extent) the same drawbacks of these extremes. But it provides a middle ground--more scatter (less smoothing) than 7PIPA (vice versa for SSA)--and thus seems best suited for present purposes. For this reason, the 3PDDA was used to calculate dc/dN.

STRESS INTENSITY FACTOR SOLUTIONS - Consider now measures of the driving force for cracking. For situations where the plasticity is reasonably confined, LEFM is appropriate.

Through Cracks - The K solution for the finite width CNP with symmetric through cracks of length c has the form

$$K = S\sqrt{\pi c}\; F\left(\frac{C}{W}\right) \quad , \tag{1}$$

where c is the surface length of the crack measured from the edge of the notch.

Values of F(c/W) must include the influence of the notch root free surface, the notch gradient, and the finite width of the specimen. For physically very small cracks, K is reasonably approximated by

$$K \cong 1.12\; K_t\; S_N\sqrt{\pi c} \quad . \tag{2}$$

Here K_t is the net section stress concentration factor, equal to the ratio of the maximum principal stress, denoted as σ, to the net section stress denoted as S_N. For two physically short cracks in notched plates, if the K_t's are the same and the gradient in σ as a function of distance across each plate is geometrically similar, K will provide similitude in the crack driving force for the same degree of through thickness constraint (same plate-thickness/hole diameter, t/ϕ). In the absence of the same constraint, K will not provide similitude based on numerical results generated for cracks in the absence of the notch gradient for values of c measured on the plate's surface. Likewise, if the diameter of the hole is the same but the width varies, Equation (2) indicates K will not provide similitude unless (F(c/W)) is geometry specific. That is, solutions for K (F(c/W)) that represent the geometry in Figure 2 are required.

A number of different forms of F(c/W) were considered. Included were numerical solutions for similar geometries, and solutions adapted to the geometry considered (for details see Appendix B of Reference 5). That of Newman [6] for an almost identical plan-form has been selected.

This solution is plotted in Figure 2. It matches closely the solution discussed by Karlsson and Backlund [7] which is similar to Equation (2), but valid for somewhat longer cracks. Correction for asymmetric cracking follows from factors developed for infinite plates normalized with respect to crack length for the approximate degree of asymmetry developed. Such corrections are discussed in the work of Newman [8].

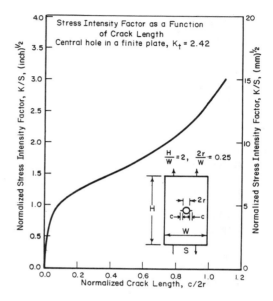

Fig. 2 - Stress intensity factor solution for through cracks [6]

Part Through Cracks - The K solutions discussed above are valid only for through cracks. If surface or corner cracks develop, the solutions must be modified to account for the 3D effects of crack shape and for the biaxial, through-thickness stress that develops as a function of t/ϕ in the CNP.

There are many analyses of part through (corner) cracks at notch roots. These solutions range from somewhat general empirical equations to geometry specific trends generated via 3D finite element analyses. A broad range of K/S values are observed for quarter circular cracks, the extent to which is shown, for example, in Figure 3, reproduced from [9]. True 3D solutions (stress field and crack configuration) exhibit a t/2r dependence. These trends are apparent in results presented in detailed summaries and reviews of 3D K solutions (e.g., [8-14].

Newman [14] has compared results of numerical 3D solutions (near the hole) for otherwise infinite plates to the Bowie [15] result. (Note that Tweed and Rooke [16] have been quite critical of the accuracy of the Bowie solution. Nevertheless, it remains popular and is widely used in comparing results for infinite plates.) Examination of this comparison reproduced in Figure 4 shows that the 3D value of K/S, for the same surface crack length, is less than that for a through crack the extent to which strongly

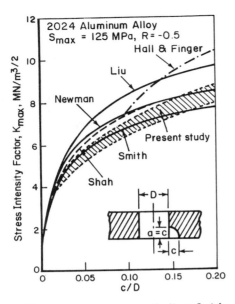

Fig. 3 - Various corner crack K-solutions [9]

depends on the aspect ratio, a/c. Values of a/c > 1 are of interest for the present analysis. In this case, Figure 4 indicates K_{3D} (the 3D value of K) is bounded above by the through-crack value and below by about 0.7K, depending on a/t. Other results examined showed that K_{3D} is less than the corresponding through-crack value as summarized in Appendix B of Reference 5.

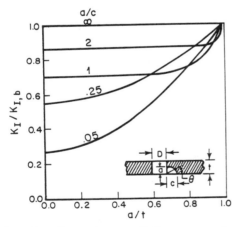

Fig. 4 - Dependence of K on aspect ratio for a corner crack

In view of Figure 4 and the general trends in the literature, the driving force for growth of physically small cracks is reduced by as much as 30 percent (i.e., to about 0.7K) in the presence of circular corner cracks. As the crack propagates across the thickness, the aspect ratio tends toward a plane-fronted through crack, and the surface length increases so that $K_{3D} \rightarrow K$ as evident in Figure 4. Where corner cracks developed, the results presented in Figure 4 have been used to modify K/S in Figure 2, in accordance with the aspect ratios observed until transition. Thereafter, Equation (2) (Figure 2, modified if so indicated by Figure 4) has been used.

The 3D corrected K solution embodies the mechanics of the 3D situation. It remains to account for the crack growth rate dependence on local stress state (e.g., [17]). Analysis performed to assess the significance of this aspect is alluded referenced in Appendix A of Reference 5 and leads to the results reproduced in Figure 5. For the CNP geometry examined, the value of t/φ is 0.18. Three-dimensional elasticity analysis [18] indicates the local biaxiality ratio $\mu \equiv \sigma_2/\sigma_1$, corresponding to t/φ = 0.18, is $\mu \doteq 0.012$, for Poisson's ratio of 0.3. For fatigue crack initiation, the literature [19] indicates the influence of local biaxiality may be significant. For the CNP used in this investigation, the literature suggests that local biaxiality results in a slightly decreased life to develop small cracks and indicates there is a preference to form corner cracks in the absence of artificial preflaws. Figure 5 indicates that the limited local biaxiality developed in the CNP used ($\mu = 0.012$) causes an almost negligible decrease (<2%) in growth rate over the entire range of stress levels imposed in this study. Thus, it can be concluded that notch-induced stress state effects do not affect the growth rate of corner cracks as compared to that observed in the usual long crack uniaxial geometries, beyond the effect already embedded in K as function of crack geometry.

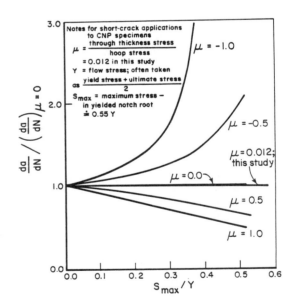

Fig. 5 - Dependence of uniaxial growth rate on biaxiality ratio

RESULTS

REFERENCE LONG CRACK DATA - Results from several specimens have been used to develop a long-crack reference data base. Raw data have been analyzed using the 7PIPA with a view to reduce data scatter as much as possible. Results representative of the long crack finite-crack-growth rate behavior of this material have been developed from five separate cracks for R = 0.01, from two cracks for R = -0.6, and from three cracks for R = -1.0. Typical data show scatter of about a factor of 2 on coordinates of dc/dN and K_{mx}. Trends shown for R = 0.01, R = -0.6 and R = -1.0 in Figure 6 form the long crack reference data base. The trend for R = 0.01 matches unpublished data independently developed by GE for a fine-grained Inconel 718 under otherwise identical loading conditions (except T = 148 C), [5]. Note that all GE data have been developed for surface-cracked geometries.

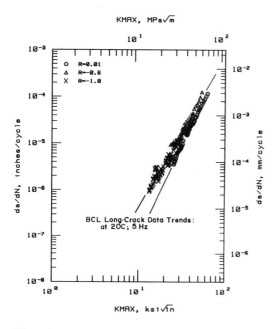

Fig. 6 - Long crack reference crack growth rate behavior

SHORT CRACK DATA - The results are presented for raw data analyzed as just detailed. When c-N data are discussed, Part (a) of a given figure presents edited data points on coordinates of c and N for each specimen and Part (b) presents data on coordinates of log dc/dN and log K_{mx}. Where reference is made to the long crack trend, the results presented in Figure 6 for the appropriate stress ratio have been used for the 20 C comparisons. Results of 5 specimens provide useful information. The specimens are CH1, CH2, CH3, CH6, and CH20. Raw crack length versus cycles data edited as detailed earlier are presented in Appendix E of Reference 5, as are the c-N plots which are not discussed herein. Likewise, dc/dN-ΔK plots are included in Reference 5 for cracks not used to illustrate the present discussion.

The range of results developed using CNP specimens is constrained by the maximum stress that can be imposed. The maximum stress is set by the materials fracture toughness and the crack size. Accordingly, the normalized nominal stresses ranged from 0.20 to 0.40. Because the notch field is local, this geometry permits

testing local levels of the ratio at plastic zone size, r_p, to surface crack length from the notch root, c higher than in unnotched samples. Values of r_p/c, are tied to values of peak stress and to the depth of the notch plastic field along the transverse net section, denoted as x_p. At low levels of S_{mx}, because the notch field is elastic, the crack's plastic zone is the dominant (only) plastic zone. Values of r_p/c of 0.21, 0.35, and 0.31 have been developed in samples CH6, crack 1; CH20; and CH2; respectively for local elastic response. At higher values of S_{mx} the crack grows (at least over part of its length) through an inelastic notch field. One depth of notch field has been explored--460 μm. The value of r_p/c developed in this case is 0.85. This value of r_p/c is much larger than the 0.1 generally associated with valid applications of LEFM. But this level of r_p/c is very small in comparison to values often associated with short crack effects. Likewise, the depths of the notch plastic fields possible in this high-strength material are small compared to those found associated with short crack effects in the literature.

Tests on CNP samples have been used to explore whether or not the initial growth of short cracks occurs at rates in excess of LEFM predictions. Also CNP specimens have been used to assess the role of closure via the wake removal concept. The CNP samples also have been used to study the influence of natural initiation and 3D effects absolute stress level and notch stress field. Note that cracks in preflawed samples can be (and have been) considered as plane fronted while those for corner cracks have aspect ratios a/c, of about 1 for a/t up to 0.25 and between 1 and 3 at breakthrough. After breakthrough, the transition to a plane-fronted crack with nearly equal surface lengths occurs very quickly.

Influence of Stress Level and Notch Fields - Results for samples CH3, CH4, CH20, and CH6 crack 1 can be used to examine the influence of absolute stress level on crack-growth rate behavior. But since the size of the notch plastic field and the value of r_p/c increase with increases in stress, both the stress level and the depth of the plastic field must be considered together.

Results for long cracks indicated that, by itself, r_p/a does not lead to differences in growth behavior over the range of values investigated. A similar result can be seen for notches in the absence of notch plasticity effects. This can be seen by comparing the results for CH6 crack 1 and CH20. These data, plotted on c-N coordinates are similar to results for long cracks [5]. Trends for inelastic action at the notch, developed in CH3 and CH4, are similar to that for elastic behavior in regard to CH6 crack 1 and CH20.

The c-N data for the CNP samples do not indicate trends that could point to even a modest short crack effect. But because the influence of the notch field masks the trend evident when only √a drives K, the data must be examined on coordinates of dc/dN and K_{mx} before conclusions can be drawn. Data showing growth rate as a function of K_{mx} in the absence of inelastic notch action are plotted in Figure 7 for CH6 crack 1 and CH20. In comparison to the long crack trend, the results for CH6, crack 1,1 (the only results captured for small cracks) show growth rates for small cracks in excess of the long crack trend by more than an order of magnitude! Similar although less dramatic results are shown for CH20 crack 1,2, while somewhat scattered results develop for CH20 at crack 1,1, and crack 2,2 [5]. Data for crack 2,1 are not available at small crack sizes.

The increased values of dc/dN just examined for growth through elastic fields becomes more complex for growth through an inelastic field, as evident in Figure 7 for CH3 and CH4. As with cracks in the elastic notch field, a decrease in growth rate with increasing crack length is evident particularly for CH3 crack 1,1 and CH4 crack 1,2. However, in contrast with the results for the elastic field, this decrease ceases at a trend which lies a factor of 5 to 10 above the long crack trend (Figure 6). The data for an elastic notch field quickly approaches the long crack trend; however, following initially high growth rates in CH3 and CH4, growth rates 5 to 10 times that of the long crack trend are evident so long as the cracks grow in the inelastic field.

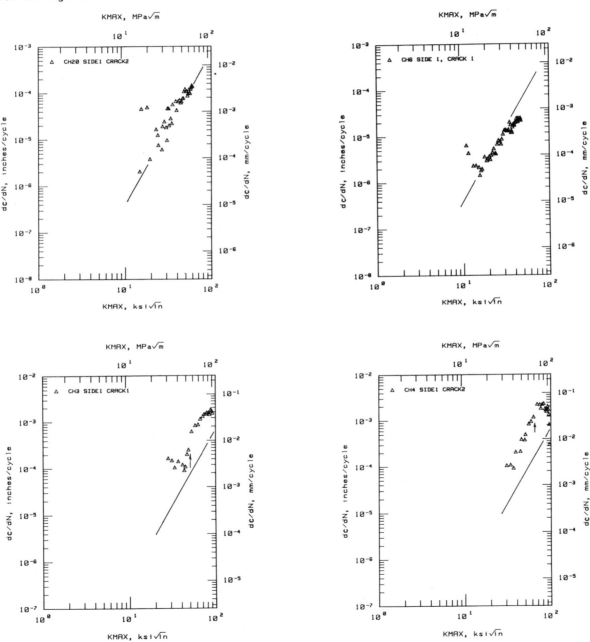

Fig. 7 Crack growth rate behavior for preflawed specimens

220

As apparent, at least for CH3 crack 2,1 and CH4 crack 1,2 and crack 2,2, growth continues at rates 5 to 10 times greater than the long crack trend at crack lengths greater than 575 μm and less than 1 mm [5]. Beyond this, there is a decrease in growth rate evident toward the long crack trend. But, only for CH4, crack 1,2, is this decrease steep enough to indicate that the growth rate would actually meet the long crack results.* This decrease in growth rate as the crack tip passes into the elastic field in CH3 and CH4 has also been observed in more ductile materials. References 20 and 21 present such trends for steels, whereas References 43 and 44 show such results for 2024 Aluminum. Plausible explanations for such behavior are considered in References 21 and 22.

For the situation just considered, approximate lower bound calculations indicate that $x_p = 460$ μm. Given the accuracy of this approximation, it is reasonable to conclude that errors in the LEFM calculation of crack driving force due to local inelastic action are responsible for the 5 to 10 times increase in growth rates. Analysis for cases where cracks grow in an inelastic notch field have been presented in the literature. Several authors have discussed the use of detailed inelastic analysis for this problem (e.g., [23]).

Influence of Natural Initiation (Free Surface/Initiation Transients) and 3D Crack Geometry - Consider now the results for CH2 presented in Figure 8. This test involves natural initiation under conditions bounded below

and above by test conditions for preflawed samples CH6 and CH20, respectively (Figure 7). Observe from Figure 8(a) that two cracks developed on side 1 whereas only one crack was observed on side 2. Crack 1,1 was first to initiate followed by crack 1,2. Both initiated and grew as circular corner cracks based on stereo macrofractography (35X) until $a/t \doteq 0.25$. At this point, a/c began to increase from a value of 1 to about 3 at breakthrough. Growth of crack 1 to beyond the camera field occurred before 2,2 appeared. Growth of crack 1 outpaced that of crack 2 leading to separation well before crack 2 grew appreciably. Several points concerning the behavior evident in Figure 8(a) are noteworthy as follows.

First, there is an interplay between cracks on adjacent faces of the plate at the same notch root. Crack 1,1 initiated first and grew quickly as a corner crack. Then, because the driving force for c is decreasing as $F'(c/W)$ takes the crack out of the dominance of the notch field (e.g., see Appendices B and F in Reference 5), growth across the transverse net section slows radically. Growth across the thickness is still in the dominance of the notch field so that, even though surface growth has slowed, the crack continues to propagate through thickness. As discussed earlier in regard to Figure 4, continued growth in the thickness direction is toward increasing stress intensity factor (as $\sqrt{a}$ increases) so that the through-thickness growth process accelerates as a increases. As $a \to t$, breakthrough occurs and the shorter "just

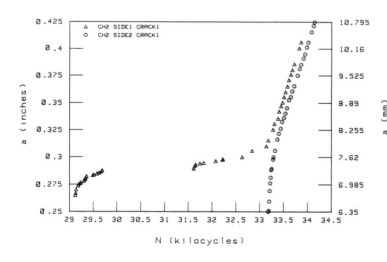

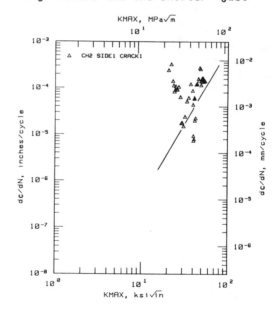

Fig. 8 Crack growth behavior for natural initiation

---

* Given the critical crack size operative at the stress level causing inelastic action, stable tearing is expected to intervene long before growth rates approaching the long crack trend could develop. Further testing beyond this exploratory study are necessary to confirm the apparent very significant short crack effect due to inelastic notch action.

initiated" crack, being tied to its longer counterpart on the other face, propagates rapidly. The now almost-plane-fronted through crack continues to grow as any other long crack.

In view of the above, the rapid change in growth rate of crack 1,1 is due to the fact that crack initiation generates crack configurations that are not stable as the crack grows longer. Had the value of $t/2r$ been larger, the results of work cited in Reference 19 indicate corner crack initiation could occur along with multiple through thickness initiation. The nearly-plane-fronted through-thickness crack initiated would not exhibit the initial high growth rate and the ensuing transient evident for crack 1,1.

Another facet of the corner cracking process is that corner crack morphologies tend to involve extensive crystallographic cracking. While initially this leads to high rates (perhaps due to the absence of closure), the transition from a mixed Mode I/Mode II to Mode I cracking and the associated increased closure could lead to a continued reduction in rates with continued growth toward a stable crack geometry. In these respects, the result for crack 1,1 is interpreted as an initiation transient due to free surface effects admitted by natural initiation. It is due, for the most part, to the mechanics controlling crack initiation in gradients at a doubly free surface. For the case in point, this initiation transient influenced the first 800 μm of surface crack growth. However, the crack length over which this effect can occur is a function of the gradient, the peak stress, the notch geometry, the plate thickness, and other factors.

Another interesting feature evident in the data of Figure 8(a) is the stepped nature of the crack growth rate, even for longer cracks. Such steps are commonly observed in the growth of cracks during the transition from corner to plane-fronted cracks. To some extent, this can be ascribed to the interaction of separate cracks. But it is more likely attributed to the fact that c is growing in field where $d(K/S)/dc$ is decreasing, whereas a is growing in a field where $d(K/S)/da$ is increasing. This results in an unsteady balance wherein the growth along the crack front has to satisfy counteracting driving forces at the extreme tips of the crack front.

Whether or not the largely mechanics-controlled initiation transient is responsible for a significant portion of the short crack effects is not clear in the literature. But, it is certain that the shape of an initiated crack often differs from that associated with its steady-state "long crack" geometry. Some data in the literature for another engine material [24], IN 100, attest to this fact. In that case, on coordinates of da/dN and ∆K, the results indicated that the growth rate first decreased and then increased becoming nearly coincident with the long crack trend as crack length increased. This is exactly the tendency for crack 1,1--as shown in Figure 8(b). As evident in the figure, the first crack to initiate also shows initial growth rates in excess of the long crack trend based on LEFM analyses. In summary, the results for this test show two features often observed in what are called short crack effects. These are initially higher growth rate and a decreasing, then increasing, growth rate.

Comparison of the results in Figure 8 with the corresponding c-N data in Figure 7 shows trends for natural initiation similar to those observed for preflawed specimens. However, there are several differences. First, the trend for

natural initiation involves several data points over which the growth rate is very much higher than for subsequent growth while the trend for preflawed geometries (through cracks) does not. As just discussed, this behavior is rationalized in terms of the double free surface which leads to initially high rates of cracking under locally large r_p/c. Decreases in growth rate to the long crack trend follow as a result of the 3D nature of the crack and the associated transient shapes leading to a stable value of a/c.

Another major difference is that natural cracks show significant scatter in initiation and consequent asymmetric crack growth, particularly at low stress levels. In some cases, multiple cracks initiate at the same notch root. Again this tendency is often accentuated at high stresses, as evident in the literature (e.g., [20]). Multiple cracking is evident in Figure 9 in the results of CH6, crack 2. While planned for elastic local stress behavior, crack 1 initiated well before crack 2 in this specimen. Crack 1 therefore had grown well across the plate before crack 2 initiated. The asymmetric cracking and the related loss of section caused yield at the notch where crack 2 initiated. For this reason, this multiple initiation developed in an inelastic field. Yet another feature unique to corner cracks is the periods of dormancy, such as evident in CH1, crack 1,1 [5].

The final feature unique to natural initiation is the development of Mode II cracks, along with the Mode I cracks observed when preflawed. As detailed fractographically in Reference 5, the fracture surface near the initiation site for the natural initiation is very stepped and coarse compared to its preflawed counterpart. Related roughness-induced closure is expected for natural initiation as compared to preflawed cracks, and may cause a significant reduction in growth rate for cracks grown beyond the effect of the doubly free surface. In contrast, the locally enhanced plasticity associated with the doubly-free surface and the presence of Mode II cracks for the natural initiation may lead to decreased closure, particularly for negative values of R. Natural initiation thus may represent a fine balance between competing mechanisms.

Results for CH2, crack 1,1, have already been discussed in regard to Figure 8. The essential difference between the data for natural initiation in CH2 and the precrack results that bound it (CH1 and CH6, crack 1) is the much larger growth rates for the natural initiation case. Obviously, if the desire is to study the growth of small cracks or to screen for short crack effects, the natural initiation process is significant in the presence of a free surface. When double free surfaces occur, natural initiation transients may be even more important. Another major difference is the driving force for an initiated crack to find the equilibrium crack shape under mechanics conditions which differ as it grows away from the notch. As this process depends on component geometry, it is difficult to predict when it will be most significant. But, earlier discussion of results for CH2 suggest

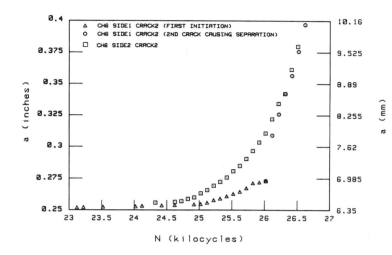

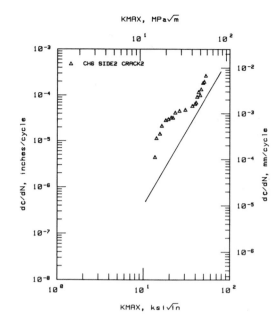

Fig. 9 Corner crack initiation after asymmetric crack growth and multiple initiation

that changes in crack shape can have a major influence on the growth rate of a part-through crack.

The significance of initiation transients and 3D crack geometry effects on the behavior of small cracks growing in an inelastic notch field can be extracted by study of data for CH1, crack 1,2, CH6, crack 2, and CH16, crack 1,1 [5]. These three sets of data are representative of 3D crack effects and initiation transients; only these results are discussed. In contrast to the c-N data for corner crack growth in an elastic notch field, the data for samples CH1 and CH16 do not show the initially high growth rate to the extent found for CH2 [5]. The analysis of these corner crack data on dc/dN-K_{mx} coordinates has been done in two ways [5]. First, the cracks have been considered as through cracks with length equal to that measured on the surface. In the second analysis they have been treated as corner cracks. For this second analysis, the aspect ratio has been changed as a function of surface crack length based on fractographic results and surface crack length data. Values of corner crack K determined in this manner show that, for a given surface crack length, K is less than that for the through crack case up to about a/t = 0.5. That is, for the same surface crack length, the value of K is reduced for corner cracks as compared to through cracks. Since a/t ≤ 0.5 lies in the small crack domain these results suggest that an inappropriate through crack K underestimates the short crack effect that would be evident were a more appropriate corner crack K solution used for data analysis. Analyzed as corner cracks, the results are similar to the trend for CH2. As anticipated in view of Figure 4, the short crack effect is enhanced by the use of a corner crack versus (an inappropriate) through crack K solution.

As has been noted for preflawed CNP data, the inelastic notch field operative in CH2 also is associated with growth rate trends in excess of the long crack data for growth through the notch field. Also, as was noted for preflawed data, the inelastic action of the notch seems to "wash out" the initially higher growth rates associated with small cracks in the CCP samples. Therefore, one could speculate that inelastic notch effects swamp the initiation transients and 3D effects observed to dominate the behavior for locally elastic conditions. In this respect, when local inelastic action occurs the value of r_p/a would appear to be the key driver for the short crack effect.

SUMMARY AND CONCLUSIONS

Data developed for preflawed CNP samples showed only a limited short-crack effect, consistent with the results for the CCP samples. Notch inelastic action was observed to significantly elevate the crack-growth rate as compared to that expected based on LEFM analysis. Natural initiation was associated with initially high growth rates attributed to the doubly free surface at corner-crack initiation sites. Fracture surfaces developed in these corner initiations showed mixed growth modes, and were very stepped and coarse. Closure and the development of constraint were asserted as the cause for reductions in growth rate as the corner crack grew. Three-dimensional crack configuration was also noted to be a factor when the configuration at initiation changed with crack advance to some other steady-state configuration.

The results developed for the precracked samples are not inconsistent with the use of r_p/a as a criterion to assess the possible extent of a short-crack effect. But results developed at nominally low values of r_p/a for both center cracked and center notched geometries indicate the significance of free surfaces--a micro-mechanics contribution to r_p/a (or r_p/c). The CNP samples also showed transients in crack geometry from initiation through the development of a stable configuration are important. Likewise the CNP samples showed inelastic action at a notch field may by itself be responsible for a short-crack effect.

The major conclusions of the experiments done in this study are summarized below.

o In the absence of the elevated temperature, corner cracking, and inelastic notch fields, LEFM anlaysis is appropriate for small cracks in fine-grained Inconel 718. Tests meant to simulate the mechanical service conditions showed an elevation of crack growth rate less than five times for short cracks. This increase operated over a very small crack growth increment.

o Inability to measure and/or analyze corner-initiating cracks can cause LEFM to become practically invalid for short cracks of that type. Consolidation of crack growth data by LEFM requires the accurate calculation of the stress intensity factor (SIF). The discussion of K solutions establishes the widely varying solutions which are available in the literature. If the appropriate SIF cannot be determined or if the measurements needed to use the appropriate SIF cannot be made, then there is little hope that LEFM can consistently consolidate corner-initiating short cracks. Corner initiation can lead to elevated growth behavior caused by stress-state and transient-crack geometry effects.

o Notch root plasticity can elevate crack growth rates above those predicted by LEFM. Growth rates 5 to 10 times the rates expected from LEFM anlaysis have been observed in tests.

o Free surface and the lack of constraint to flow normal to free surfaces enhances inelastic action for crack growth in the first few grains. This noncontinuum behavior may confound attempts to apply LEFM, and to characterize the closure behavior in the wake of plasticity in the first several grains.

REFERENCES

1. Frost, N. E., Pook, L. P., and Denton, K., "A Fracture Mechanics Analysis of Fatigue Crack Growth Data for Various Materials", Engineering Fracture Mechanics, Vol. 3, pp. 109-126, 1971.

2. Ohuchida, H., Nishioka, A., and Usami, S., "Elastic-Plastic Approach to Fatigue Crack Propagation and Fatigue Limit of Materials with Crack", Vol. V., ICF3, Munich, 1973; see also "Fatigue Limit of Steel with Cracks", Bulletin of the JSME, Vol. 18, No. 125, November, 1975.

3. Kitagawa, H., and Takahashi, S., "Applicability of Fracture Mechanics to Very Small Cracks or the Cracks in the Early Stages", Proc. Second International Conference on Mechanical Behavior of Materials, Boston, pp 627-631, 1976.

4. Leis, B. N., Kanninen, M. F., Hopper, A. T., Ahamd, J., and Broek, D., "A Critical Review of the Short Crack Problem in Fatigue", AFWAL-TR-83-4019, January, 1983.

5. Leis, B. N., Goetz, D. P., Ahmad, J., Hopper, A. T., and Kanninen, M. F., "Mechanics Aspects of Microcrack Growth in INCONEL 718--Implications for Engine Retirement for Cause Analysis, AFML/AFWAL, AFWAL-TR-4041, March 1985.

6. Newman, J. C., Jr., "An Improved Method of Collocation for the Stress Analysis of Cracked Plates With Various Shaped Boundaries", NASA TN D-6376, 1971.

7. Karlsson, A., and Backlund, J., "Summary of SIF Design Graphs for Cracks Emanating from Circular Holes", International Journal of Fracture, Vol. 14, No. 6, December 1978, pp 585-596.

8. Newman, J. C., Jr., and Raju, I. S., "Stress-Intensity Factor Equations for Cracks in Three-Dimensional Finite Bodies", NASA TM 83200, August 1981.

9. Sova, J. A., Crews, J. H., Jr., and Exton, R. J., "Fatigue Crack Initiation and Growth in Notched 2024-T3 Specimens Monitored by a Video Tape System", NASA Technical Note No. D-8224, August, 1976.

10. "Part-Through Crack Fatigue Life Prediction", ASTM STP 687, J. B. Chang, editor, 1979.

11. Newman, J. C., Jr., "A Review of Stress-Intensity Factors for the Surface Crack", in Part-Through Crack Fatigue Life Prediction, ASTM STP 687, 1979.

12. "The Surface Crack: Physical Problems and Computational Solutions", The American Society of Mechanical Engineers, J. L., Swedlow, editor, 1972.

13. Broek, D., Elementary Engineering Fracture Mechanics, Noordhoff, 1974.

14. Newman, J. C., Jr., "Predicting Failure of Specimens with Either Surface Cracks or Corner Cracks at Holes", NASA TN D-8244, 1976.

15. Bowie, O. L., "Analysis of an Infinite Plate Containing Radial Cracks Originating at the Boundary of an Internal Circular Hole", J. Math. and Physic., 25 (1956), pp 60-71.

16. Tweed, J., and Rooke, D. P., "The Distribution of Stress Near the Tip of a Radial Crack at the Edge of a Circular Hole", Int. J. Engng Sci., Vol. 11, pp 1185-1195.

17. Proceedings, International Conference on Multiaxial Fatigue, December, 1982, ASTM STP 853, edited by K. J. Miller and M. Brown, 1985.

18. Sternberg, E., and Sadowsky, M. A., "Three-Dimensional Solution for Stress Concentration Around a Circular Hole in a Plate of Arbitrary Thickness", J. App. Mech., Trans. ASME, March, 1949, pp. 27-38.

19. Leis, B. N., and Topper, T. H., "Long-Life Notch Strength Reduction in the Presence of Local Biaxial Stress", Journal of Engineering Materials and Technology, Trans. ASME, Vol. 99, No. 3, July, 1977, pp 215-221.

20. Leis, B. N., "Microcrack Initiation and Growth in a Pearlitic Steel", 15th National Fracture Symposium, June, 1982, ASTM STP 833, 1984, pp 449-474.

21. Leis, B. N., "Fatigue Crack Propagation Through Inelastic Gradient Fields", Int. J. Pres. Ves. & Piping, 10, pp 141-158, 1982; see also "Displacement Controlled Fatigue Crack Growth in Elastic-Plastic Notch Fields and the Short Crack Effect", Engineering Fracture Mechanics, Vol. 22, 1985, pp 279-293.

22. Leis, B. N., and Galliher, R. D., "Growth of Physically Short Center Cracks at Circular Notches", Low Cycle Fatigue and Life Prediction, ASTM STP 770, pp 399-421, 1982.

23. Hammouda, M. M., and Miller, K. J., "Elastic-Plastic Fracture Mechanics Analyses of Notches", ASTM STP 668, pp 703-719, 1979.

24. Lankford, J., Cook, T. S., and Sheldon, G. P., "Fatigue Microcrack Growth in a Nickel-Base Superalloy", Int. J. of Fracture, in press.

LOW CYCLE THERMAL FATIGUE AND FRACTURE OF REINFORCED PIPING

W.J. O'Donnell, J.M. Watson
O'Donnell & Associates, Inc.
Pittsburgh, Pennsylvania, USA

W.B. Mallin, J.R. Kenrick
Eckert, Seamans, Cherin & Mellott
Pittsburgh, Pennsylvania, USA

ABSTRACT

A large diameter steel pipe reinforced by stiffening rings with saddle supports was subjected to thermal cycling as the system was started up, operated and shut down. The pipe sustained local buckling and cracking, then fractured during the first five months' operation. Failure was due to low cycle fatigue and fast fracture caused by differential thermal expansion stresses. Thermal lag between the stiffening rings welded to the outside of the pipe and the pipe wall itself resulted in large radial and axial thermal stresses at the welds. Redundant tied down saddle supports in each segment of pipe between expansion joints restrained pipe arching due to circumferential temperature variations, producing large axial thermal bending stresses. Thermal cycling of the system initiated fatigue cracks at the stiffener rings. When the critical crack size was reached, fast fracture occurred. The system was redesigned by eliminating the redundant restraints which prevented axial bowing, and by modifying the stiffener rings to permit free radial thermal breathing of the pipe. Expert testimony was also provided in litigation resulting in a court decision requiring the designers of the original system to pay damages to the furnace owner.

THE PIPE WHICH FAILED was an exhaust duct in an emission control system at a plant in Washington state. There are a pair of submerged arc electric furnaces at the site, with parallel exhaust systems running generally east/west from the furnaces to a baghouse, as shown on Figure 1. These furnaces burn a mixture of wood chips, bituminous coal and coke. In each system, exhaust gases are drawn from the furnace up through three stacks which converge into a single duct. The gases pass through the duct to a spark box, then through loop coolers to the baghouse. There, particulate material is removed from the gas which is then vented to the atmosphere.

The north furnace is a silicon metal furnace, the south is a ferrosilicon furnace. The ferrosilicon furnace was operated at increasing power levels for almost five months; at which time large cracks occurred at many locations on the south exhaust duct, creating a safety hazard and forcing a shutdown of the furnace.

At the time of the failure of the south exhaust duct, the north duct hadn't yet been placed into service. It was necessary to determine what modifications should be made to the north duct so that it could be operated without experiencing failure similar to that of the south duct. It was also necessary to determine what repairs or modifications to the south duct would be required to put it safely back into service. A failure analysis of the south duct was performed consisting of an analytical fatigue and fracture evaluation, combined with visual and fractographic examination of the duct. The results showed that failure was caused by low cycle thermal fatigue. The thermal stresses were caused by the duct stiffening rings and redundant saddle supports which did not allow for thermal expansion of the duct. In the duct design, stiffening rings, welded to the outside of the duct, prevented free thermal expansion of the duct in the radial and axial directions. The design included saddle supports between the end supports at the expansion joints of each segment of duct. These redundant supports prevented free thermal bending of the duct.

Since high thermal stresses were caused by these improper constraints, they were removed in modifying the north duct which had not yet been put into service. The south duct had undergone very extensive low cycle thermal fatigue damage and cracking. It was therefore necessary to have it torn down and rebuilt without rings or redundant saddle supports. Both ducts have since operated eight years without any cracking

Fig. 1 - North duct viewed from just north of baghouse. Damaged south duct can be seen in background.

problems. Moreover, this operation has been above the power levels and temperatures which caused the failure in five months.

Detailed failure analyses of the south duct were performed to quantify the stress levels and failure mode evaluation. Dead weight and thermal stress analyses and low cycle fatigue analyses of the duct with its support structure were carried out. Metallurgical and fracture studies were performed to determine whether there were any material deficiencies, corrosion problems, fabrication defects or abnormal operating temperatures which may have contributed to the failures. Operating data were examined in order to determine the temperatures at which the duct had been operated. Finally, the original design calculations were reviewed to determine why the thermal stress problems were not recognized at the time of the original design.

The evaluations, examinations, and calculations which were performed are discussed in more detail in the remainder of this paper. The results showed that the duct was operated well within the anticipated temperature ranges, and that there were no fabrication defects or corrosion problems which were of significance in causing the duct failure. The failure was caused by low cycle thermal fatigue directly attributable to the use of stiffening rings welded to the duct, and to the use of intermediate redundant saddle supports in the duct segments between saddle supports at the expansion joints. When the critical crack size was reached, fast fracture occurred, ultimately producing crack lengths comparable to the duct diameter.

BACKGROUND

The emission control system for the silicon metal and ferrosilicon furnaces at the Washington state plant was designed and built over a period of about three years. The ferrosilicon south duct furnace was first operated at low power on January 10, 1976. Even at low power levels which were used in the early weeks of the system operation, the duct reportedly arched in response to circumferentially nonuniform thermal expansion. As the power increased, the temperature differences and thermal bending also increased. Because of the thermal bending, saddle support holddown bolts began breaking in February. The nonuniform temperatures were caused by several factors including flue gas impingement, weather conditions, nonuniform heat convection from the outside surface of the duct, and a build-up of insulating dust on the upper inside surface of the duct.

In early March, an insulation blanket was placed atop a portion of the duct in an attempt to raise the temperature at the top of the duct to a value comparable to that at the bottom. This reduced the arching but caused local buckling. Even with reduced arching, by early April some additional holddown bolts had failed and keepers had been bent at the duct supports. In late May, a large crack was discovered under the insulation midway between two supports at a local buckle in the duct wall. This crack was repaired by cutting out the cracked region and welding on a patch.

The system was restarted and major cracks developed in many sections of the duct: one on May 30, and several more on the morning of June 2, 1976. By June 2, cracks had propagated to

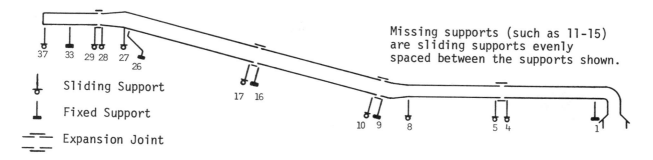

Fig. 2 - Schematic representation of the south duct showing supports and expansion joints.

the extent that the system could not safely be operated and required major repairs and modifications.

DESIGN CONDITIONS - The ductwork was designed for 100,000 hours operation (equivalent to 11.4 years continuous operation). Normal furnace operation was defined for forty-eight weeks per year as producing gas temperatures which would not exceed 704°C (1300°F) and duct wall temperatures which would not exceed 482°C (900°F) at the stack area.

An upset furnace operation was also defined for the remaining four weeks per year which would produce gas temperatures which would not exceed 934°C (1714°F) and duct wall temperatures which would not exceed 632°C (1170°F) at the stack area. This upset condition was expected to occur twice a year on the average and it was judged that the condition might persist for as long as two weeks each time.

The maximum internal vacuum was defined as 15 cm (6 in.) of water at 649°C (1200°F) and 41 cm (16 in.) of water at room temperature based on full-speed fan operation with two stacks plugged or all furnace doors closed. The duct was to be assumed to operate half full of dust and was to meet ASA standards for wind, snow, and seismic loading.

ORIGINAL DESIGN - The duct was originally designed using lengths of rolled and welded COR-TEN steel plate which were butt welded together on site. As seen on Figure 2, the duct was supported by thirty-seven saddles numbered consecutively from the baghouse (cool) end of the duct. Welded to each saddle was a stiffening ring which in turn was welded to and completely encircled the duct. The saddles were mounted on a structural steel truss.

The upper inlet (hot) portion of the duct is horizontal, with an inside diameter (ID) of 2.2 meters (7.25 feet). Each of the three furnace stacks has a refractory lined pipe with an ID of 1.8 meters (6 feet) which leads into this part of the duct. The duct has a 1.9 cm (0.75 in.) wall thickness over the first 5.2 meters (16.9 feet), a 1.3 cm (0.5 in.) wall thickness over the next 4.9 meters (16 feet) and in the first expansion joint, which is 1.3 meters (4.3 feet) long. Then there is a section 4 meters (13 feet) long with a 1.9 cm (0.75 in.) wall thickness in which the duct ID expands to

2.7 meters (9 feet).

The sloped portion of the duct coming off the roof of the furnace building has an ID of 2.7 meters (9 feet). The initial section is a bend 1.3 meters (4.4 feet) long with a 1.9 cm (0.75 in.) wall thickness which provides the transition from the upper horizontal portion of the duct to the sloped portion. The initial straight section is 23.3 meters (76.4 feet) long followed by an expansion joint 1.8 meters (6 feet) long and another section 22.5 meters (73.6 feet) long. The duct wall is 1.3 cm (0.5 in.) COR-TEN in all of these sections. Then there is an expansion joint 1.8 meters (6 feet) long followed by another bend section 4.3 meters (14 feet) long which tapers down to an ID of 2.4 meters (8 feet) and provides the transition from the sloped portion of the duct to the lower horizontal portion. Both of these sections have 1.9 cm (0.75 in.) wall thicknesses.

The lower portion of the duct is horizontal, with an ID of 2.4 meters (8 feet) and a 0.95 cm (0.375 in.) wall thickness. There is one section 16.5 meters (54 feet) long, an expansion joint 2.1 meters (7 feet) long, and another section 16.5 meters (54 feet) long.

OPERATIONAL HISTORY - Early operation of any complex system involves numerous brief shutdowns to make adjustments and correct minor problems. The furnace operating data disclose that the ferrosilicon furnace was shut down on a number of occasions for periods of time ranging from ten minutes to several hours. Twenty of these shutdowns were routine (twelve because electric power was off, seven for scheduled maintenance, and one because coke was not available), and would have occurred independent of the newness of the system

By the time that insulation was placed atop the duct in early March to try to reduce arching, the highest duct wall temperature reading obtained from the thermocouples in the upper section of the duct was 205°C (400°F). With the insulation on, the furnace had been operated at gradually increasing power levels until keeper damage and additional holddown bolt failures were noted in early April. By April 1, the recorded upper section duct wall temperature had not exceeded 300°C (570°F).

Prior to the June 2, 1976 shutdown, the normal duct wall design operating temperature of 482°C had been recorded only three times. The design upset condition, which produces a wall temperature of 649°C had never been attained. The maximum recorded wall temperature throughout the entire operating history was 515°C (960°F).

The accuracy of the duct wall temperature thermocouple readings was confirmed by instrumented thermal tests of April 14-16, 1976. Additionally, metallurgical examinations described later in this paper, and tests on duct samples confirmed that the duct wall temperature would have been in the temperature range recorded by the thermocouples. During fabrication and construction, large paint markings had been placed upon the duct sections to aid in erection. At the time of failure, it was observed that these paint markings remained visible. It was believed that if high temperatures had been experienced by the duct wall, the paint markings would have been evaporated and no longer visible. An experiment was performed to test this belief. A paint-bearing sample was taken from the hot end of the duct (between saddles 34 and 35), upstream from the thermocouple locations. The sample was heated to the upset duct wall temperature of 1200°F to see how long the paint could withstand elevated temperatures. In less than three hours at this inside surface temperature, no trace of paint remained. The test conclusively demonstrated that the duct walls had not spent any significant time at elevated temperature because the amount of paint loss depends upon the cumulative time at that temperature.

Operating personnel were also questioned to determine whether there had been any excursions above normal operating temperatures. It was noted that in late May 1976 the duct had been operated at 22 megawatts (29,500 horsepower) for several hours with two of the three exhaust stacks plugged. An analysis of the air flow and temperature conditions with two stacks plugged was conducted and the results showed that the total gas flow from the furnace would be reduced by less than ten percent. The resulting increase in gas temperature was small. This conclusion was verified by field measurements taken the first time the stacks plugged. A pitot tube traverse taken at the horizontal port before the dropout box showed that the gas temperature was 570°C (1058°F), more than 360°C below the design upset condition. The traverse also showed that the volume flow rate through the system with two stacks plugged was 2,034 cubic meters per minute at standard temperature and pressure (71,800 SCFM). The most important effect of the plugged stacks was therefore to increase the gas velocity in the unplugged stack which would have increased the heat transfer at local gas impingement points, thereby raising the local duct wall temperatures to a value closer to the inlet gas temperature of about 590°C (1100°F). While this would have caused some local "hot spots" near the manifold at the top of the duct, there was little effect after

expansion from the stack into the main duct where the velocity was reduced to normal.

The duct had also been observed glowing at night. Since the glowing region was localized, the duct would not have been much above the minimum temperature at which it can be seen to glow. Steels are known to glow at 480°C (900°F) with an intensity which depends very strongly upon how much light is present. A test was conducted on a COR-TEN A sample from the duct which showed that the apparent color also strongly depends upon the lighting conditions. The observers judged that the COR-TEN sample first began to glow somewhere between 480°C and 495°C (900°F and 925°F) in semidarkness. Color descriptions are subjective, but each observer described the color as "orange" when viewed in semidarkness. As the amount of light was slowly increased while holding the temperature constant, each observer found that the color became more "red," until eventually it appeared to have become that color altogether. In a fully lighted room, the glow became visible around 675°C (1250°F).

REDESIGN - In modifying the silicon metal furnace (north) duct, the welded stiffener rings were cut away, and the number of saddle supports reduced from thirty-seven to ten by eliminating all redundant supports between expansion joints. Each of the ten remaining saddles was modified so that, although it supported the duct, it was not welded to the duct and did not constrict its diametral thermal expansion. In this configuration, the spans between supports were fixed by the locations of the existing expansion joints. These long spans introduced a potential problem with material creep during high temperature operation. Dead weight stresses increase with the square of the span length, and removing the redundant supports greatly increased the span length. The resulting stresses would have exceeded the elevated temperature allowable stress values, which are limited by creep effects. Therefore, a refractory lining insulation was added to keep the material at a temperature below the creep regime.

It would also have been possible to use more expansion joints in order to reduce the spans so that the dead weight stresses remained below the elevated temperature allowable stresses without using refractory lining. However, it was concluded that this option would not be as cost effective in view of the already existing hardware.

High prior fatigue damage required that the existing COR-TEN in the ferrosilicon furnace (south) duct be scrapped. The redesign required either more expansion joints or refractory lining. Cost benefit analyses were carried out, taking into account safety, reliability and maintenance factors as well as differences in fabrication costs. Such factors as outside duct temperature, duct movement, the combined effects of the small amounts of air leakage at each expansion joint, and the details of the support structure design in the sloped region were considered. Moreover, there were distinct advantages to having identical maintenance procedures for both ducts, and benefiting from lessons learned from operating experience. Hence, it was concluded that twin duct systems would be the better design.

Fig. 3 - Bottom view of typical shell crack at stiffener ring

EXAMINATION OF FAILED DUCT

Although thermal stresses could not be measured during furnace operation because the system could not safely be operated in the severely cracked condition, the consequences of thermal expansion were quite evident, as can be seen on Figures 3-8. Holddown bolts attaching the fixed saddles to the support truss had failed because thermal bending of the duct was restricted by redundant restraints. Keepers which hold the sliding saddles down had also failed because of the thermal bending of the duct. The many cracks and buckles in the duct

Fig. 4 - Bottom view of shell crack at stiffener ring which has propagated away from the ring

itself also attested to the presence of high thermal stresses.

The source of the high thermal stresses was readily determined by examination of the failures and the geometry of the structure. Support rings on the outside of the duct were welded to the duct preventing free diametral thermal breathing of the duct. Large temperature differences between the rings and the duct produced large differential thermal expansions between the ring and the duct in both the radial and axial directions, resulting in large thermal stresses in the duct at the welds. Weather conditions (wind, rain, snow, darkness) caused changes in heat transfer from the outside surface of the duct, resulting in cyclic thermal stresses. Predictable variations in the gas temperature due to startup, normal operation, and shutdown of the furnace also caused cyclic thermal stresses. These cyclic stress conditions produced fatigue cracks in the duct. The fatigue cracks initiated near the stiffener rings, and propagated through the duct under the influence of the thermal stresses caused by the rings and the thermal bending stresses in the duct caused by the redundant supports.

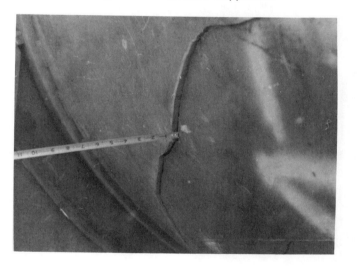

Fig. 5 - Crack propagated through the duct wall

The crack propagation was aggravated by the use of redundant supports in each segment of duct between expansion joints. Basic design to accommodate thermal expansion consists of supporting each segment of the duct system only at the end expansion joints. This allows the duct to flex without introducing large axial bending moments when the temperature distribution is nonuniform around the circumference of the duct. There are many causes of such nonuniformities including impinging gas at duct intersections or bends, weather conditions (the duct is exposed to the weather), and nonuniform dust buildup inside the duct. In the original design, several intermediate redundant supports were used in each segment of the duct. These additional redundant supports introduced significant thermal bending loads in the duct. These loads were evident from the arching of the duct. Not only did they break holddown bolts and keepers, they also tended to open up the fatigue cracks which had been initiated at the support rings, causing them to propagate to dangerous lengths.

The midspan buckle which occurred near the circumferential butt weld midway between saddles 7 and 8 was due to locally higher temperatures which produced axial compression in the duct wall where insulation was placed on the top of the duct. Eventually, buckling produced cracking in that region.

Fig. 6 - Open Crack at stiffener ring

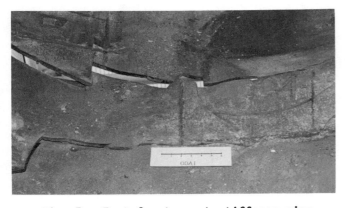

Fig. 7 - Fast fracture at stiffener ring

Fig. 8 - Midspan buckle

MATERIAL EVALUATION

Tests of the duct material were conducted, including various mechanical tests, metallurgical assessments, and chemical analyses. Metallurgical, scanning electron microscopy and electron-excited, energy-dispersive x-ray analysis examination of fracture surfaces and cross section samples were made. Such analyses were made on pieces of COR-TEN removed from the duct and also on virgin samples. Macroscopic examinations of material from the failed duct showed numerous cracks in close proximity to the intermittent welds and arc strikes. Photomicrographs indicated that most of the areas of the fracture surfaces examined were mixed mode, i.e., ductile and brittle fractures. Low cycle fatigue cracks initiated at locally high strain points and subsequently propagated as fast fractures once they had reached a critical size.

The microstructure at the hot end of the duct differed from that in unused material in that it contained spheroidized carbides. This indicates that the duct became hot enough to cause this microstructural change. Such changes are, of course, a function of both the time and temperature of exposure. The temperature had to exceed 445°C (833°F) for some period of time to cause this particular microstructural change in COR-TEN steel. The spheroidized carbides could have resulted from a long time exposure at 454°C (850°F), or from a very brief exposure at 689°C (1200°F), or from an intermediate time at a temperature between 454°C and 689°C .

According to the operating data for the main 2.7 meter diameter duct, the duct skin temperature exceeded 445°C for varying lengths of time during a two week period. The observed microstructure for a sample taken from the sloped region of the duct is consistent with the cumulative effect anticipated for these time-temperature conditions. Hotter temperatures would have been experienced in the 2.2 meters diameter duct where stack gases impinge directly against the wall. This region was observed glowing at night on several occasions. The microstructure of the sample taken from this region contains more spheroidized carbides than the other sample, consistent with the greater cumulative exposure to temperatures above 445°C in the manifold.

The fatigue properties of COR-TEN steel used in the failure analysis were verified by tests. Material which had originally been purchased for use in the duct but which was never used was soaked for 100 hours at 540°C (1000°F). This material had seen no prior strain damage. Low cycle fatigue tests were performed on this material in air at 1000°F using hourglass-shaped specimens. The axial strain was controlled in the tests to give strain ranges of 1.5%, 2.0%, 4.0% and 5.0%, respectively, based on diametral strain measurements. In addition, six hold-period tests were performed with hold periods of three or fifteen minutes. Hold periods in tension only were employed in five of these tests while one was performed using a hold period in compression only. The fatigue tests included only very short hold times compared to the service conditions. It was not practical to run laboratory fatigue tests with hold times comparable to service conditions because it takes too long to generate the data. During hold times at elevated temperature, elastic strains are converted into creep strains, which produce much more fatigue damage in materials such as COR-TEN. Thus, for a fixed total strain range such as used in these tests, the fatigue life would be expected to be reduced with increasing hold times. The data showed this to be the case. A three minute hold time reduced the fatigue life by about a factor of two. Extrapolations to the service condition hold times on the order of days indicate quite good agreement with the theoretical failure curve at 1000°F.

The resulting low cycle fatigue test data is plotted on Figure 9, along with the theoretical mean failure curves which had previously been derived using tensile test data reported by U. S. Steel. The theoretical failure curves were based on the Langer-Coffin equation which does not include consideration of creep effects. However, the use of a relatively low reduction in area value of 34 percent was believed to account for the reduced ductility due to thermal aging and creep effects at 1000°F. Note that all of the fatigue test data falls between the theoretical failure curves for 370°C and 540°C (700°F and 1000°F),

respectively. Thus, the fatigue test results indicate that the fatigue properties of COR-TEN steel are much lower at 1000°F than at 600, 700 or 800°F. Moreover, the data substantiates the validity of the fatigue properties used.

FAILURE AND REDESIGN ANALYSES

Failure analyses and redesign analyses were performed to evaluate the failure of the south duct and to assure the structural integrity and reliability of the redesigned ducts. This section summarizes those analyses.

FINITE ELEMENT ANALYSES - Temperature distributions in the duct stiffener ring were calculated based on a reference 538°C (1000°F) inlet gas temperature and corrected for other operating inlet gas temperatures. Gas temperature measurements taken at various times during furnace operation indicate that the actual gas inlet temperature at the hot end of the duct was in the range from 540°C TO 595°C (1000°F to 1100°F) during the period of south duct operation at 22 megawatts.

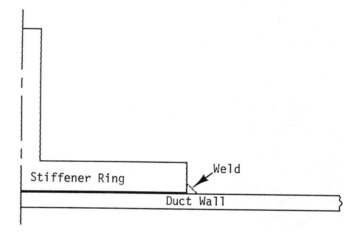

Fig. 10 - Stiffener ring welded to duct

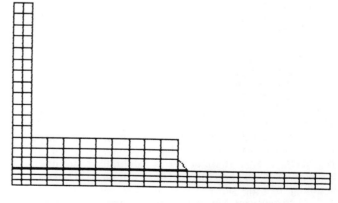

Fig. 11 - Finite element model of ring/duct interface

The computer model illustrated on Figures 10 and 11 shows a gap between the duct and the stiffener ring. Field measurements show that

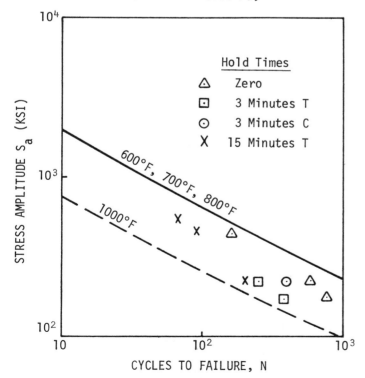

Fig. 9 - Low cycle fatigue test data

this gap, which resulted from the fabrication process, varied from a minimum (usually direct contact) near the bottom of the duct to a maximum near the top of the duct. The gap is also larger at the welded edges of the ring than at the center of the ring due to the permanent anticlastic curvature introduced when the rings were fabricated. The maximum gap varied from ring to ring. A series of thermal problems were run with the gap thickness varied from zero to 2.54 cm (1 in.). The reference problem has a 0.119 cm (0.0469 in.) gap considered representative of the "typical" gap thickness around the circumference.

Thermal stresses in the duct due to the differential thermal expansion between the duct and the stiffener ring were also evaluated. Here, as in the thermal analysis, a series of problems were run with various gap thicknesses. The reference detailed analysis corresponded to the reference thermal problem. Elastically calculated stresses were far above yield, demonstrating the detrimental effect of using stiffener rings welded to the duct.

STRUCTURAL ANALYSES USING BEAM ELEMENTS - Structural analyses were performed using beam elements, employing both hand and computer calculations. The computer program employed uses linear temperature variations across a beam cross-section from top to bottom and/or from side to side. Hence, initial calculations were performed to determine an equivalent linear temperature distribution for the actual nonlinear temperature variation. The computer analyses provided the thermal bending stresses in the complex duct and support system resulting from these equivalent linear temperature variations. Again, the elastically calculated stresses were far above yield, demonstrating the detrimental effect of having redundant supports preventing thermal bending of the duct.

Thus, the computer structural model provided support reactions and thermal stresses in the duct due to the basic geometry of the duct and its support locations. Detailed local finite element stress analysis accounted for the effect of putting stiffening rings around the duct and welding them to the duct. The effects of the saddle support reactions on the duct shell were included using stress solutions available in the literature.

The results of the structural analyses of the original design showed that the flexibility of the truss increased dead weight stresses and reduced thermal bending stresses in the duct. This result was due to the improper design of the duct which incorporated redundant supports between the expansion joints. Hence, thermal bending of the duct imposed high cyclic loads and stresses on the support truss, buckling several truss members. With properly designed duct supports, thermal bending of the duct does not bend the supporting truss. Moreover, the flexibility of the truss would have no effect on either the thermal or dead weight stresses in the duct. This is important since the typical

design sequence involves first designing the duct, and then using the resulting weight to design the truss.

FATIGUE ANALYSES - The calculated loads and stresses were used to perform a low cycle fatigue evaluation using the actual operating data. The operating temperatures were lower than the temperatures anticipated in the design specifications. There were no known cycles corresponding to the anticipated 934°C gas inlet temperature upset condition, and only three cycles to the normal operating temperature. Hence, the actual operating cycles were used to evaluate the fatigue damage.

Since local stresses were far in excess of the yield strength of the materials at the ring to duct stitch welds, elastic-plastic analyses were used to obtain the strain ranges needed to perform a low cycle fatigue evaluation. The evaluation was made using fatigue design curves obtained by applying a factor of twenty on cycles to the theoretical failure curves shown on Figure 9. Based on the design curves, a cumulative fatigue usage of 13 had been reached when failure occurred. This is consistent with the knowledge that cracks had initiated and grown to a critical size and propagated as fast fractures at this usage. Actual failure occurred between the design fatigue curve and the theoretical mean failure data for small polished laboratory test specimens. Failure below the mean laboratory failure curve is expected due to size effects, surface finish effects, environmental effects and scatter in the data.

FRACTURE MECHANICS ANALYSES - Exhaust gases and dust were blowing through cracks up to three meters long by the time the ferrosilicon furnace was shut down in June 1976. In order to understand why the cracks had propagated so far, fracture mechanics analyses were employed to assess crack propagation into regions well away from the stiffener rings. This behavior is a function of the stress field and the fracture toughness of the material. The elastically calculated stresses varied from far above yield at and near the stiffener rings to well below yield at midspan. Temper embrittlement associated with elevated temperature operation caused the fracture toughness K_{IC} to vary from a minimum of 27.5 MPa$\sqrt{m}$ (25 ksi $\sqrt{in.}$) at ambient temperatures near the furnace up to five times that value for uneffected material near the duct outlet where it was not exposed to elevated temperatures. These values were obtained by applying standard correlations to Charpy V-notch test data obtained on the duct material. The cracks ran during the shutdown transient when the duct wall temperature dropped, reducing the toughness below the critical value for the existing fatigue cracks. Fast fractures moving at the speed of sound in the material were heard by the system operators.

Linear fracture mechanics is applicable to the elastic stress regions away from the stiffener rings, but not to the plastic stress regions near the rings. Initiation of fatigue cracks occurred at the 7.6 cm (3 in.) long stitch welds attaching the stiffener rings to the duct wall. Critical crack sizes for various observed crack configurations were evaluated over the K_{Ic} range for stress intensities up to yield. These evaluations showed that cracks propagating from the stiffener welds into regions at these lower stresses would have grown to critical sizes even for the unembrittled material. The critical crack sizes at yield were such that even a shallow crack the length of the stitch weld would exceed the critical size and propagate through the duct.

REVIEW OF ORIGINAL DESIGN CALCULATIONS

When the crossover duct failure became the subject of litigation, the original design calculations were reviewed to determine why the stiffening rings had been welded to the saddles and duct and why redundant supports had been used between the expansion joints. The original design was for an operating condition with a continuous gas inlet temperature of 934°C (1714°F) and a duct wall temperature of 649°C (1200°F). This wall temperature comes from the upset condition with a 17°C (30°F) margin. Axial temperature profiles for both the gas and duct wall corresponding to this operating condition were determined.

It is fundamental in designing ducts or large diameter thin-walled pipes for elevated temperature service to allow for thermal expansion. In this ductwork, expansion joints were provided to accommodate the axial thermal expansion of the duct. However, the radial thermal expansion of the duct was improperly restrained by welding stiff reinforcing rings on the outside of the duct as shown on Figure 10. Moreover, the rings were 0.26 meter (10.275 in.) wide and welded on both sides so that the thermal axial expansion under the rings also introduced large stresses at the welds. The rings were intended to keep the duct from creeping out of round due to the small vacuum pulled by the exhaust fan. However, they could have served this function as well had a small radial clearance for thermal expansion been provided between the duct and had the rings not been welded to the duct. As designed, the welded rings restrained the thermal expansion of the duct since the duct wall operates at a much higher temperature than the rings. The rings were not insulated and were exposed to atmospheric cooling. Very high thermal discontinuity stresses were therefore created in the duct material at the reinforcing rings. The thermal structural interaction between the duct and the stiffening ring and the resulting fatigue were not considered in the original design calculations.

When the stiffener rings were welded to their saddle supports to keep the duct from sliding down the slope, the problems created by the rings were amplified. Over the arc subtended by the saddle, the stiffness of the ring was increased while the effective temperature was decreased. This accentuated the radial constraint on the duct and increased the thermal stresses in the duct at the reinforcing rings. The original design calculations did not consider the effect of the saddles on the temperature distribution or on the thermal stresses in the duct, stiffening rings or saddles.

Each segment of ductwork between expansion joints should have been supported only at its ends so that the duct would have been free to bend thermally. This is a basic design consideration for such ductwork. Thermal bending is caused by impinging gases, dust build-up and nonuniform cooling on the outside wall of the duct which is exposed to the atmosphere. None of these factors were considered in the original design. When the duct is tied down between expansion joints, it is not free to arch when the top and bottom of the duct are at different temperatures. Moreover, the additional improper redundant supports impose loads on the duct by restraining the free thermal bending. The resulting loads can be many times higher than the dead weight loads. These loads tended to propagate the low

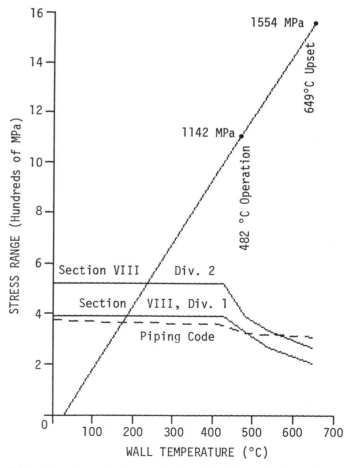

Fig. 12 - Comparison of elastically calculated stress range to limits imposed by Codes

235

cycle fatigue cracks which had been initiated by the welded reinforcing rings, yet had never been considered in the design calculations.

COR-TEN steel is not an ASME Boiler and Pressure Vessel Code material. However, the basis for establishing Sm described in Section VIII, Division 1 and that described in Section VIII, Division 2 of the Code were used to determine stress range limits for thermal stresses in this application. The allowable stress range formula for expansion stresses given in paragraph 102.3.2C of the ANSI B31.1 Piping Code was also considered. As can be seen on Figure 12, by the time the duct wall temperature goes above 233°C (452°F), the allowable stress range has been exceeded for all three of these approaches. Normal anticipated operation and the occasional upset condition produced elastically calculated stresses far above these limits.

Unrealistic simplifying assumptions were made in performing the original design calculations. For example, it was assumed that there would be no circumferential temperature variations around the duct even though a specified design condition included having the duct half filled with dust. Dust acts as an insulator, thereby creating temperature differences between the top and bottom of the duct. Note that there are also significant temperature differences across the duct at the bends due to gas impingement and from top to bottom due to weather conditions even if there were no dust buildup.

Commencing shortly after system shutdown, the engineers performing the failure analysis, the redesign analysis, and the material evaluation worked closely with the attorneys prosecuting the claim on behalf of the plant owner against the system designer. This team approach greatly facilitated the ultimate utilization of the analyses and material evaluation at the trial. It also assured that an appropriate and judicially admissible record supporting the engineering decisions which were made was maintained as work progressed. This minimized the potential for "second guessing" by persons in an adversary position who were challenging the failure and redesign analyses and the cost of the redesign work. The attorneys (three of whom have engineering degrees) benefited from this team approach by acquiring a better understanding of the complex engineering issues involved in the failure analysis and redesign effort. This better understanding in turn enhanced the attorneys' ability to present the expert testimony at the trial in a readily comprehensible manner.

The outcome of the lawsuit was a court determination that the duct failure was caused by defective design and multimillion dollar compensation was awarded to the plant owner.

IMPROPER FABRICATION OF ROTATING BLADES RESULTS IN PREMATURE FAILURE

Fred W. Tatar

Factory Mutual Research Corporation
Norwood, Massachusetts, USA

ABSTRACT

An investigation was conducted into the failure of a number of rotating blades in a diffuser at a sugar beet processing plant. Several of the blades, which had been fabricated from rectangular bars removed from rolled steel plate, fractured brittlely. However, otherwise apparently identical blades underwent significant plastic deformation without fracture. Inspection of both fractured and bent blades revealed similar pre-existing cracks at the toes of bar attachment welds. Metallographic examination of the bent and the fractured bars revealed that they had been cut parallel and transverse, respectively, to the rolling direction of the steel plate. Due to the combined effects of the low fracture toughness of the plate on planes parallel in the rolling direction, the presence of the preexisting cracks, and the relatively large section thickness of the bars, the bars whose lengths were transverse to the rolling direction fractured brittlely when subjected to impact loads. Had the poor transverse properties of thick-section plate been recognized, and all the bars properly cut with respect to the rolling direction, the premature fractures would not have occurred.

MECHANICAL DAMAGE occurred to the internal components of a sugar beet diffuser at a midwestern United States processing plant. The slope type diffuser extracts raw beet juice from sliced beets, as highlighted in the simplified diagram of the vessel shown in Figure 1. Sliced beets are dropped from a conveyor belt into the bottom of the vessel. The beets are carried from the low to high end of the diffuser by two intermeshed screw-type conveyor assemblies that have opposite rotations. Along their routes of travel, any solid plugs of beets are separated by stationary breaking bars welded to the vessel shell. As the beets are conveyed upward, they are intermixed with hot water which enters the high end of the diffuser and flows downward by gravity. Sugar is leached from the beets, forming raw beet juice which flows out of the bottom of the diffuser. The leached beet pulp is then expelled from the vessel top.

The two intermeshed screw-type rotating components, known as scrolls, contain 32 flight arm assemblies. As shown in Figure 2, the assemblies consist of three pie-shaped flight arms. The flight arms are curved similar to turbine blades, but are not solid, consisting rather of a number of components. Two 4-1/2 in. width x 2-5/8 in. thickness carbon steel bars cut from rolled plate are welded to the scroll shaft approximately 70º from each other and at slightly different elevations. Three curved concentric steel stiffeners are welded to the bars, forming a skeleton similar to a turbine blade. Finally, aluminum screens, manufactured from 1/4 in. thick perforated sheet, are bolted to the bars and stiffeners. Beet juice can therefore pass through the flight arms while the pulp is carried upward.

INCIDENT

During operation, a noise was heard coming from the sugar beet diffuser. Although the scrolls were still rotating, opening of the inspection doors showed that some of the flight arms had fractured. After the diffuser had been shut down, internal examination revealed that thirteen flight arms were damaged. A number were bent, while the remainder had fractured 4 to 6 in. from the scroll shaft at the rectangular bar/aluminum screen attachment welds. All of the aluminum screens were found to be torn; some were rolled into balls, indicating that they had been dragged along by the rotating scrolls.

EXAMINATION

A number of the bent and fractured bars were removed from the diffuser and shipped to independent laboratories for examination. A photomacrograph showing the fracture face of one of the bars removed from the diffuser is presented in Figure 3. The fracture face exhibits significant smearing of metal, the result of mechanical damage during the incident. However, other than the smeared regions, little plastic deformation is apparent, either on or at the edges of the fracture face. Since the fracture face is also almost entirely on one plane, the failure is clearly brittle rather than ductile.

Subsequent examination revealed that the initiating area of the fracture was a preexisting crack at the toe of the attachment weld. Figure 4 is a photomacrograph showing the attachment weld region; an arrow highlights the preexisting crack. Stereomicroscopic examination indicated that the crack face was smooth with no apparent branching, consistent with corrosion fatigue cracking. Due to the presence of the preexisting cracks, and the large section thickness of the bars (4-1/2 in. width x 2-5/8 in. thickness), fractures might be expected if the bars were subjected to impact loads. Apparently, one of the aluminum screens became dislodged, jamming one of the blades and setting off a chain reaction of fractures. As shown in Figure 5 (Ref. 1), the resistance to crack propagation is very low as the section size increases and plane strain conditions are approached.

However, a number of flight arm bars exhibited almost identical pre-existing cracks, but had significantly deformed during the incident. Figure 6 highlights one of the bent bars; evident is a crack at the toe of the attachment weld. Clearly, the fracture toughness of the bent bars was sufficient to withstand the incident impact loads. It was therefore believed that the

fractured and bent bars had been fabricated from plate steels with significantly different fracture toughnesses. To confirm this initial conclusion, Charpy impact specimens were removed from a number of the bars with the notch aligned on a plane parallel to the fracture face. Results of the impact testing, presented in Table 1, showed that the fractured blades had average impact strengths less than one-third those of the bent blades.

At the sugar beet processing plant's request, chemical analyses of the bar samples were also conducted. The results of the analyses, presented in Table 2, showed that the average chemical compositions of the bent and fractured bars were almost identical. Due to the vastly different impact properties of the bars, the possibility of significant heat treatment variations between the bent and fractured bars appeared likely.

Metallographic specimens were therefore removed parallel to the pre-existing cracks in both bent and fractured bars. The representative microstructures of the bent and fractured bars are presented in Figures 7 and 8, respectively. Rather than indicating varying heat treatments, the microstructures show that the bent and fractured bars had been cut parallel and perpendicular, respectively, to the rolling direction. Similar to wood logs, which split easily lengthwise along the wood grains, the fracture toughness of steel is typically far greater on planes oriented across the steel grains (transverse to the rolling direction). A close-up photomacrograph concentrating on one of the bar fracture faces is shown in Figure 9. The terraced topography, which indeed indicates that the fracture occurred along the rolling direction of the plate, is similar to the wood-grained structure of a split log.

CONCLUSIONS

The combined evidence indicates that the fracture of the flight arm bars was the result of three factors, namely, the presence of a preexisting crack at the toe of an attachment weld, the large section thickness of the bars, and the removal of the bars from plates with their lengths transverse to the steel plate rolling direction. When these bars were subjected to impact loads, the fractures occurred. Had these bars been cut parallel to the rolling direction, as had the bent bars, the premature fractures would not have occurred.

ACKNOWLEDGEMENT

The author wishes to thank Mark H. Gilkey of Factory Mutual Research Corporation for his extensive assistance in the analysis of these flight arm failures.

Reference 1   -   J. I. Bluhm, "Failure Analysis, Theory and Practice," Failure Analysis, Edited by J. A. Fellows, ASM, pp. 61-124, 1969.

	Impact Strength ft-lb
Bent Bars	130
Fractured Bars	42

Table 1. Average Charpy V-notch Impact Strengths of Flight Arm Bars

	Carbon %	Manganese %	Phosphorus %	Sulfur %
Bent Bars	0.18	1.10	0.022	0.020
Fractured Bars	0.20	1.04	0.006	0.022

Table 2. Average Chemical Compositions of Flight Arm Bars

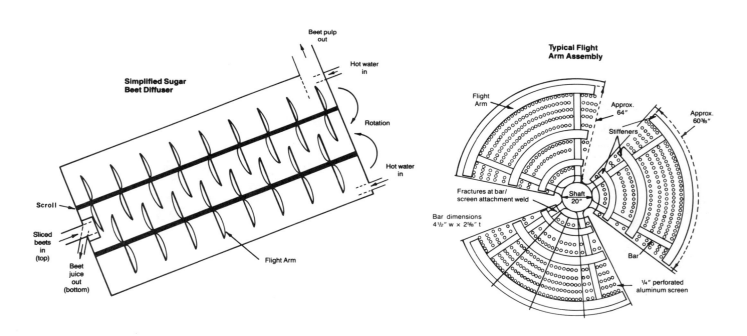

Figure 1. Simplified sugar beet diffuser diagram.

Figure 2. Detailed drawing of typical flight arm assembly.

Figure 3. Photograph of one of the bar fracture faces. A brittle fracture is indicated.

Figure 4. Photomacrograph highlighting preexisting crack (arrow) at toe of attachment weld of fractured bar.

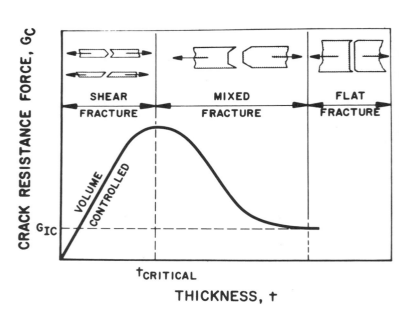

Figure 5. Effect of section size on resistance to crack propagation.

Figure 6. Photograph showing crack at the toe of attachment weld on one of the bent bars.

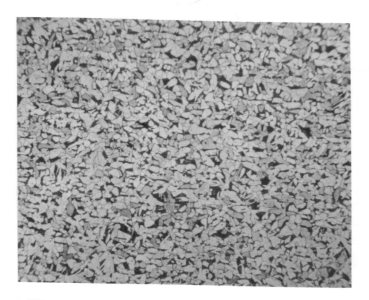

5% Nital   Mag 50X

Figure 7. Photomicrograph showing microstructure of cross section parallel to pre-
existing crack in bent bar. Equiaxed grain
structure is indicated.

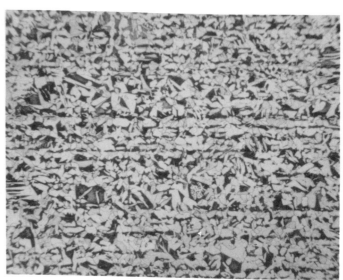

5% Nital   Mag 50X

Figure 8. Microstructure of cross section
parallel to fracture face. Fracture
occurred along the plate rolling direction.

Mag 4.0X

Figure 9. Photomacrograph highlighting area
on bar fracture face. Terraced topography
indicates a brittle failure.

FAILURE ANALYSIS OF A COLLAPSED ROOF

Walter L. Bradley
Mechanical Engineering
Texas A&M University
College Station, Texas, USA

ABSTRACT

A portion of the roof of a single story building collapsed during a thunder storm. A failure analysis was conducted to determine whether this structural failure was due to inproper design, substandard construction materials, faulty erection, or extreme weather conditions. The failure analysis indicated substandard structural components in combination with faulty construction was responsible for this service failure.

A PORTION OF THE ROOF of a one year old single story building collapsed during a thunder storm which "dumped" twelve inches of rain in less than two hours. Winds approaching 100mph were recorded at the height of the storm. Tornadoes were seen in the vicinity of the building during the storm, and it was speculated that the roof collapse might have been precipitated by one of the tornadoes. A failure analysis was undertaken to determine whether the roof failure was the result of poor design or workmanship or the unavoidable consequence of severe weather.

The building with partially collapsed roof is seen in an aerial photograph in Figure 1 and from the inside of the building in Figure 2. Several horizontal girders are seen to have broken free from vertical I-beams during the roof failure. Numerous horizontal trusses broke during the roof collapse, as seen in Figure 3. Some broken trusses were also noted in sections of the roof adjacent to the section that failed during the storm.

The failure analysis consisted of (1) an onsight inspection, (2) macrofractographic examination of the fractures where the girders were welded to the columns, (3) macrofractographic examination of the fractured trusses, (4) metallographic examination of the girder and truss materials, (5) chemical analysis of the girder and truss materials, and (6) mechanical testing of the truss material. Experimental results of the investigation will presented next followed by discussion and conclusions.

EXPERIMENTAL RESULTS

ONSIGHT INSPECTION - The initial onsight inspection was used to survey the damage and determine what failed first, triggering the general collapse. It was noted during the onsight inspection that the roof had a slight pitch toward the center of the building, with downspouts placed in the interior columns to allow water drainage. The collapsed section was in the center section of the roof where water loading would be greatest if a heavy rain released water at a faster rate than the downspouts, or drains could handle. The fact that the damage was localized where water loading would be heaviest with minimal damage around the perimeter suggested that rainwater accumulation on the roof was more responsible for the failure than wind, including tornadoes.

The total amount of water loading was limited in design by the pitch of the roof and the height of the small lip erected along the perimeter of the roof to minimize loss of gravel. However, any reduction in stiffness in construction or service would allow a greater depth of water ponding on the roof during a heavy rain storm, with the increased loading giving further roof deflection and additional ponding. Thus, a load sufficient to collapse the entire structure may be reached under such circumstances.

Approximately thirty trusses were removed from a section adjacent to the section that collapsed in service. These trusses were found to be either uncambered or undercambered. The purpose of the camber is to anticipate the deflection in the trusses that will result from the load of the roofing material and camber to such a degree that the trusses under normal roofing material loading will be essentially horizontal. This avoids unnecessary water ponding due to truss deflections.

A second observation concerning the trusses was that they frequently contained splices. In the roof section that collapsed, all broken trusses failed at splices. Thus, the onsight inspection included X-ray inspection of the splices in the thirty trusses removed from the section adjacent to the section that collapsed in service. This X-ray inspection indicated poor penetration welding in the splices. However, the code does not require full penetration welds if splice bars are used for reinforcement, as shown in Figure 4. The only code requirement is that the splice must be at least as strong as the remainder of the truss; i.e., failure should not occur at splices.

Since it was not possible to calculate the strength of the poor penetration welds combined with the splice bars, plans were made to cut out sections of thirty trusses containing splices to be mechanically tested. The results of this portion of the investigation will be reported later.

In summary the onsight inspection seemed to suggest that water ponding due to premature truss failure at splices was the primary cause of failure of the roof. Lack of proper camber in the trusses further aggravated the problem.

MACROSCOPIC EXAMINATION OF GIRDER WELD FRACTURES - A closer inspection of the girder structure which failed in service indicated bolt holes were enlarged by torch cutting to accommodate misfited joints, as seen in Figure 5. This has the adverse effect of reducing structural stiffness, which aggravates the water ponding problem during a heavy rain storm. An examination of the frac-

tured surface suggested that the weld failure which caused the girder collapse seemed to be a tensile overload fracture of a basically sound weld.

MACROFRACTOGRAPHIC EXAMINATION OF FRACTURED TRUSSES - Many trusses were found to have fractured in the roof section that collapsed. Additional trusses fractured in sections of the roof adjacent to the collapsed one. All fractured trusses were at splices in the angle iron, as shown in Figure 6. Approximately thirty sections containing such fractures were cut from the trusses that failed in service. These were subsequently cleaned and photographed. Typical results are seen in Figure 7. The raised lip around the perimeter of the fracture surface is the fractured weld metal. The recessed area in the center is the base metal and also represents the area of lack of full penetration in the weld. Typically the weld penetration was about 60% of the total cross-sectional area of the angle iron. This lack of penetration in welding if not fully compensated by the splice bar would explain why the truss failures were all at the splices. Tensile testing of the spliced angle iron with splice bars was conducted to quantify the strength of the trusses at the various splices.

TENSILE TESTING OF SPLICED ANGLE IRON USED IN TRUSSES - Approximately thirty sections of angle iron used in the trusses and containing splices were cut from trusses which did not fail in service. Fixtures were welded to the end of each 30cm length of angle iron to allow for gripping in the tensile test machine. Each spliced angle iron contained a splice bar for reinforcement. In some cases, both angle irons and their common splice bar were test simultaneously. (Note in Figure 3 that the trusses are constructed using a pair of angle iron bars at the top and a pair at the bottom, with the angle iron being joined by welding to the paddle bars and splice bars.)

Using strength of materials equations based on linear beam theory and knowing the rated loads and physical dimensions of the trusses, it is a straight forward exercise to calculate the load and stress in each angle iron at the rated service load. The calculated load was 134KN for a single angle iron or 268KN for a pair of angle iron sections. These calculated loads were used to evaluate the tensile test results on the spliced angle iron sections.

A specimen being tensile tested is shown in Figure 8. This particular specimen included the pair of angle iron sections containing a splice and the splice bar. Of the thirty specimen tested, twenty-five were single specimens and only five were pairs of angle iron as shown in Figure 8. The average load at fracture for the thirty tests was only 60% of the calculated load at the rated service load for the trusses. There were no specimens that reached the calculated load in the angle iron that would be experienced when the truss structure was loaded to its rated load. Thus, it may be concluded that the trusses would have failed on the average at service loads of only 60% of the rated load limit. It is interesting to note that conventional design practice is to design for a maximum service load of approximately 60% of rated truss load. The maximum service load would only be experienced when a heavy rain caused water ponding of sufficient depth that excess water would run off the perimeter of the building. This was indeed the case at the time that the roof collapse occurred.

METALLURGICAL AND CHEMICAL ANALYSIS OF THE TRUSS AND GIRDER MATERIAL - The investigation was concluded with a metallurgical and chemical analysis of the girder and truss materials. The chemistry and microstructure of both materials were consistent with specified values for the low carbons steels used in such components.

DISCUSSION

The experimental results may be summarized as follows. The steel used in the girders and trusses were acceptable. The fabrication of the trusses included poor penetration welding with poor welding of the splice bars, giving a splice joint with only sixty percent of the required strength. The first trusses to fail was apparently one that contained adjacent splices midway across the truss span. It should be noted that adjacent splices are prohibited by code as are splices near the midspan where loads are greatest.

The failure of this truss will locally reduced the structural stiffness, allowing more deflection of the roof and more water ponding. The additional ponding increases the load on trusses adjacent to the first failed truss, causing a second one to break. Additional ponding results until a third truss breaks and the processes repeats itself. Eventually, the ponding becomes sufficiently great that the joints between girders and columns begin to fail, at which point a whole section of the roof will collapse.

Lack of camber in the trusses and the oversized bolt holes used in the girder erection would have increased somewhat the water pondering, but are not considered the primary cause of failure.

CONCLUSION

The primary cause of the roof collapse was found to be poorly fabricated trusses.

Figure 1. Aerial view of building showing one section of roof collapsed.

Figure 2. View of roof collapse from inside the building.

Figure 3. Fractured truss in roof section adjacent to one that collapsed.

Figure 4. Two sections of angle iron with splice bar across splice in one
of angle iron sections.

Figure 5. Holes enlarged by torch cutting to accommodate beam misalignment in construction.

Figure 6. Section of angle iron in truss which failed at splice.

Figure 7. Fractured surface of angle iron which failed at splice. Raised section around perimeter is fractured weld metal. Recessed area is base metal and indicates area which was not penetrated during welding.

Figure 8. Section of angle iron with splice and splice bar being loaded tensile test.

ELASTIC PLASTIC FRACTURE EVALUATION FOR PART-THROUGH FLAWS IN STAINLESS STEEL PIPE WELDS

Akram Zahoor, Ronald M. Gamble
Novetech Corporation
Rockville, Maryland, USA

Abstract

Elastic-plastic fracture mechanics and limit load analyses were performed to evaluate the load carrying capacity of austenitic stainless steel pipe weldments that contain circumferential throughwall or finite length, semi-elliptical, part-throughwall flaws. The welds included shielded metal arc weld (SMAW), and submerged arc weld (SAW), which are often used to fabricate stainless steel reactor piping systems in the United States. The purpose of the analyses was to assess the difference in load carrying capacity predicted by stability and limit load analyses. Flaw instability was determined using J-Integral/tearing modulus (J/T) analyses.

The results for part-throughwall flaws were obtained from newly developed elastic-plastic fracture mechanics computational techniques for finite length, circumferentially oriented, part-throughwall flaws in piping under combined axial and bending loads.

The results indicate that instability loads for SMAW are between 62 and 88 percent of the limit load, while the instability loads for SAW are between 59 and 78 percent of the limit load. The specific instability to limit load ratio depends on pipe size, and flaw type and size.

THE INCIDENCE OF INTERGRANULAR STRESS CORROSION CRACKING (IGSCC) discovered in stainless steel piping at operating boiling water reactor (BWR) facilities (1,2) motivated placement of allowable flaw sizes for stainless steel piping into Paragraph IWB 3640 of Section XI to the ASME Code (3). The allowable flaw sizes were determined from limit load analysis. The bases for using limit load analysis were the results from previous analyses (2, 4-9) and experimentation (10) that indicated cracked wrought stainless steel piping would fail by plastic collapse prior to unstable crack extension.

However, subsequent experimental data from small test specimens (11, 12) indicate that SAW and SMAW stainless steel welds can have lower resistance to first extension and subsequent growth of a crack compared to wrought stainless steel base metal. This reduced crack extension resistance may preclude the cracked weld section of the pipe from reaching limit load prior to instability and, consequently, result in load carrying capacity less than that predicted from limit load analysis.

The objective of the work presented in this paper was to determine if the apparent lower weld toughness would result in reduced load carrying capacity of the cracked pipe section relative to that predicted from limit load analysis. Pipe size and throughwall and part-throughwall flaw size were included as parameters in the analyses.

ANALYSIS

LIMIT LOAD - The limit load equations used to develop the ASME, Section XI relationship between applied stress and circumferential flaw length and depth for wrought austenitic stainless steel piping are (10,13-15)

$$P'_b = 2\sigma_f [2\mathrm{Sin}\,\beta - a(\mathrm{Sin}\,\Theta)/t]/\pi, \qquad (1)$$

where

$$\beta = [(\pi - \Theta a/t) - (P'_m/\sigma_f)\pi]/2,$$

or if $\Theta + \beta > \pi$, then

$$P'_b = 2\sigma_f[(2 - a/t)\,\mathrm{Sin}\,\beta]/\pi, \qquad (2)$$

where

$$\beta = \pi[1 - a/t - P'_m/\sigma_f]/[2 - a/t],$$

and

P'_b = bending stress at plastic collapse, with

P'_m = membrane stress, assumed equal to $0.5\ S_m$,

σ_f = Material flow stress, equal to one-half the sum of the yield and ultimate strengths,

a = part-throughwall radial flaw depth,

Θ = throughwall or part-throughwall circumferential half crack angle in radians,

t = pipe wall thickness.

In IWB 3640 of Section XI, allowable flaw depths are presented as a function of flaw length and stress ratio (SR), which is the total applied bending and membrane stress (P_m + P_b) normalized with respect to the design stress intensity, S_m. The value of S_m used in IWB 3640 was 16.8; values for P_m and P_b are obtained from the pipe design stress reports.

Two sets of allowable flaw size tables were developed from Eqs. (1) and (2) and are presented in the code. One set of tables has been developed for normal (including upset) operation and includes a safety factor (SF) of 2.77 on the total plastic collapse load (P'_m + P'_b); a second set of tables is available for emergency and faulted operating conditions and includes a SF = 1.39. The tables are constructed so that the safety factor is included in the stress ratio, ie, the stress ratio corresponding to any allowable flaw size in Tables IWB 3641-1 thru 4 of Section XI is

$$SR = (P'_m + P'_b)/\ SF \cdot S_m \qquad (3)$$

The following computation procedure was used to determine the difference in load carrying capacity between cracked weld pipe sections assumed to fail by ductile crack instability versus plastic collapse.

For part-through flaws, Eq. (3) was used (with $P'_m = 0.5S_m$, SR = 1.0, and SF = 2.77) to determine the bending stress P'_b at plastic collapse. Elastic plastic fracture mechanics analyses were then performed as a function of flaw length to determine the part-throughwall flaw depths that result in flaw instability at the loads determined from Eq. (3). Next, the instability flaw sizes were used with Eqs. (1) or (2) to determine the collapse load for the instability flaw sizes. Finally, the load carrying capacity relative to limit load was determined by the ratio of the instability load to limit load at the corresponding instability flaw size.

The throughwall flaw evaluation was performed for bending loads alone. In these analyses, throughwall flaw lengths ranging from 10 to 40 percent of circumference were used to determine the limit moments (from Eq. (1) or (2) with $P'_m = 0$ and $a/t = 1$) and

instability moments.

The following section describes the fracture mechanics analysis procedures used to determine the instability moments for pipe sections containing either throughwall or part-throughwall circumferential flaws.

FRACTURE MECHANICS ANALYSIS - The elastic plastic fracture mechanics analyses are based on the J integral (16), which is a measure of the intensity of the stress-strain field around the crack tip (17, 18).

Two important aspects should be considered when evaluating ductile crack extension; namely, initiation or first extension of the existing flaw and stability of a growing flaw subsequent to initiation. The value of J associated with initiation of crack extension is denoted as J_{IC}. If the applied value of J is less than J_{IC}, first crack extension will not occur and stability of the existing crack is ensured automatically. When extension of the existing crack is predicted, the crack extension must be evaluated to determine if it occurs in a stable manner or if the crack will grow unstably and result in a predicted full break.

Crack instability is evaluated using the tearing modulus instability criterion originally defined (19) as

$$T = (dJ/da) \cdot (E/\sigma_f 2) \qquad (4)$$

where

dJ/da indicates the increment of J needed to produce a specified increment of crack extension at any specified load and crack state,

E is the material elastic modulus,

σ_f is the material flow stress defined as one half the sum of material yield and ultimate strengths.

To determine the instability conditions, values of J and T are first determined for the structure using the applied loads and crack geometry. These J and T values are denoted as $J_{applied}$ or J_{app}, and $T_{applied}$ or T_{app}.

The material resistance to unstable crack extension is determined experimentally from test data that relate J and material crack extension (J-R Curve). The material tearing modulus can be determined as a function of J from Eq. (4) and the slope at various points along the J-R Curve. The tearing modulus obtained from the materials data is denoted as T_{mat}. At any specified J level, where J_{app} is greater than J_{IC}, stable crack extension will occur when

$$T_{mat} > T_{app} \qquad (5)$$

A convenient means now commonly used to

illustrate the instability conditions involves plotting J as a function of T for the applied loads and material resistance as shown schematically in Figure 1, where material crack growth resistance is plotted along with the crack growth potential associated with the load and crack in the component. The intersection of the curves defines the instability point.

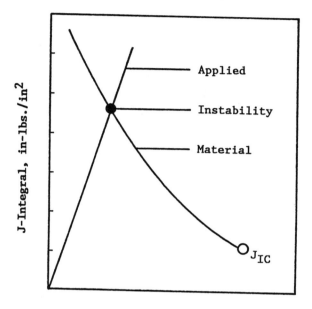

Fig. 1 - Illustration of J/T Diagram

<u>Computation of Japp and Tapp</u> - The computation of J presented here follows the method described in (7) and (20) for throughwall flaws. The same general procedure also was used in this study to obtain J for part-through, finite length semi-elliptical surface flaws. The part-throughwall and throughwall flaw geometries are illustrated in Figure 2.

In the generalized elastic-plastic computation scheme the J integral is separated into elastic and plastic components, or

$$J = J_e + J_P \qquad (6)$$

where:

J_e is the plasticity adjusted elastic contribution to J,

and

J_P is the plastic contribution to J.

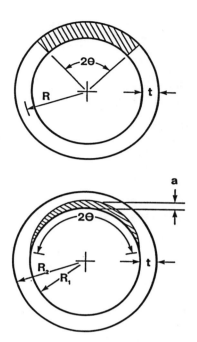

Fig. 2 - Illustration of Part-throughwall and Throughwall Flaw Geometries

The elastic portion of J is directly related to the elastic stress intensity factor, K_I, by the relationship

$$J_e = K^2_I/E'. \qquad (7)$$

where:

the purely elastic solutions are modified for plastic zone size effects, $a = a_{eff}$.

The plastic component of J is expressed in the form

$$J_P = f(\alpha, n, \sigma_0, \varepsilon_0) \cdot g(a, R/t, c) \cdot (L/L_0)^{n+1} \qquad (8)$$

where:

σ_0 and ε_0 are the reference yield stress and strain, and α and n are material constants determined from the material true stress strain curve fit to the Ramberg-Osgood stress strain relationship

$$\varepsilon/\varepsilon_0 = \sigma/\sigma_0 + \alpha(\sigma/\sigma_0)^n, \qquad (9)$$

c is the remaining circumferential ligament of the cracked portion of the pipe,

L_0 is the applied load required to develop an average stress of magnitude σ_0 in the cracked section,

L is the applied load.

Elastic K solutions are available from (21,22) for throughwall flaws, and from (22,23) for part-throughwall flaws with combined axial and bending loads. In all cases, the method for adjusting the elastic solutions for plastic zone size is that described in (20). Plastic solutions are available for throughwall flaws from (20); the plastic solutions for part-throughwall flaws were developed for combined axial and bending loads using procedures presented in (4,23-25).

The applied tearing modulus can be expressed in general terms (for bending) (26) as

$$T_{app} = [E/\sigma_f^2] \cdot [(\partial J/\partial a)_M - (\partial J/\partial M)_a \cdot (\partial \varphi_c/\partial a) \cdot$$
$$\{C + (\partial \varphi/\partial M)_a\}^{-1}] \qquad (10)$$

where:

a = crack size,
M = moment,
φ = total angle of rotation,
φ_c = angle of rotation of cracked section, and
C = system compliance.

The system compliance ranges between two extremes; C = 0 for fixed displacement boundary conditions and C = ∞ for dead weight type loading. A conservative estimate of T_{app} can be obtained for any load distribution by placing C = ∞ into equation (10) so that it becomes:

$$T_{app} = (dJ_{app}/da) \cdot (E/\sigma_f^2) \qquad (11)$$

Stability Computations - To determine instability crack size for part-throughwall flaws, applied values of J and T were calculated from Eqs. (6) and (11), respectively, as a function of part-throughwall flaw depth, with combined stress determined from Eq. (3), and circumferential crack length held constant. The results were placed on a J/T plot (see Figure 1) to obtain the crack depth at the defined combined load and crack length that corresponds to the intersection of the applied and material J/T curves.

The value of J at instability was then used to define the instability crack depth using the J versus crack depth relationship determined from Eq. (6). When the instability flaw sizes were defined, the values of P'$_b$ (with P'$_m$ = 8.5 KSI) that would produce limit load at the cracked section were determined for the instability part-throughwall flaws using Eq. (1) or (2).

Analyses were performed for four pipe diameters: 4 inch (t = 0.35 inch), 12 inch (t = 0.56 inch), 16 inch (t = 0.85 inch), and 28 inch (t = 1.4 inch). Instability part-through-wall crack depths were determined for crack lengths corresponding to 10, 20, 30, 40, 50, and 75 percent of circumference for each pipe size.

For throughwall flaws, the instability moment was determined for bending only using flaw lengths ranging from 10 to 40 percent of circumference, and pipe diameters ranging from 4 to 28 inches. In these analyses, crack length was held constant and J and T were calculated as a function of moment from Eqs. (6) and (11), respectively. The instability moment was determined from the intersection of the applied and material J/T curves and the J versus moment relationship calculated from Eq. 6.

Materials Data - The Ramberg-Osgood stress strain parameters for both the SAW and SMAW welds at 550F are presented in Table 1.

Table 1

Ramberg-Osgood Stress Strain Parameters for SMAW and SAW Stainless Steel Welds, 550 F

Property	Value	
	SMAW	SAW
Reference Stress, σ_o, ksi	53.9	33.7
Elastic Modulus, E, 10^3 ksi	25.5	25.5
α	2.83	11.0
N	11.8	6.9

The J-R curves for the two welds (11) are presented in Figure 3. The values of J in Figure 3 were calculated using the modified J (J_m) methodology (11). All data were obtained from 1-inch thick, face-grooved compact tension type (1TCT) test specimens, which is the small specimen configuration that appears to best simulate the crack tip condition associated with cracked piping. The values of J_m at which first extension of the existing crack would occur (J_{IC}) are 650 in-lb/in^2 for the SAW and 990 in-lb/in^2 for the SMAW weld.

RESULTS

The results from the J/T computation are illustrated in Figure 4. The dashed portion of the curve in Figure 4 indicates the region in which the curves were extrapolated beyond the range of available data.

The curves in the figure illustrate that instability occurs at relatively high J and low T values for throughwall cracks, and low J and high T values for part-throughwall cracks. These trends are applicable to both SMAW and SAW welds.

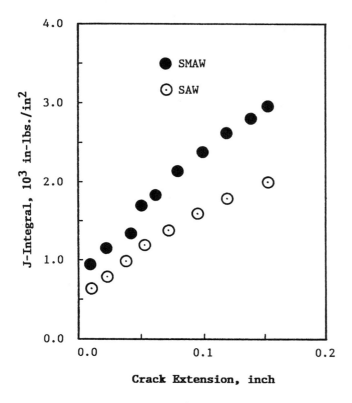

Fig. 3 - J-R Curves for SAW and SMAW, 550

The loads at instability relative to the load at net section collapse, and the instability part-throughwall flaw depths are presented in Table 2 as a function of weld type and circumferential crack length. The computational results indicate that the instability to limit moment ratio is less than one, and is not sensitive to pipe size for the part-throughwall flaw geometry. The results in Table 2 for 12 inch diameter pipe represent within a few percent the results for the other pipe sizes.

The results for throughwall flaws are presented in Table 3, and again indicate instability to limit moment ratios less than one. Because a more significant pipe size effect was indicated for piping with throughwall flaws and pure bending loads, data are presented in Table 3 for both 4 and 28 inch diameter piping.

CONCLUSIONS

The results indicate that the load carrying capacities of SAW and SMAW welded austenitic stainless steel piping containing throughwall or part-throughwall flaws are reduced relative to limit load as a result of lower material toughness.

The indicated load carrying capacities for cracked pipe sections are between 62 and 88 percent of the limit for SMAW load, and between 59 and 78 percent of the limit load for SAW. The specific instability to limit load ratio depends on pipe size, and flaw type and size.

The throughwall crack analysis procedure can be used for most applications to quickly estimate reduced load carrying capacity relative to limit load because the method is easy to use and generally will provide conservative results.

The part-throughwall analysis procedure can be used for when a more realistic assessment is desired for the actual flaw shape, and adequate time and data are available for a more detailed analysis.

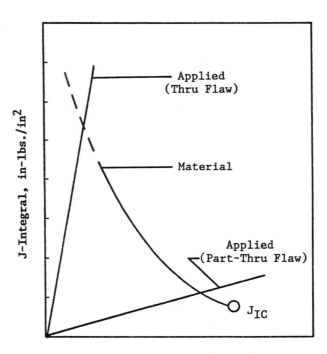

Fig. 4 - Illustration of J/T Results for Part-throughwall and Throughwall Flaws

Table 2

Instability Part-throughwall Flaw Depth and Instability to
Limit Load Ratio for SAW and SMAW Welds,
12 inch Diameter Pipe, 550F, Stress Ratio = 1.0
Combined Axial/Bending Loading

Relative Flaw Length (Θ/π)	SAW		SMAW	
	Relative Instability Flaw Depth (a/t)	Instability to Limit Load Ratio	Relative Instability Flaw Depth (a/t)	Instability to Limit Load Ratio
0.1	0.14	0.68	0.41	0.72
0.2	0.12	0.69	0.36	0.76
0.3	0.11	0.70	0.34	0.80
0.4	0.10	0.70	0.32	0.83
0.5	0.095	0.71	0.31	0.85
0.75	0.095	0.71	0.31	0.88

Table 3

Instability to Limit Load Ratio for Throughwall
Flaws, SAW and SMAW Welds, 4 and 28 inch
Diameter Pipe, 550F, Bending Loading

Relative Flaw Length (Θ/π)	SAW		SMAW	
	Pipe Diameter (inch)	Instability to Limit Load Ratio	Pipe Diameter (inch)	Instability to Limit Load Ratio
0.1	4	0.78	4	0.84
0.2	4	0.76	4	0.82
0.3	4	0.76	4	0.84
0.4	4	0.78	4	0.85
0.1	28	0.63	28	0.70
0.2	28	0.59	28	0.65
0.3	28	0.59	28	0.62
0.4	28	0.61	28	0.64

REFERENCES

1. Pipe Crack Study Group, 1975: Technical Report-Investigation and Evaluation of Cracking in Austenitic Stainless Steel Piping of Boiling Water Reactor Plants, NUREG-75-067, Nuclear Regulatory Commission, Washington, D.C.

2. Pipe Crack Study Group, 1979: Investigation and Evaluation of Stress-Corrosion Cracking in Piping of Light Water Reactor Plants, NUREG-0531, Nuclear Regulatory Commission, Washington, D.C., February 1979.

3. ASME Boiler and Pressure Vessel Code, Section XI, Paragraph IWB-3640, American Society of Mechanical Engineers, Winter, 1983 Addenda.

4. Tada, H., Paris, P.C., and Gamble, R.M., Stability Analysis of Circumferential Cracks in Reactor Piping Systems, NUREG/CR-0838, U.S. Nuclear Regulatory Commission, Washington, D.C., June 1979.

5. PWR Pipe Crack Study Group, 1980: Investigation and Evaluation of Cracking Incidents in Piping in Pressurized Water Reactors, NUREG-0691, U.S. Nuclear Regulatory Commission, Washington D.C., September 1980.

6. Cotter, K.H., Chang, H.Y., and Zahoor, A., Application of Tearing Modulus Stability Concepts to Nuclear Piping, EPRI NP-2261, Electric Power Research Institute, Palo Alto, California, February 1982.

7. German, M. D. and Kumar, V. July 1982. Elastic-Plastic Analysis of Crack Opening, Stable Growth and Instability

Behavior in Flawed 304 SS Piping. In Aspects of Fracture Mechanics in Pressure Vessels and Piping PVP-Vol. 48, ASME.

8. Paris, P. C., Tada, H., and Macek, R., Fracture Proof Design and Analysis of Nuclear Piping, in NUREG/CR 3465, U.S. Nuclear Regulatory Commission, September 1983.

9. Zahoor, A., and Gamble, R.M., Leak Before Break Analysis for BWR Recirculation Piping Having Cracks at Multiple Weld Locations, EPRI NP-3522-LD, Electric Power Research Institute, Palo Alto, CA, April 1984.

10. Zahoor, A., Kanninen, M. F., Wilkowski, G., Abou-Sayed, I., Marschall, C., Broek, D., Sampath, S., Rhee, H., and Ahmad, J., Instability Predictions for Circumferentially Cracked Type-304 Stainless Steel Pipes Under Dynamic Loading," Vol. 2, EPRI NP-2347, April 1982.

11. Landes, J. D., McCabe, D. E., Ernest, H. E., Pryle, W. H., and Liaw, P. K., Elastic Plastic Fracture Methodology to Establish R-Curves and Instability Criteria, Sixth Semi-annual Report for EPRI RP 1238-2, August 1982.

12. NRC Program FIN B-6290, David Taylor Naval Ship R. D. Center, In Progress.

13. Kanninen, M.F., Toward an Elastic Plastic Fracture Mechanics Predictive Capacity for Reactor Piping, Nuclear Engineering and Design, 48, 117-134, 1978.

14. Horn, R.M., Ed., The Growth and Stability of Stress Corrosion Cracks in Large Diameter BWR piping, Electric Power Research Institute, Report NP-2472, July 1982.

15. Ranganath, S., and Metha, H.S., Engineering Methods for the Assessment of Ductile Fracture margin in Nuclear Power Plant Piping, Elastic Plastic Fracture Symposium, Vol. 2, ASTM STP 803, American Society for Testing and Materials, Philadelphia, PA, 1983.

16. Rice, J. R. 1968. Fracture Vol. 2, Academic Press, New York.

17. Hutchinson, J. W. January 1968. Singular Behavior at End of Tensile Cracks in Hardening Material. Journ. of Mech. and Phys. of Sol., Vol. 16, No. 1.

18. Rice, J. R. and Rosengren, G. F. January 1968. Plane Strain Deformation Near a Crack Tip in Power-Law Hardening Material. Journal of Mech. and Phys. of Sol., Vol. 16, No. 1.

19. Paris, P. C. Tada, H., Zahoor, A., and Ernst, H., August 1977. A Treatment of the Subject of Tearing Instability. NRC. NUREG-0311.

20. Kumar, V., German, M. D., and Shih, C. F. July 1981. An Engineering Approach for Elastic Plastic Fracture Analysis. EPRI NP-1931, Research Project 1237-1.

Electric Power Research Institute.

21. Sanders, J. L. Jr. 1983. Circumferential Through-Cracks in a Cylindrical Shell Under Combined Bending and Tension. ASME Journal of Applied Mechanics, Vol. 50, No. 1, p. 221.

22. Zahoor, A.,Closed Form Expressions for Fracture Mechanics Analysis of Cracked Pipes, ASME Journal of Pressure Vessel Technology, Vol. 107, No. 2, pp 203-205.

23. Zahoor, A., Fracture of Circumferentially Cracked Pipes, To Appear in the ASME Journal of Pressure Vessel Technology, 1986.

24. Zahoor, A. and Kanninen, M. F., A Plastic Fracture Instability Analysis of Wall Breakthrough in a Circumferentially Cracked Pipe Subjected to Bending Loads," ASME Journal of Engineering Materials and Technology, 103, pp. 194-200, July 1981.

25. Zahoor, A. and Norris, D. M., Ductile Fracture of Circumferentially Cracked Type 304 Stainless Steel Pipes in Tension, ASME Journal of Pressure Vessel Technology 106, pp. 399-405, November 1984.

26. Hutchinson, J. W. and Paris, P. C. 1979. Stability Analysis of J-Controlled Crack Growth. Elastic-Plastic Fracture, ASTM STP 668, pp. 37-64.

HEAVY SECTION TYPE 316H SS PIPE AND ITS NARROW-GAP GTA WELDMENT WITH REDUCED SUSCEPTIBILITY TO SIGMA-PHASE EMBRITTLEMENT FOR ADVANCED FOSSIL PLANTS

T. Koyama, K. Tamura, T. Takuwa, Y. Sakaguchi
Babcock-Hitachi K.K.
Kure, Hiroshima, Japan

DEVELOPMENT of high thermal efficiency power plants is strongly desired from the viewpoint of energy saving. For this reason, the authors have studied materials used under high pressure and high temperature steam conditions. The final target for steam conditions is 5,000psi/1,200°F (34.5MPa/649°C), though the predominant steam conditions are 3,500psi/1,000°F (24.1MPa/538°C). In the case of high pressure and high temperature steam conditions, heavy section stainless steels have to be used for main steam pipes and headers instead of ferritic steels. (Ref. 1-5). Type 316H austenitic stainless steel was selected from some austenitic steels with consideration of high temperature strength, weldability and actual working experience at Eddystone No. 1 of Philadelphia Electric Company. A heavy section Type 316H stainless steel pipe, 546mm OD x 106mm thick, was produced. Special attention was paid to the nickel balance in order to reduce susceptibility to sigma phase embrittlement. Several tests were performed to establish the fabrication technique of this pipe. At first, variation in mechanical properties and chemical composition through the wall was examined to confirm homogeneity which is a important property in a thick wall product. As to welding, a special narrow gap gas tungsten arc welding process using 16Cr-8Ni-2Mo welding wire was used. High temperature tensile properties and creep rupture strength of base metal and welded joint were examined, and aging properties were also investigated in order to confirm stability after long term exposure at high temperatures.

MATERIAL SELECTION

Type 304H, 316H, 321H and 347H are the most popular steels for high temperature use.

Strength, weldability, resistance to steam oxidation and experience in use as main steam pipes of these four steels were compared and the results are shown in Table 1. For strength, Type 316H and 347H show higher allowable stresses than the others. Weldability and weld cracking susceptibility of these four steels were compared. Stress relief cracking may occur in Type 321H and 347H, and Type 347H is sensitive to hot cracking. In this respect, Type 304H and 316H have good weldability. Resistances to steam oxidation of these four steels are almost the same, because these resistances are determined by the grain size of each steel and shot blasting can make any steel fine grained. Type 316H and 347H have been used as main steam pipes in utility power plants. (Ref. 1-5)

From the above results, Type 316H is considered the most suitable steel for main steam pipes and headers, and we selected Type 316H for this study.

Table 1 Comparison of properties

Steel type	Allowable stress at 600 °C(MPa)	Weldability	Resistance to steam oxidation	Experience in use	Evaluation
304H	64	good	good (1)	—	X
316H	81	good	good (1)	Eddystone	O
321H	59	SR cracking	good (1)	—	X
347H	86	SR cracking Hot cracking	good (1)	Drakelow	△

(1) With inside shot blasting

CHEMICAL COMPOSITION

The heavy section Type 316H austenitic stainless steel pipe, 546mm OD x 106mm thick, was produced by Sumitomo Metal Industries, LTD. in Japan. Table 2 shows the chemical composition. This steel contains 16% chromium, 12% nickel and 2% molybdenum.

The main steam pipe of Eddystone No. 1, made of Type 316H, leaked in March, 1983 as a result of sigma phase embrittlement. Because delta ferrite will rapidly change to sigma phase at elevated temperatures, a fully austenitic microstructure is demanded in order to prevent sigma phase formation. A fully austenitic microstructure can be obtained by setting the nickel balance at a 2.5 minimum and the nickel balance is determined from the following. (Ref. 6)

$$\text{Nickel balance} = \text{Nieq} + 11.6 - 1.36 \times \text{Creq}$$

where

$$\text{Nieq} = 30\,(\%C + \%N) + 0.5\,\%Mn + \%Ni$$

and

$$\text{Creq} = 1.5\%Si + \%Cr + \%Mo + 0.5\%Nb$$

In this equation, if the nickel balance is zero, the equation will be represented by the zero ferrite line on Fig. 1 that presents a De Long diagram. The 2.5 nickel balance line will be a line located 2.5 nickel equivalents above this. Also the point plotted on Fig. 1 represents the pipe used for this study. The nickel balance of this pipe is 2.7 and a fully austenitic microstructure can be obtained.

Table 2   Chemical composition

										(wt.%)
	C	Sl	Mn	P	S	Ni	Cr	Mo	N	Ni balance[1]
	0.08	0.50	1.45	0.022	0.001	12.85	16.40	2.24	0.0497	2.70
Spec.	0.04 ― 0.10	0.75 max.	2.00 max.	0.040 max.	0.030 max.	11.00 ― 14.00	16.00 ― 18.00	2.00 ― 3.00	―	―

(1) Ni balance = Ni eq + 11.6 − 1.36 Cr eq
Ni eq = 30 (%C + %N) + 0.5%Mn + %Ni
Cr eq = 1.5%Si + %Cr + %Mo + 0.5%Nb

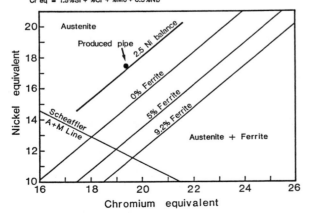

Fig. 1   De Long diagram

HOMOGENEITY THROUGH THICKNESS

Compositional segregation often occurs in thick wall products such as main steam pipes and headers. This segregation leads partially to sigma phase formation and reduction of strength, and reduces the utility of the pipes. For this reason, homogeneity through the wall is one of the most important properties and the following items concerning variation through the wall's thickness were studied.

(1) Chemical composition (Mainly C, Cr, Ni, Mo)

(2) Tensile properties at room temperature and at 600°C (Tensile strength, 0.2% yield strength, elongation and reduction of area)

(3) Impact property

(4) Hardness

(5) Microstructure

CHEMICAL COMPOSITION − Fig. 2 shows the variation of chemical contents, such as carbon, chromium, nickel and molybdenum. These are the main elements of Type 316H stainless steel. The variation of these elements through the wall is very slight and in every case the points fall within the specified ranges indicated by the two long and short dashed lines in Fig. 2. Also the other elements, silicon, manganese, phosphorus and sulfur, show no compositional segregation through the wall.

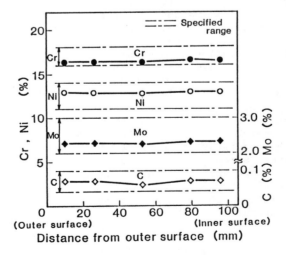

Fig. 2   Variation of elements through the wall

TENSILE PROPERTIES - Fig. 3 shows the result of the tensile test at room temperature. The specimens, taken from various locations through the wall were 10mm in diameter and 50mm in gauge length and were taken in a longitudinal direction. Tensile strength, 0.2% yield strength and elongation of all specimens satisfy the minimum specification indicated by the alternate long and short dashed lines in Fig. 3. These tensile properties, including reduction of area, show a tendency to decrease near the inner surface, however the variation is very small and the absolute values are high enough.

Fig. 4 shows the result of the tensile test at 600°C. The data band presented by the National Research Institute for Metals (NRIM) in Japan (Ref. 7) is shown in the figure. The variation of all tensile properties through the wall is very small and most of the points fall within the data band.

At a position of 1/4 thickness from the outer surface, the effect of the specimen's direction on the tensile properties was examined at room temperature and 600°C. It became clear that there was no difference in tensile properties through the specimen's direction.

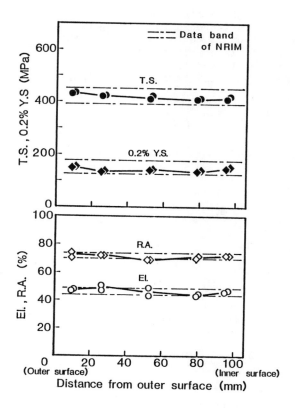

Fig. 4   Variation of tensile properties through the wall (600°C)

IMPACT PROPERTY - Fig. 5 shows the result of the V-notch charpy impact test at 0°C. The dimension of the specimen was 5mm x 10mm x 55mm and the specimens were taken from various locations through the wall. Absorbed energy was uniform through the whole thickness and all values were very high, above 115J. Also no difference of absorbed energy by the specimen's direction were confirmed.

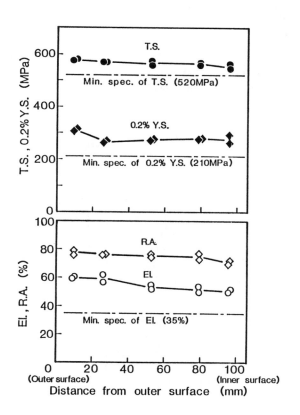

Fig. 3   Variation of tensile properties through the wall (Room temperature)

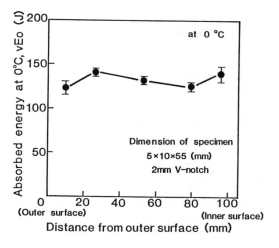

Fig. 5   Variation of absorbed energy at 0°C through the wall

HARDNESS - Fig. 6 shows the variation of the hardness through the wall. The hardness changes from 73 to 78 in the Rockwell B scale. It is considered that this variation is negligible and the pipe used for this study has uniform hardness distribution.

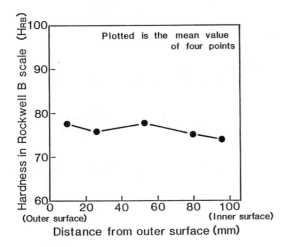

Fig. 6   Variation of hardness through wall

MICROSTRUCTURE - Fig. 7 shows the microstructure under 100 x magnification at various points through the wall. These photos show that each grain size is almost the same. (From 2.7 to 3.3 in ASTM grain size number)

It became clear that this pipe had homogeneous properties through the wall from a viewpoint of chemical composition, tensile and impact properties, hardness and microstructure.

WELDING

WELDING PROCESS - A special narrow gap gas tungsten arc (GTA) welding process was tried for welding. The welding process was developed by Babcock-Hitachi K.K. and is called The Hot Wire Switching TIG (HST) welding process. (Ref. 8) The conceptional circuit of the electric power source for the HST welding process is shown in Fig. 8. Generally a high quality welded joint can be obtained by a GTAW process. However the GTAW process is not so efficient due to its low deposition rate. The hot wire system was developed by A.F. Manz in 1964 (Ref. 9) instead of the cold wire system to improve the deposition rate. In the hot wire system, the filler wire is heated by an electric current before it enters the molten weld pool. Therefore the wire current and the arc current repulse each other and the arc deflects.

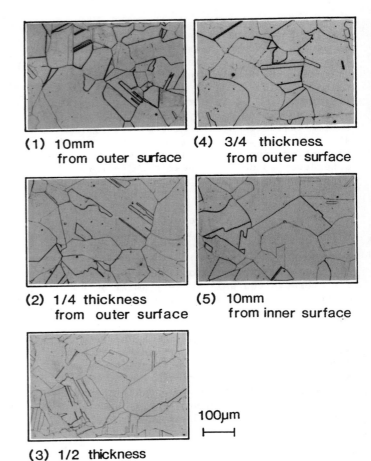

(1) 10mm from outer surface

(4) 3/4 thickness from outer surface

(2) 1/4 thickness from outer surface

(5) 10mm from inner surface

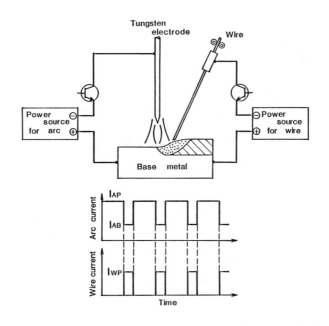

100μm

(3) 1/2 thickness from outer surface

Fig. 7   Microstructure at several locations through the wall

Fig. 8   Principles of HST welding process

The arc deflection leads to difficulties in welding work. This problem was solved by alternate switching between wire and arc current. (See Fig. 8) Namely, two power sources which are used for heating the filter wire and welding are employed, and the wire current is zero when the arc current is high (I_{AP}), while the arc current is low (I_{AB}) maintaining the arc when the wire current flows (I_{WP}). Using this system, the arc deflection can be almost completely prevented.

WELDING WIRE - For HST welding, 16Cr-8Ni-2Mo welding wire was used. The chemical composition is shown in Table 3. This wire has been used in Babcock-Hitachi Kure Works for the welding of austenitic stainless steels. As the weld metal, by means of this wire, will show a fully austenitic microstructure, it is considered that sigma phase will not be formed easily. Because sigma phase formation will be caused by the presence of delta ferrite, the weld metal contains hardly any delta ferrite.

Table 3 Chemical composition of welding wire

(wt.%)

C	Si	Mn	P	S	Ni	Cr	Mo	V
0.072	0.29	1.67	0.025	0.006	9.03	15.98	1.69	0.27

WELDING CONDITIONS - Table 4 shows the welding conditions. The switching frequency between the arc and wire current is 70Hz and the duty - this means the percentage of peak current time to total current time - is 70%. The deposition rate in this process is from 15 to 25g/min., while in the cold wire GTAW process it is about 5g/min.. So the welding efficiency increases by a factor of 3 - 5. The root gap is narrow about 10mm at the inner surface, and 16.8mm at the outer surface.

Table 4   Welding conditions

Arc current (A)			Wire current (A)	Welding speed (cm/min)	Deposition rate (g/min)	Frequency (HZ)	Duty (%)
Peak	Base	Mean					
230		180	40	11	15		
	50					70	70
400		300	60	17	25		

WELDABILITY - Fig. 9 is a photo that shows the weldment after butt welding by means of the HST welding process as mentioned above. As this pipe has a thick wall (106mm), restraint strength applied to the weld is very high. Therefore there is the possibility of weld defects occuring during or after welding. The cross sections were examined to determine whether there are welding defects or not. Fig. 10 shows typical cross section and Fig. 11 shows the microstructure under 100 x magnification.

No microscopic defect could be found. Also a side bend test using a specimen of full thickness was performed to detect embedded defects. Fig. 12 shows the specimen after the side bend test and no defect was found.

As described above, this pipe has good weldability and sound welded joints were obtained.

Fig. 9   Appearance of welded joint

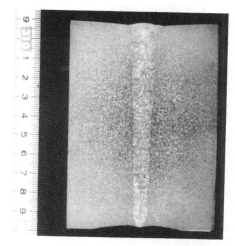

Fig. 10   Cross section of weldment

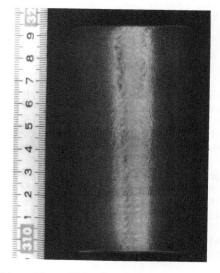

Fig. 12   Side bended specimen

263

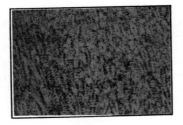

(1) Weld metal

(2) Weld metal
     interface

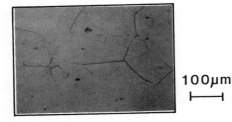

100μm
⊢————⊣

(3) Base metal

Fig. 11   Microstructure of weldment

the properties of both base and weld metals are evaluated at the same time in a tensile test of a welded joint. However it is considered that the elongation in the welded joint is allowable. Also the reduction of area is high enough though the weld metal broke.

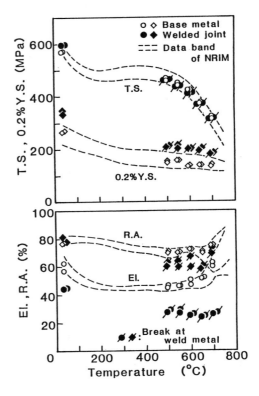

Fig. 13   Tensile properties

HIGH TEMPERATURE STRENGTH, AGING PROPERTIES

TENSILE PROPERTIES - Tensile tests of the base metal and the welded joint were performed at room temperature and at 500 - 700°C. The specimens were taken at a position of 1/4 thickness from the outer surface, and they were 10mm in diameter and 50mm in gauge length. The result is shown in Fig. 13 and the data band presented by NRIM is also shown in the figure.

For the base metal, all tensile properties fall within the data band and are considered reasonable. Concerning the welded joint, specimens were fractured at the weld metal at high temperatures. The tensile strength is almost the same as that of the base metal in spite of the weld metal breaking. The 0.2% yield strength is higher and the elongation is lower than these properties of the base metal. These phenomena are common in a tensile test of a welded joint because weld metal generally has high 0.2% yield strength and low elongation, and

CREEP RUPTURE STRENGTH - Creep rupture tests were performed at 600, 650 and 700°C for the base metal and 650°C for the welded joint. The specimens were taken from the same position as those of the tensile test and they were 6mm in diameter and 30mm in gauge length. Fig. 14 shows the result and the data band presented by NRIM is also shown in Fig. 14.

The creep rupture strength of the base metal falls within the data band. In the welded joint, the creep rupture strength is higher than that of the base metal and it is located at the upper limit of the data band in spite of the base metal breaking. It is considered that this increase of rupture strength is caused by the restraint with the weld metal. If the ductility, especially reduction of area, shows a low value, this welded joint is of no practical use. However, the reduction of area of the ruptured welded joint is almost the same value as that of the ruptured base metal.

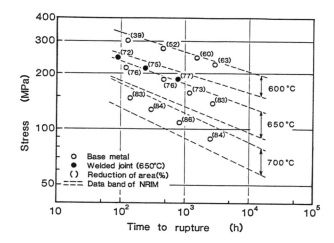

Fig. 14   Creep rupture strength

AGING PROPERTIES - The main steam pipes and headers are exposed to a high temperature environment over a long time. With this in mind, stability in microstructure and strength of a steel after aging is required. The tensile and charpy impact tests and microstructural examination of base and weld metals were performed with samples aged at 650°C and 700°C for 1,000 hours. These accelerated aging conditions are equivalent to about 20,000 hours and 300,000 hours at 605°C calculated by the Larson-Miller parameter and this parameter (P) is given by the following equation.

$$P = T(c + \log t) \times 10^{-3}$$

where

    T ; absolute temperature (K)
    t ; time (h)
    c ; material constant (= 20)

The temperature of 605°C is the expected metal temperature when the main steam temperature is 593°C (1,100°F).

Tensile Properties - Tensile tests of the base metal and the welded joint were performed at room temperature and at 600°C after aging. Fig. 15 shows the result of the base metal tests and the horizontal axis is presented by the Larson-Miller parameter. In room temperature tests, the tensile and 0.2% yield strengths become high, and the elongation and the reduction of area are reduced as the parameter increases. Also these properties show a similar tendency at 600°C with the exception of the tensile strength. The tensile strength is almost the same as that at room temperature. In this way, the elongation and the reduction of area are reduced, but these values are considered high

enough because the elongation satisfies the minimum specification (35%) in the room temperature tests and in the 600°C tests the value is higher than 20%. Fig. 16 shows the result of the welded joint tests. Variation of tensile properties by aging is similar to the result of the base metal test, however, the degree of variation is smaller.

Impact Properties - Fig. 17 shows the result of the V-notch charpy impact test at 0°C after aging. A full size specimen (10mm x 10mm x 55mm, 2mm V-notch) was used. The base metal has 120J absorbed energy after aging, though it has very high absorbed energy (about 400J) before aging. Also weld metal has about 90J absorbed energy after aging. This value is high enough because, in comparison, there are a few materials, like 321 and 347 weld metals, which show extremely low values (about 20 - 30J) after aging. (See Fig. 17) Though absorbed energies of the base and weld metals are reduced by aging, the change of absorbed energies from 21.2 (650°C x 1,000h) to 22.4 (700°C x 1,000h) within the LarsonMiller parameter is very small. This tendency suggests that no extra drop of absorbed energy is expected after longer aging.

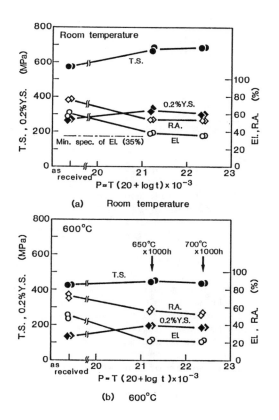

(a)   Room temperature

(b)   600°C

Fig. 15   Change of tensile properties after aging (Base metal)

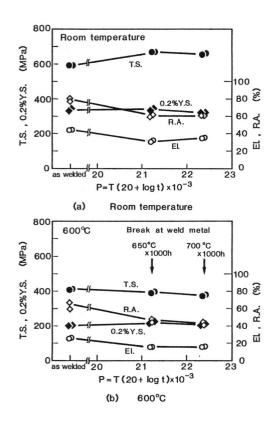

(a)　　Room temperature

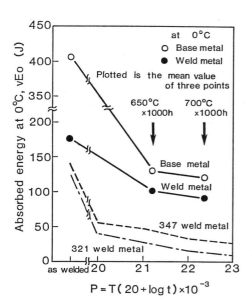

(b)　　600°C

Fig. 16　Change of tensile properties
after aging (Welded joint)

Fig. 17　Change of absorbed energy
at 0°C after aging

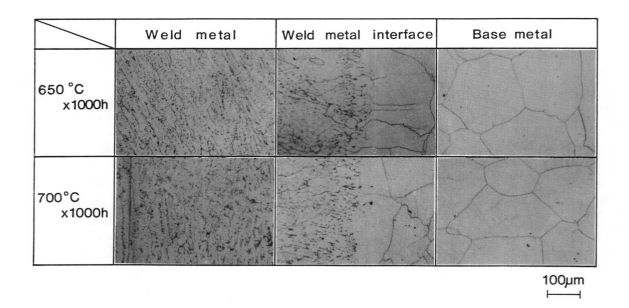

100μm

Fig. 18　Microstructure after aging

266

Microstructure – Fig. 18 shows the microstructure of the samples after aging. By aging, carbides precipitate at the grain boundary, but sigma phase does not seem to form. No sigma phase formation was confirmed by KOH etching which can color the sigma phase orange. The carbide precipitation makes the tensile strength rise and the absorbed energy drop as above mentioned. This is called 'age-hardening'. Though the sigma phase formation leads to similar phenomena, the absorbed energy is reduced to under 50 J.

It is considered that no sigma phase formation is caused for the following reason. In the base metal, the nickel balance is considered in the production of this pipe and a fully austenitic microstructure is obtained. In the welding, 16Cr-8Ni-2Mo welding wire is used and a fully austenitic microstructure is also obtained. The absence of delta ferrite ensures a low possibility of conversion to sigma phase.

Generally a small percentage of delta ferrite will prevent weld cracking, especially hot cracking, from occuring on welding of austenitic steel. These welding wires, like 321 and 347, are used frequently. However, 16Cr-8Ni-2Mo welding wire used for this study has a low susceptibility to weld cracking though it has a fully austenitic microstructure. Also the special narrow gap GTAW process prevents weld cracking because of its intrinsic low heat input.

CONCLUSION

A heavy section type 316H SS pipe, 546mm OD x 106mm thick, for advanced fossil plants was produced and ; weldability, strength and aging properties of this pipe were examined.
The following are the results of this study.

(1) This pipe, a thick wall product, had homogeneous properties in cross section from a viewpoint of chemical composition, tensile and impact properties, hardness and microstructure.

(2) A special narrow gap GTA welding process (Hot wire switching TIG welding process) using 16Cr-8Ni-2Mo welding wire was tried and sound welded joints, which had no defects, were obtained.

(3) High temperature strengths of both the base metal and the welded joints satisfied the requirements.

(4) By aging, which is equivalent to 300,000 hours at 605°C, ductility and toughness were reduced but these reduced

values were still high enough to allow continued use for a long period. Also, in microstructure, sigma phase formation after aging did not occur.

REFERENCE

1. R. H. Caughey and W. G. Benz, Jr., Trans. ASME 82, 293-313 (1960)

2. R. A. Baker and H. M. Solden, Combustion, Nov., 24-28 (1963)

3. R. A. Baker and H. M. Solden, Combustion, Dec., 42-45 (1963)

4. Harry S. Blumberg, Combustion, May, 33-39 (1965)

5. F. E. Asbuzy, B. Mitchell and L. F. Toft, British Welding Journal, Nov., 667-678 (1960)

6. T. C. Mcgough, J. V. Pigford, P. A. Lafferty, S. Tomasevich, H. E. Zielke, C. T. Ward and J. F. De long, Welding Journal 64, 29-36 (1985)

7. National Research Institute for Metals, "NRIM Creep Data Sheet No. 6A" (1978)

8. The Japan Welding society, "Narrow Gap Welding (NGW), The State-of-the-Art in Japan" p.119-122, Kuroki Publishing Company, Osaka, Japan (1984)

9. Manz, A. F., U. S. Patent 3,122,629

THE EFFECT OF RECOVERY AND RECRYSTALLIZATION ON THE HIGH-TEMPERATURE FATIGUE CRACK GROWTH BEHAVIOR OF ALLOY 800H IN CONTROLLED-IMPURITY HELIUM

J. R. Foulds
GA Technologies Inc.
San Diego, California, USA

Abstract

Alloy 800H has gained major importance as a high-temperature engineering material. In service, the alloy may experience cyclic loading (e.g. in a steam generator) as well as changes in microstructure due to temperature and/or environment. The paper describes the effect of cold-work, recovery and recrystallization on the fatigue crack growth behavior of Alloy 800H at 677°C (1250°F) and 732°C (1350°F) in controlled-impurity helium. Results at 677°C indicate that cold-working increases the crack growth rate over that of the solution-annealed material. Increasing recovery and recrystallization after cold-working decreases the crack growth rate from the cold-worked condition, despite decreases in material yield strength. This effect of material condition is not significant at the higher test temperature of 732°C.

Fatigue crack growth rate was observed to increase with temperature for all material states. This is attributed to an environmental effect from low level helium impurities. The increased environmental effect with increasing temperature may explain the accompanying increase in crack growth rate.

The paper points to the need for considering the effect of in-service time-dependent changes in microstructure on the fatigue crack growth behavior of Alloy 800H, where appropriate.

ALLOY 800 AND ITS VARIANTS, such as Alloy 800H, are ductile austenitic iron - nickel - chromium alloys that have found widespread applications as structural materials. The applications have emerged as a consequence of their good corrosion, carburization and oxidation resistance in aggressive environments and their relatively high strength at elevated temperatures, particularly under creep conditions. For this reason, the alloy group has found use as a heat exchanger/steam generator material in a variety of energy conversion systems; for example, as a superheater/reheater tubing material in fossil-fired power plants[1] and as a steam generator tubing material for liquid-metal fast breeder reactors (Super Phenix), high temperature gas cooled reactors (Thorium High Temperature Reactor-THTR, Peach Bottom #1) and a few pressurized water reactors outside the United States[1].

Alloy 800H is a controlled grain size (solution-annealed), high carbon (0.05-0.10 wt.% C) version of Alloy 800 (C≤0.10 wt.%). When used as steam generator tubing material, the alloy is subject to elevated temperature fatigue, creep and creep-fatigue loading primarily from tube-wall thermal gradients, thermal expansion restraint and steam pressure. A flaw (crack) in the tube, having been initiated in service or pre-existing from fabrication, may grow as a result of these loadings. It is noted that the present ASME Code for Nuclear Power Plant Components[2] does, in fact, address the issue of flaw acceptability under fatigue loading conditions, but only for lower temperature ferritic components. In general, flaw growth rates depend on the loading conditions as well as the material properties. Material properties often change significantly during service since microstructures evolve with time at temperature. For example, cold-worked regions of the tube, such as cold-bends in a steam generator, may experience recovery and recrystallization during high temperature service.

The results reported here constitute one part of a study to model the crack growth behavior of Alloy 800H in various material states under fatigue, creep and creep-fatigue loading conditions. In particular, the effect of time- and temperature-dependent changes in material microstructure (as a result of cold-working) on the pure fatigue crack growth behavior of Alloy 800H in a typical high temperature gas-cooled reactor (HTGR) steam generator atmosphere is being reported here.

EXPERIMENTAL

MATERIAL - The material was procured as solution-annealed tubing, 25.4 mm (1.0 in) O.D., 5.1 mm (0.2 in) thick wall with an average grain intercept of 60 µm (ASTM grain size no.≈4). Following is the vendor ladle analysis of the Heat:

Table 1 - Vendor Ladle Analysis (wt.%)

C	Si	Mn	P	S	Cu
0.039	0.61	1.10	0.008	0.004	0.01

Cr	Ni	Ti	Al	N	Fe
20.78	33.70	0.51	0.52	0.024	Rem.

A vendor check analysis for carbon indicated 0.046 wt.% C. It is expected that the material behavior studied here represents that of Alloy 800H fabricated to existing ASME SB-163 specifications[3]. The as-received tubing was in a slightly cold-worked state as a result of post-anneal tube straightening. The material was processed to final states of 20% cold-worked (CW), 15%, 40% and 80% recrystallized (RX), where percent recrystallized represents the volume fraction recrystallized. The actual processing is described below.

SPECIMEN FABRICATION - Figure 1 is a schematic of the various stages in the center-cracked tension (CCT) specimen fabrication process.

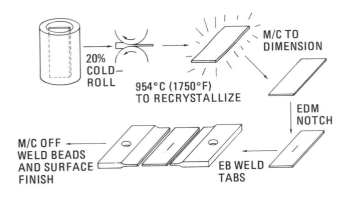

Figure 1 - Schematic of the center-crack tension (CCT) specimen fabrication process

Specimen blanks, 102 mm (4 in) long × 12.7 mm (0.5 in) wide × 3.3 mm (0.130 in) thick were machined out of the tube wall with the long end of the blank parallel to the tube axis. These were then rolled in near plane strain and 4 passes to a final thickness reduction of 20%. This represents a 20% CW condition. Other material states were obtained by heat treating similarly cold-worked blanks at 954°C (1750°F) for 2.6, 13.2 and 187 hours to produce material recrystallized 15, 40 and 80%, respectively. Volume fractions recrystallized were estimated optically from two-dimensional area fractions of material recrystallized, identified through grain refinement and measured using a point count method[4]. The heat treating times were estimated from a recrystallization kinetics equation experimentally determined for this lot of material[5]. Each blank was then machined to dimensions of 76 mm (3 in) × 12.7 mm (0.5 in) × 2.5 mm (0.10 in). A through-thickness central notch approximately 9.5 mm (0.375 in) long and 0.5 mm (0.020 in) wide was electro-discharge machined into the blank with the notch long side parallel to the long end of the blank. Gripping tabs machined from rolled Alloy 800, 12.7 mm (0.5 in) thick were then precision electron-beam (EB) welded to the blank as shown in figure 1 such that the long side of the blank is the CCT specimen width. The gage section (reduced thickness section) of the specimen was then machined and ground to a 15 µm surface finish. Figure 2 is a drawing of the final specimen configuration. This final configuration does not meet the present ASTM standard test method requirement for constant-load-amplitude fatigue crack growth rate testing, ASTM E 647-83[6], but was arrived at as a result of practical constraints of specimen strength and size, and tube and furnace dimensions available.

TESTING - Fatigue crack growth tests were run in tension-tension under constant load amplitude on four different material conditions (20% CW, 15% RX, 40% RX and 80% RX) at temperatures of 677°C (1250°F) and 732°C (1350°F) in a typical HTGR helium environment at 1.35 atmospheres (0.14 MPa). In addition, one test was run on 20% CW material at room temperature and one on as-received material at 677°C (1250°F). Table 2 shows the nominal helium environment chemistry.

Table 2 - Nominal Impurity Concentration in the helium test environment

	Impurity in Helium				
	H_2	CH_4	CO	CO_2	H_2O
Partial Pressure µatm(Pa)	150(15)	10(1)	50(5)	10(1)	10(1)

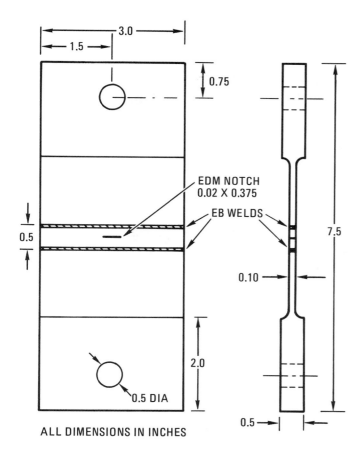

ALL DIMENSIONS IN INCHES

Figure 2 - Schematic of final test specimen configuration

Tests were run on a servo-hydraulic MTS machine with the specimen heated in a three-zone resistance furnace enclosed in a stainless steel environmental chamber. All tests were run at a frequency of 10 Hz, a sawtooth waveform and a load ratio, R(=min.load/max.load) of 0.05. Specimens were fatigue precracked at room temperature with the same load ratio and tested in accordance with ASTM Standard E 647-83[6]. Crack length was continuously monitored during tesing using a direct current (DC) electric potential (EP) technique[3,8]. A DC current of 10 amperes was applied centrally and symmetrically across the crack. The output EP was measured to microvolt resolution by 0.25 mm (0.10 in) dia. Nichrome wires spot-welded symmetrically to the specimen on each side of the notch centerline at a distance of 4.6 mm (0.18 in) from it. The Johnson[7] formula:

$$a = \frac{W}{\pi} Cos^{-1}\left[\frac{Cosh(\pi y/W)}{Cosh\left[\frac{V}{V_o} Cosh^{-1}\left\{\frac{Cosh(\pi y/W)}{Cos(\pi a_o/W)}\right\}\right]}\right] \quad (1)$$

where a = half-crack length at any instant,
     V = output DC EP at that instant,
a_o, V_o = half-crack length and DC EP at a reference time, $t=t_o$,
     W = specimen width
and   y = distance of output lead location from notch centerline,
was found to work satisfactorily without need for a thermal EMF correction.

The half-crack length (a) vs. cycles (N) data were reduced using a seven-point incremental polynomial for determination of the fatigue crack growth rate (da/dN). The applied stress intensity factor range, ΔK, was calculated from the load- and geometry-dependent expression for a CCT specimen given in reference (6). For convenience and ease of comparison, the logarithmic da/dN vs. ΔK fatigue crack growth data are in most cases represented by least-square-fit straight lines or line segments to data that appear to follow the linear Paris[9] behavior. In all cases, the regression coefficient of logarithmic data exceeded 0.95. A comparison of actual data with the regression line is graphically shown for the results on the effect of temperature on crack growth of cold-worked material.

RESULTS

TENSILE PROPERTIES - Tensile properties for the various material conditions (20% CW, 15%, 40% and 80% RX) were derived from tests in air at 677°C (1250°F) and 732°C (1350°F) on flat specimens machined from the tube wall and stressed parallel to the tube axis. In addition, a specimen in the as-received condition was tensile-tested at 677°C (1250°F). Tables 3 and 4 summarize the results obtained at 677°C (1250°F) and 732°C (1350°F) respectively.

Recovery and recrystallization from cold-working effects a significant decrease in the yield and ultimate tensile strength upto the 40% recrystallized condition. This is accompanied by an increased elongation (ductility) and strain hardenability. Further recrystallization above 40%, however, appears to have a negligible effect on tensile behavior.

FATIGUE CRACK GROWTH AT 677°C (1250°F) - Figure 3 is a graphical representation of the fatigue crack growth results obtained on the various material conditions tested at 677°C (1250°F). The "as-received" condition represents, as mentioned earlier, a "solution-annealed + slightly cold-worked" state. The straight line segments in the da/dN vs. ΔK plot of figure 3 are obtained from linear regression of the data (regression coefficient ≥ 0.95) and are used to represent the results in preference to actual data

Table 3 – Tensile Properties of Alloy 800H at 677°C (1250°F)

Material Condition	0.2% Yield Strength σ_{YS}, MPa (ksi)	Ultimate Strength σ_{UTS}, MPa (ksi)	Percent Elongation	Percent Redn. Area
20% CW	417.1 (60.5)	459.9 (66.7)	18	27.0
15% RX	288.9 (41.9)	375.1 (54.4)	30	36.7
40% RX	131.0 (19.0)	319.2 (46.3)	55	49.5
80% RX	124.8 (18.1)	320.6 (46.5)	50	38.0
As-recd.	158.5 (23.0)	323.9 (47.0)	38	Not measured

Table 4 – Tensile Properties of Alloy 800H at 732°C (1350°F)

Material Condition	0.2% Yield Strength σ_{YS}, MPa (ksi)	Ultimate Strength σ_{UTS}, MPa (ksi)	Percent Elongation	Percent Redn. Area
20% CW	346.1 (50.2)	368.9 (53.5)	27	29.3
15% RX	254.4 (36.9)	298.6 (43.3)	43	47.0
40% RX	128.2 (18.6)	246.8 (35.8)	64	54.9
80% RX	120.7 (17.5)	255.8 (37.1)	64	46.7

points for convenience and ease of comparison. Some published results on Incoloy (Alloy) 800 tested in air[10,11] have been superimposed for comparison. These are also linear representations of actual data of references (10) and (11). Note the significantly lower crack growth rate observed here compared with James' data of reference (10) obtained on Alloy 800 in air at the higher test temperature of 649°C (1200°F). This is discussed in a subsequent section.

At 677°C (1250°F), as seen in figure 3, there is a definite increase in crack growth rate going from the as-received to the cold-worked state. Recovery and recrystallization after cold-working generally results in a decrease in crack growth rate compared with the cold-worked material data. However, the significantly lower sensitivity of crack growth rate to ΔK (shallow slope of curve) for the 20% CW material results in a "crossover" or intersection of the curve with the recrystallized and as-received material data curves. At ΔK values above this "crossover", recrystallized material actually shows a higher fatigue crack growth rate than does the 20% CW material. The "crossover" ΔK increases from approximately 25 MPa√m (22.8 ksi√in) for the 15% RX material to > 50 MPa√m (45.5 ksi√in) for the 80% RX material. The regime of practical interest is usually below "crossover", so that it may be stated that recovery and recrystallization results in an improved fatigue crack growth resistance over the cold-worked condition. Note that the 80% RX material showed fatigue crack growth rates even lower than for the as-received material. This is attributable to the slightly cold-worked condition of the as-received material from tube straightening.

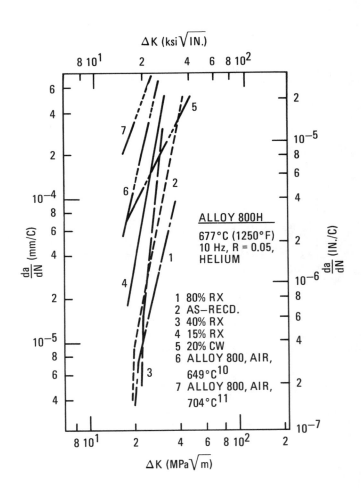

Figure 3 – Fatigue crack growth data for various material states at 677°C (1250°F)

James'[12] results on 316 stainless steel indicate that cold-working increases the fatigue crack growth resistance (lower fatigue crack growth rate at a given applied ΔK) at 649°C (1250°F). This effect of increasing cyclic or monotonic yield strength on decreasing the fatigue crack growth rate is predictable from models such as that of Donahue et al.[13] Indeed, temperature effects on fatigue crack growth rate are sometimes explained on the basis of varying yield strength, such as Ellison and Walton's[14] results on low alloy steel. The data obtained here on Alloy 800H tested at 677°C (1250°F), however, indicate quite the opposite, if any, effect of yield strength. Increasing recovery and recrystallization from cold-working decreases the material monotonic yield strength but results in an improved fatigue crack growth resistance or lower fatigue crack growth rate. In fact, highly recrystallized material exhibited some crack retardation and self-arrest behavior below approximately 18 MPa√m (16 ksi√in) similar to that observed by James[10] on solution-annealed Alloy 800 at 649°C (1200°F). It is evident that ductility and strain hardenability, not just yield strength, contribute significantly to fatigue crack growth resistance in this case. Table 5 summarizes the fatigue crack growth results in terms of the constants C,m in the Paris[9] equation:

$$da/dN = C(\Delta K)^m \qquad (2)$$

where a = half-crack length, N = cycles
ΔK = applied stress intensity amplitude and C,m are constants.
In addition, the table contains pertinent tensile data that could influence crack growth behavior.

The two significant observations to be made from Figure 3 and Table 5 are that overall fatigue crack growth resistance is improved and the crack growth rate sensitivity to the applied stress amplitude (exponent m in eq.(2)) is increased with recovery and recrystallization from the cold-worked

condition. These are discussed later on the basis of pertinent tensile parameters − strength, ductility and strain hardenability.

FATIGUE CRACK GROWTH AT 732°C (1350°F) − Figure 4 is a graphical representation of the fatigue crack growth results obtained on the various material conditions tested at 732°C (1350°F). As with the data reported for tests at 677°C (1250°F), linear segments are being used to represent the results. Again, some published results on Alloy 800 tested in air[10,11] have been superimposed for comparison.

Figure 4 shows a definite trend toward increasing fatigue crack growth resistance with increasing recovery and recrystallization as represented by the "mean" behavior line segments. The differences in "mean" behavior at this temperature however, are too small to be significant, especially in comparison with the effects of material state observed at the lower test temperature of 677°C (1250°F) in figure 3. Evidently, increasing temperature appears to wash out the effect of material condition on fatigue crack growth resistance. Post-test room temperature hardness measurements on 20% CW material were no different than those on untested 20% CW material, ensuring that little or no recovery occurs during testing at this temperature. Examination of the sensitivity of crack growth rate to the applied stress intensity amplitude shows an increasing sensitivity (m in eq.(2)) with increasing recovery and recrystallization. Table 6 summarizes the crack growth data along with the relevant tensile properties that could influence this sensitivity.

As with the results at 677°C (1250°F), comparison with published data on Alloy 800 tested in air[10,11] shows a superior fatigue crack growth resistance of Alloy 800H tested here in helium. Also, from table 6 it is immediately obvious that the sensitivity of crack growth rate to ΔK increases as the yield strength decreases and the ductility and strain hardenability increase with recovery and recrystallization. In summary, at 732°C

Table 5 − Fatigue crack growth* and relevant tensile results at 677°C (1250°F)

Material Condition	Yield Strength σ_{ys}, MPa (ksi)	Ductility % Elong.	Strain Hardening Exponent***, n	Crack Growth Data** C	m
20% CW	417.1 (60.5)	18	0.0521	2.46 E-07	1.92
15% RX	288.9 (41.9)	30	0.0918	1.52 E-12	5.68
40% RX	131.0 (19.0)	55	0.1843	3.04 E-19	10.09
80% RX	124.8 (18.1)	50	0.1723	9.02 E-11	4.43
As recd.	158.5 (23.0)	38	0.1334	2.63 E-12	5.07

* Assumed Paris[9] equation, eq.(2)
** C,m determined for da/dN in mm/cycle and ΔK in MPa√m
*** Strain hardening exponent, n = slope of logarithmic true stress-true strain curve from 0.1 to 2.0% strain

(1350°F), recovery and recrystallization does not significantly influence the overall fatigue crack growth resistance but does increase the sensitivity of crack growth rate to the applied stress intensity amplitude.

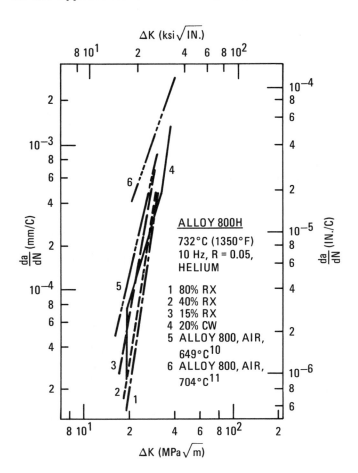

Figure 4 – Fatigue crack growth data for various material states at 732°C (1350°F)

EFFECT OF TEMPERATURE ON FATIGUE CRACK GROWTH BEHAVIOR – A comparison of the crack growth data of figure 3 (677°C) and figure 4 (732°C) indicates that for each material condition, increasing temperature has the effect of decreasing the crack growth resistance. Figure 5 is a comparison of the fatigue crack growth data obtained on 20% CW Alloy 800H in helium at room temperature, 677°C (1250°F) and 732°C (1350°F). The straight line segments drawn through the elevated temperature data are typical regression line representations of the actual data as seen in figures 3 and 4. The crack growth rates are observed to increase with temperature. This temperature effect may be attributed to changes in tensile properties (Elastic Modulus, Yield Strength) and/or an increasing environmental effect as discussed below. Note that creep effects are expected to be negligible at the high test frequency of 10 Hz.

DISCUSSION

EFFECT OF MATERIAL CONDITION ON THE FATIGUE CRACK GROWTH BEHAVIOR – Two obvious effects of material condition on crack growth are discussed: (1) Recovery and recrystallization from cold-working improves the fatigue crack growth resistance, the effect decreasing with increasing temperature and (2) recovery and recrystallization from cold-working appears to increase the sensitivity of crack growth rate, da/dN, to the applied stress intensity amplitude, ΔK, as measured by an increase in "m" of equation (2).

The effects may be exaggerated by use of an improper descriptive parameter for crack growth; for example, surface roughness and/or plasticity induced crack closure effects[15] may increase as a result of recovery and recrystallization (note the reduction in yield

Table 6 – Fatigue crack growth* and relevant tensile results at 732°C (1350°F)

Material Condition	Yield Strength σ_{YS}, MPa (ksi)	Ductility % Elong.	Strain Hardening Exponent***, n	Crack Growth Data**	
				C	m
20% CW	346.1 (50.2)	27	0.0693	6.09 E-10	3.95
15% RX	254.4 (36.9)	43	0.0691	4.47 E-12	5.56
40% RX	128.2 (18.6)	64	0.1418	2.13 E-13	6.32
80% RX	120.7 (17.5)	64	0.1579	6.19 E-13	8.78

*   Assumed Paris[9] equation, eq.(2)
**   C,m determined for da/dN in mm/cycle and ΔK in MPa√m
*** Strain hardening exponent, n = slope of logarithmic true stress-true strain curve from 0.1 to 2.0% strain

strength and increase in ductility and strain hardenability with recrystallization - tables 3 and 4). An increasing closure load (load at which opposing crack surfaces behind the crack tip make contact to effect a zero net crack tip stress intensity factor) will cause a decrease in the stress intensity amplitude effective in driving the crack, ΔK_{eff}. ΔK_{eff} may be written as: $\Delta K_{eff} = K_{max} - K_{cl}$, where K_{max} is the applied stress intensity amplitude at the peak load and K_{cl} is the applied stress intensity amplitude at the closure load. Since K_{cl} increases in proportion to an increasing closure load, logarithmic da/dN - applied ΔK plots will indicate an overall reduction in crack growth rate as well as an increased sensitivity of da/dN to ΔK (higher "m").

Since actual closure load measurements were not made, a simple exercise was attempted to estimate closure stress intensities (or loads) to be subtracted from K_{max} in order that "m" in da/dN = $C(\Delta K_{eff})^m$ equals that for cold-worked material. It is assumed that

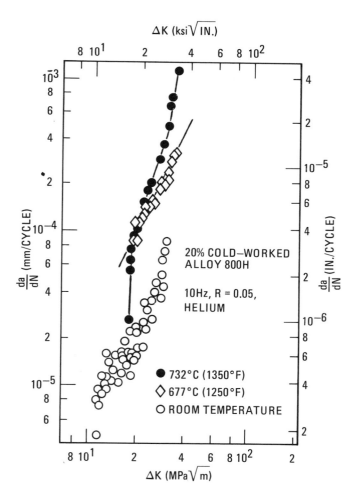

Figure 5 - Effect of temperature on the fatigue crack growth behavior of 20% cold-worked Alloy 800H

closure effects on cold-worked material are negligible. If changes in "m" arise from closure effects alone, then the closure stress intensities, K_{cl}, determined as above should provide reasonable estimates of the closure load for the various recrystallized material conditions. Actually, the estimates provided unreasonably high values of K_{cl}, varying from 0.5K_{max} for 15% RX material tested at 732°C (1350°F) to 0.93K_{max} for 80% RX material tested at 732°C (1350°F) at a K_{max} level of $\approx$ 16.5 MPa√m (15.0 ksi√in). These results suggest that closure effects do not entirely explain the variation in observed crack growth behavior with recrystallization. Testing at higher R ratios, not possible within the experimental constraints of specimen size and geometry in this study, will enable determination of significant closure effects, if any. Following is a discussion on the role of material state in influencing the intrinsic fatigue crack growth behavior based on tensile properties.

All the proposed models for second stage "Paris" crack growth predict a reduction in crack growth resistance with decreasing monotonic or cyclic yield strength. The results obtained here are quite the opposite, crack growth resistance increasing with a decrease in yield strength as a result of recovery and recrystallization following cold-working. Evidently, other tensile properties such as the ductility and strain hardenability play a role in influencing crack growth behavior. Examination of the fracture surface of the two extreme case test specimens, 20% CW and 80% RX material tested at 677°C (1250°F) shows no difference in the mechanism of crack advance. Figures 6a and 6b are Scanning Electron Micrographs of the crack surfaces of 20% CW and 80% RX material test specimens respectively, taken at crack lengths corresponding to an applied stress intensity amplitude of 29 MPa√m (26.4 ksi√in).

From figure 6 it is evident that, for both material states, cyclic crack advance occurs by shear sliding displacements resulting in the formation of ductile striations. The geometric model representations of Laird and Smith[16] and of Neumann[17] adequately describe the fractographic ductile striation appearance observed. The model leads to a prediction of cyclic crack growth rate, da/dN, proportional to the crack tip opening displacement, CTOD. The effect of plasticity on CTOD is normally excluded for small scale yielding and da/dN predicted to vary as $(\Delta K)^2/E\sigma_{ys}$, where E is the elastic modulus. Results of this study indicate a need to incorporate ductility and/or strain hardenability (in addition to yield strength) into the cyclic crack growth prediction even when conventional Linear Elastic Fracture Mechanics (LEFM) criteria are

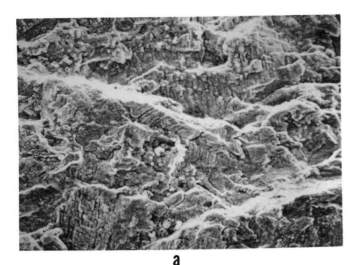

a

50 μm   **crack advance** ←

b

Figure 6 - Scanning Electron Micrographs of the fracture surface of (a) 20% CW and (b) 80% RX material at applied ΔK≈29 MPa√m

met. Weertman's[18] theoretical "accumulated plastic strain criterion for fatigue crack growth" incorporates material ductility and strain hardenability into the crack growth equation (eq.(31) of reference 18):

$$da/dN = (\{\Delta K\}^2/2\pi n\varepsilon_c\sigma_{ys}\mu)(1+n)^{n/(n+1)} \quad (3)$$

where da/dN, ΔK, σ_{ys} and n are as mentioned above, ε_c is the critical strain required for crack advance (proportional to ductility) and μ is the elastic shear modulus. The changes in

overall fatigue crack growth resistance observed here with changes in material tensile properties are explainable on the basis of this strain exhaustion criterion. The varying sensitivity of da/dN to ΔK, however, needs further investigation.

It should be noted that in the presence of significant plasticity and crack tip blunting the logarithmic linear da/dN - applied ΔK correlation seen here is unexpected. Indeed, highly recrystallized material did exhibit some crack retardation and arrest behavior at low ΔK values below approximately 18 MPa√m (16 ksi√in) which may be attributed to crack tip blunting. The effect manifests itself as significant scatter on the da/dN - ΔK plot as a result of alternating growth - arrest behavior. However, these blunting effects were not observed over the higher ΔK (>18 MPa√m) region of the test, where good linearity was obtained. In view of the correlation obtained, no attempt was made to experimentally measure ΔJ (amplitude of the J-integral) or use it as a fatigue crack growth rate descriptive parameter.

The effect of material condition on the overall fatigue crack growth resistance seen at 677°C (1250°F) appears to wash out at 732°C (1350°F) - compare figures 3 and 4. This may be attributed to an increasing environmental effect with increasing temperature that controls the crack growth rate. The "low-level impurity" helium environment has been shown to be weakly oxidizing[19] to Alloy 800H. Also, the fatigue life of Alloy 800H at 649°C (1200°F) has been shown[19] to be sensitive to the test environment. The specimens tested here showed some post-test oxide discoloration and oxidation of the fracture surfaces (see figure 6). Although the crack growth rates for the various material conditions were comparable at 732°C (1350°F), their sensitivity to ΔK (see table 6) increased with recovery and recrystallization. Again, this variation in sensitivity, also seen at 677°C (1250°F) is not understood.

EFFECT OF TEMPERATURE - Temperature is shown to increase the fatigue crack growth rate for a given material condition and applied ΔK (see figure 5). This effect has been observed on Alloy 800 for tests in air[10,11] and may be attributed to an increasing environmental effect, probably oxidation, as explained above. Creep effects are expected to be negligible at the high test frequency of 10 Hz. It should be pointed out that the environment here is significantly weaker than air in its oxidation potential and this may account for the superior crack growth resistance of these results compared with air data on Alloy 800 of references (10) and (11). The observed variation in crack growth rate with temperature cannot be explained only on the basis of changing tensile properties, E and/or σ_{ys}, as has sometimes been demonstrated.[11,14,20,21] Normalization of the data using $\Delta K/E.\sigma_{ys}$ instead of ΔK does not consolidate the data and entirely eliminate the temperature effect. The effect of temperature on fatigue crack growth resistance seen here is in agreement with the published results of references (10) and (11) for Alloy 800 tested in air.

CONCLUSIONS

Following are the main conclusions of the study:

(1) Increasing recovery and recrystallization from cold-working results in an increasing material fatigue crack growth resistance at 677°C (1250°F), but has a negligible effect at 732°C (1350°F). The effect at 677°C (1250°F) cannot be entirely explained on the basis of crack closure and has been at least partially attributed to an increasing tensile ductility and strain hardenability with increasing recrystallization. The disappearance of this effect at the higher test temperature of 732°C (1350°F) may be explained on the basis of an increasingly dominating environmental effect with increasing temperature.

(2) Recovery and recrystallization causes an increase in the sensitivity of cyclic crack growth rate to ΔK (exponent in Paris law). Reasons for this effect are not understood.

(3) For a given material state and applied ΔK, fatigue crack growth rates are observed to increase with increasing temperature. The weakly oxidizing helium environment contributes to this phenomenon.

(4) Comparison with "air" data on Alloy 800 shows superior fatigue crack growth resistance measured for microstructurally similar Alloy 800H in helium, indicating the sensitivity of the alloy to the operating environment.

ACKNOWLEDGMENT

The author thanks the Electric Power Research Institute (EPRI) for the support provided this project under Contract RP-2079 and for the encouragement of R.E.Nickell and R.B.Hayman through the course of the program. The invaluable contribution of R.O.Harrington Jr., who conducted the tests, is also gratefully acknowledged. Helpful discussions with A.M.Ermi and L.A.James of Westinghouse Hanford Co. led to the specimen geometry employed in this study.

REFERENCES

1. Cordovi,M.A. in "Alloy 800", p.3, North-Holland Publishing Co., New York (1978)
2. ASME Boiler and Pressure Vessel Code, Section XI, Division I, Appendix A, The American Society of Mechanical Engineers, New York (1980)
3. ASME Boiler and Pressure Vessel Code, Section II, Part B, SB-163, The American Society of Mechanical Engineers, New York (1983)
4. Underwood,E.E., "Quantitative Stereology", p.29, Addison-Wesley Publishing Co., Reading, MA (1970)
5. Foulds,J.R., Unpublished Research, GA Technologies Inc., San Diego, CA (1984)
6. ASTM Standard E647-83, "Standard Test Method for Constant - Load - Amplitude Fatigue Crack Growth Rates Above 10^{-8} m/cycle", 1985 Annual Book of ASTM Standards, Vol. 03.01, p.739, ASTM, Philadelphia, PA (1985)
7. Johnson,H.H., Materials Research and Standards, 5, p.442 (1965)
8. Schwalbe,K.H. and D.Hellman, J.Testing and Evaluation, 9, no.3, p.218 (1981)
9. Paris,P.C. and F.Erdogan, J.Basic Engineering, 85, p.528 (1963)
10. James,L.A., J. Eng. Materials and Technology, 96, p.249 (1974)
11. Shahinian,P., Metals Technology, 5, no.11, p.372 (1978)
12. James,L.A., Nuclear Technology, 16, no.1, p.316 (1972)
13. Donahue,R.J., Clark,H.McI., Atanmo,P., Kumble,R. and A.J.McEvily, Int. J. Fracture Mechanics, 8, p.209 (1972)
14. Ellison,E.G. and D.Walton in "Proc. Int. Conf. on Creep and Fatigue in Elevated Temperature Applications, vol.13", p.173.1, Inst. of Mechanical Engineers, London (1973)
15. Elber, W. in ASTM STP-486, p.230, ASTM, Philadelphia, PA (1971)
16. Laird,C.and G.C.Smith, Phil. Mag., 7, p.847 (1962)
17. Neumann,P., Acta Met., 22, p.1155 (1974)
18. Weertman,J. in "Fatigue and Microstructure", p.279, American Society for Metals, Metals Park, OH (1979)
19. Johnson,W.R. and D.I.Roberts, ASME Paper No. 82-JPGC-NE-14, ASME, New York (1982)
20. Speidel,M.O. in "High Temperature Materials in Gas Turbines", p.207, eds. Sahm,P.R. and M.O.Speidel, Elsevier, Amsterdam (1974)
21. Shahinian,P., Smith,H.H. and H.E.Watson, Trans. ASME, B, 93, p.976 (1971)

A NONLINEAR FINITE ELEMENT ANALYSIS
OF STRESS CONCENTRATION
AT HIGH TEMPERATURE

Faysal A. Kolkailah
California Polytechnic State University
San Luis Obispo, California, USA

A.J. McPhate
Louisiana State University
Baton Rouge, Louisiana, USA

Abstract

The rotating disks that hold the turbine blades in jet aircraft engines are subjected to severe stresses and high temperatures. Local stress raisers are very often life-limiting in that the low-cycle fatigue failures generally initiate in these critical regions. Economic and reliability demands have prompted development of analytical methods, first to predict stresses and strains in these regions of complex geometry, and ultimately to predict the low cycle fatigue life.

In this paper, an elastic-plastic finite-element model incorporating the Bodner-Partom model of non linear time-dependent material behavior is presented. The parameters in the constitutive model are numerically obtained from a least-square fit to experimental data obtained from uniaxial stress-strain and creep tests at 650°C. The finite element model of a double-notched specimen, figure 1, is employed to determine the value of the elastic-plastic strain and is compared to experimental data. The constitutive model parameters evaluated in this paper are found to be in good agreement with those obtained by the other investigators. However, the numerical technique tends to agree with the response curves better than graphical methods used by the other investigations. The calculated elastic-plastic strain from the model agrees well with the experimental.

INCONEL 718 IS A HIGH TEMPERATURE SUPERALLOY specially developed for low-cycle-fatigue limited components operating under high temperatures, severe stresses and a hostile environment. It is important to determine and properly characterize its time dependent inelastic properties. A number of visco-plastic constitutive theories in which creep and plasticity effects are combined into a unified plastic strain model have recently been proposed and are still undergoing active development. One of these theories, employed in this study, is the constitutive theory of Bodner and Partom [1].

Bodner [1], utilized the constitutive equations of Bodner and Partom to represent the inelastic behavior of Rene 95 at 1200°F. Stouffer [2], used the state variable constitutive equations of Bodner and Partom to calculate the mechanical response of IN 100 at 1350°F. Hennerichs [3], estimated the material constants in IN 100 by using Bodner and Partom constitutive equations. Milly [4], represented the experimental data for Inconel 718 at 1200°F, from which the material constants were determined by the method given by Stouffer [2]. The Bodner constitutive equations were then applied. Milly compared the theory and experimental data and concluded the overall behavior is good.

As to the finite element model, there is a large number of finite elements in use today for both plane stress and plane strain analyses, such as, the constant strain triangular elements, hybrid elements and higher order isoparametric elements. For linear elasticity, the mathematical development of these elements is simple. Since the constant strain triangular elements have the advantages of being simple and economical, they have been selected to be employed in this study. The elastic-plastic code has been used in the numerical determination of the elastic stress concentration (K_t) for benchmark notched specimen under constant tensile load and for the elastic-plastic response of that specimen for cyclic loading.

Domas, Sharpe, Ward and Yau [5] in the analytical task of their study for the "Benchmark Notch Test for Life Prediction", used the finite element method for good estimation of the elastic stress concentrating factor (K_t) from calculated stress and strains at element centroids. Ahmand [6] in his report, provides a user's guide for a special purpose finite element code developed primarily for two-dimensional linear elastic analysis of test specimens. It includes some supporting programs, such as mesh generating and mesh plotting for commonly used test specimens.

Hennerichs [3], incorporated the Bodner constitutive model into a constant-strain triangular elements model to analyze creep crack growth in a nickel alloy at $1350^{\circ}F$.

THE CONSTITUTIVE THEORY OF BODNER AND PARTOM

The theory is based on the assumption that the total strain rate $\dot{\varepsilon}^t(t)$ is separated into elastic $\dot{\varepsilon}^e(t)$, and plastic $\dot{\varepsilon}^P(t)$, components, both non-zero for all loading/unloading conditions.

$$\dot{\varepsilon}^t(t) = \dot{\varepsilon}^e(t) + \dot{\varepsilon}^P(t) \qquad (1)$$

The specific representation used by Bodner and Partom [1] is given by:

$$\dot{\varepsilon}^P(t) = \frac{2}{\sqrt{3}} \frac{\sigma(t)}{|\sigma(t)|} D_o \exp\left[\frac{n+1}{2n}\left(\frac{Z(t)}{\sigma(t)}\right)^{2n}\right] (2)$$

where; $\sigma(t)$ = stress

D_o = limiting plastic strain rate in shear.
n = strain rate sensitivity parameter.
$Z(t)$ = the plastic state variable, hardness.

The evolution of the plastic state variable is in the form of a differential equation for the hardening rate, $\dot{Z}$, that depends on stress, temperature and hardness.

$$Z(t) = Z_o + \int \left[m(Z_1 - Z)\,\sigma\dot{\varepsilon} - A Z_1 \left(\frac{Z-Z_i}{Z_1}\right)^r \right] dt \qquad (3)$$

where, m = work hardening parameter.
Z_1 = the saturation value of Z.
Z_o = the initial value of Z.
A = hardening recovery parameter.
r = the exponent controlling the rate of hardening recovery.
Z_i = dead soft hardness.

The Bodner model parameters thus fall into two groups; short-time constants D_o, n, m, Z_o, Z_1; and creep constants Z_i, r, A.

NUMERICAL EVALUATION OF THE PARAMETERS

To determine the Bodner constants, the measured value of strain at a given time was compared to that modeled by a set of constants. The measured plastic strain is obtained from eq. 1 so

$$\varepsilon^t(t) = \varepsilon^t(o) + \int^t \dot{\varepsilon}^P(t)dt + \sigma(t)/E \quad (4)$$

From eq. 2 and 3, $\dot{\varepsilon}^P(t)$ can be computed for a given set of constants. $\varepsilon_{model}(t)$ is computed following eq. 4. In the numerical scheme developed, the fourth order Runge Kutta algorithm was employed for numerical time integration. A direct search algorithm was used to vary the Bodner constants in the iterative process. The error function

$$Q = \int \left(\varepsilon_{data}(t) - \varepsilon_{model}(t)\right)^2 dt \qquad (5)$$

was used as a measure of the goodness of the fit, although another real test is how well the computed stress-strain curves match the experimental.

THE ELASTIC FINITE ELEMENT MODEL

The locally written finite-element program generates constant strain triangular elements with very fine elements in the root of the notch. The 2-D Program was formulated for plane stress calculation. Only one quarter of the specimen was modelled because of symmetry, and an automatic mesh generation algorithm permitted easy variation of the total mesh size. It was thus possible to easily select the proper balance between mesh fineness and computation time. A mesh of 176 nodes and 300 elements was eventually chosen, figure 2.

RESULTS

Uniaxial tensile tests were run at strain rates of 1.6×10^{-3}, 0.67×10^{-4}, 1.0×10^{-5}, 1.1×10^{-6}, and 3.3×10^{-7} sec^{-1}. Two creep tests were run at 758 and 862 MPa. The raw data for each test was smoothed and approximately 20 data points used for each test in the fitting procedure. The modulus of elasticity was also considered as a parameter increasing the number of constants to be determined to nine. Different search starting values were tried, producing slightly different values of constants though no appreciable difference in the agreement with individual curves. A unique set of constants is not to be expected. The final parameters determined are listed in Table 1 where the values from Milly [4] are also listed. Figures 3, 4, 5, and 6 show the response curves from the model and the experimental data (only tensile tests at strain rates of 3.3×10^{-7} and 1.6×10^{-3} sec^{-1} are shown in Figures 3 and 4). The agreement is considerably better than other comparisons [2,4].

The finite-element model does a good job of evaluating the theorectical elastic stress concentration, K_t, and predicting the notch behavior under low cycle loading at high loads. The response at the notch is very complicated given the finite plastic strains generated. Figure 7 shows the experimental data and finite-element prediction for the comparison of measured notch strain with that predicted by the finite element during the first cycle of continuous cyclic loading at 650°C.

For further application, using the Bodner Partom state variable material model and model parameters developed from Uniaxial test data, creep crack growth in Inconel 718 was analyzied by using the constant strain finite element for compact tension specimen [6].

Crack growth behavior identified solely from crack opening displacement measurements made on pre-cracked specimens was analyzed by examining various parameters around the model's crack tip. The J-Integral was applied and used to predict crack growth initiation. Promising results were obtained.

In reviewing the response, figure 8, one must keep in mind the initial material parameters used were those derived from the uniaxial data. The cracked specimens were actually pre-cracked by fatigue loading and the crack propagation material state was not the same as the initial state of the uniaxial test specimen.

REFERENCES

1. Bodner, S.R. and Partom, Y., "Constitutive Equations for Elastic-Viscoplastic Strain Hardening Materials", Journal of Applied Mechanics, Trans. ASME, 42, 385-389, 1975.

2. Stouffer, D.C., "A Constitutive Representation for IN-100", Air Force Wright Aeronautical Laboratories, AFWAL-TR-81-4039, 1981.

3. Hinnerichs, T., Nicholas, T., and Pulayatlo, A., "A Hybrid Experimental-Numerical Procedure for Determining Creep Crack Growth Rates", Engineering Fracture Mechanics, 16, 265-277, 1982.

4. Milly, T.M. and Allen, D.H., "A Comparative Study of Nonlinear Rate-Dependent Mechanical Constitutive Theories for Crystalline Solids at Elevated Temperatures", Report API-3-5-82-5, Virginia Polytechnic Institute and State University, March 1982.

5. Domas, P.A., Sharpe, W.N., Jr., Ward, M., and Yau, J., "Benchmark Notch Test for Life Prediction", NASA-Lewis Research Center, NASA CR-165571, 220 pgs., 1982.

6. Jupta, Nirmalya, "The Threshold Creep Crack Prediction Using the Finite Element Method", Master Thesis in mechanical Engineering, Louisiana State University, 1984.

APPENDIX

Table 1 - Bodner Parameters for Inconel 718 at 650°C.

Constant	Units	Milly[4]	This Work
n	-	1.167	0.7374
z_o	MPa	3130	6520
z_1	MPa	4140	7030
m	MPa^{-1}	2.43×10^{-2}	6.86×10^{-1}
A	sec^{-1}	1.1×10^{-4}	6.82×10^{-4}
r	-	2.857	4.734
z_i	MPa	2760	3690
D_o	sec^{-1}	10^4	1.03×10^{-4}
E	MPa	165×10^3	172×10^3

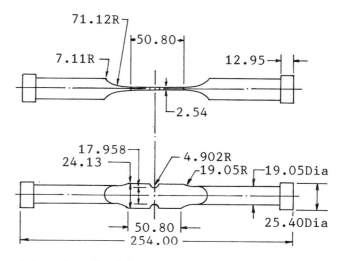

Fig. 1 - Double-notched specimen (dimensions in mm)

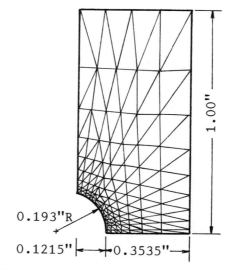

Fig. 2 - Finite element grid

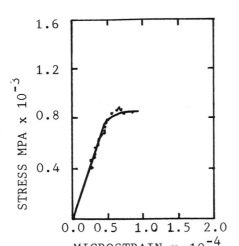

Fig. 3 - Experimental data and
numerical tensile response
($\dot{\varepsilon}$ =3.3 x 10^{-7} $_{sec}$ -1)

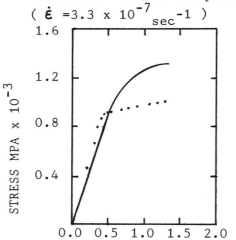

Fig. 4 - Experimental data and
numerical tensile response
($\dot{\varepsilon}$ =1.6 x 10^{-3} $_{sec}$ -1)

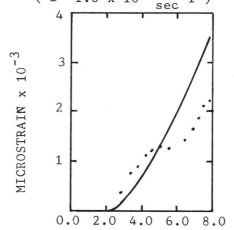

Fig. 5 - Experimental data and
creep response
(σ =758mpa)

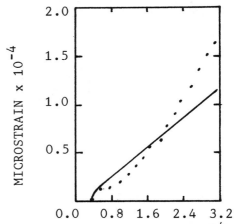

Fig. 6 - Experimental data and
creep response
(σ =862mpa)

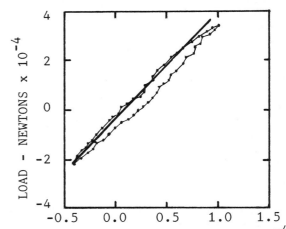

Fig. 7 - Experimental data and
finite element prediction

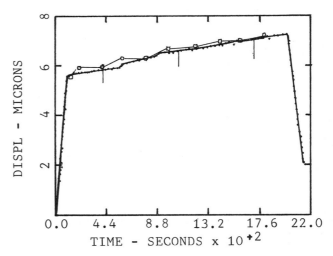

Fig. 8 - Experimental results for
creep crack growth

A STUDY OF THE TRANSITION TEMPERATURE OF COD OF PRESSURE VESSEL STEELS

Jia-Ling Jiang
Zhejiang University
Department of Chemical Engineering
HangZhou China

Abstract

In this paper a new method is introduced to obtain statistically defined reference curves of variation of COD as temperature in pressure vessel steels and to determine the characteristic temperature of ductile-brittle transition of fracture failure of these steels. The relations between the referance curve of COD and the curve of impact energy C_v or lateral expansion MLE of V-notched specimen of these steels are discussed.

THE BRITTLE FRACTURE at low stress level is one of the main failure types in pressure vessels. One of the factors affecting fracture behavior of material is temperature. At present, fracture mechanics is already widely used in analysis of of protecting pressure vessels from brittle fracture and critical crack opening displacement $_c$ is generally selected as the fracture toughness of medium or low strength steels used in manufacture of pressure vessels. In fact, the fracture failure in steel is affected by temperature, therefore it is sure that there exist close relation between fracture toughness and temperature. It is suitable to obtain statistically defined reference curve for variation of COD with temperrature. It is suitable to obtain statistically defined reference curve for variation of COD with temperature for steels.

In the ASME Code, for steel alloys used in reactor vessel construction the relation between fracture toughness and temperature has been expressed follows,

$$K_{1R}=26.77+1.233\exp[0.01449(T+160-RT_{NDT})]$$

where

K_{1R} -- the reference fracture toughness of steel,

RT_{NDT} -- the reference temperature determined from drop weight NDT and Charpy V-notch tests of steel,

T -- working temperature.

However, this is the lower bound curve, all testing points (t_j, K_{1Rj}) are scattered above the curve. For most of medium or low strength steels, this type of curve is too conservative to estimate the toughness value at the given temperature. In addition, it has not been whether RT_{NDT} can be defined as a characteristic temperature marking the behavior of ductile-brittle transition of fracture in steel. The procedure to determine RT_{NDT} for each steel is very difficult and complex.

TESTING RESULTS

The critical crack opening displacement $_c$ were determined in four steels, 16Mn, 20G, 14MnMoNb and 18MnMoNbB at temperatures from $-90^\circ C$ to $30^\circ C$. At lower temperatures, the value of inition COD (δ_i) was adopted as the value of critical COD (δ_c). In the transition region and at higher temperatures, base on the resistance curve method for each series of specimence, the value $\delta_{0.05}$ when the length of crack growth equals to 0.05mm was adopted as the value of critical COD (δ_c). The results showed that, for these steels, there exist lower and higher values of δ_c and that the transition goes within a temperatures region. In other words, at lower temperatures the toughness has a stable low value δ_L; as the temperature increases, the toughness becomes higher; finally it reaches high value δ_H. All measurement results were showed the table 1.

Table 1. Mechanical Property and
Measurement Values of δ_L and δ_H

Materials	σ_r(kgf/mm^2)	δ_L(mm)	δ_H(mm)
16Mn	34.0	0.088	0.128
20G	24.2	0.132	0.270
18MnMoNb	58.3	0.049	0.063
14MnMoNbB	86.7	0.038	0.056

REGRESSION ANALYSIS AND
CHARACTERISTIC TEMPERATURE

After we tried various mathematics models, it was found that a nonlinear function as eq.(1) is proper for regression analysis of all test data of each steel:

$$\frac{\delta_c(t) - \delta_L}{\delta_H - \delta_L} = \frac{1}{1 - e^{-(t-t_o)/b}}$$

or   (1)

$$\delta_c(t) = \tfrac{1}{2}(\delta_H + \delta_L) + \tfrac{1}{2}(\delta_H - \delta_L)\,th(\frac{t-t_o}{2b})$$

where to, b are regression coefficients, t is working temperature.

Fuction (1) possesses some desirable properties. Firstary, at low or high temperatures it tends to two constant values. Secondary, when $t=t_o$, the second derivative of δ_c with respect to t equals to zero, in other words, at point ($t=t_o$) the function has transition of curvature.

From function (1), the following expression can be deduced:

$$-\frac{t}{b} + \frac{t_o}{b} = Ln[\frac{\delta_H - \delta_c(t)}{\delta_c(t) - \delta_L}]$$ (2)

Defining a_o, a and Y(t) as

$$a_o = t_o/b$$
$$a = -1/b$$

$$Y(t) = Ln[\frac{\delta_H - \delta_c(t)}{\delta_c(t) - \delta_L}]$$ (3)

We yield a function as follows

$$Y(t) = at + a_o$$ (4)

Based on eq.(4), this transition can be used for calculation of $Y(t_j)$ from toughness

284

value, $\delta_{cj}(t_j)$ that was measured at temperature $t_j(^oC$) and for linear regression analysis of all test data. After calculating coefficients a, a_o, we by eq.(2) could calculate the coefficients b, t_o. For instance, for two steels 16Mn and 20G, we obtained all the results as showed in Table 2.

Table 2. Regression Coefficients
of two Steels

Materials	a_o	a	t_o	b
16Mn	−1.726	−0.0526	−32.8	19.0
20G	−1.402	−0.0758	−18.5	13.2

The results were verified as following

1. The standard deviation was given by

$$= \sqrt{\frac{\sum_{j=1}^{N}[\ \delta_c(t_j)\ -\ \delta_{cj}(t_j)]^2}{N-1}}$$

where $\delta_c(t_j)$ is the value of correration curve at temperature $t_j(^oC)$. N presents the number of measurement points.

It were found that the deviation in 16Mn and 20G steels were less than 5 percent of average value of toughnesses within this temperature range.

2. On the basis of formula (4), the two linear correlation coefficients equal to 0.888 and 0.953 respecting.

3. F-test at the α=0.05 level of significance

For steel 16Mn, N=32
critical value, $A_F(0.05;\ 1,\ 30)$=4.17
computed result $f=116 > A_F$
For steel 20G, N=27
critical value, $A_F(0.05;\ 1,\ 25)$=4.24
computed result $f=246 > A_F$
apparently, the regression was effective.

Two regression curves of $_c(t)$ are shown in Fig.1 and Fig.2. In addition, the confidence region at 0.95 confidence level also is indicated, its half width $d\delta_c(t)_{0.95}$ was calculated

by following formula:

$$d\delta_c(t)_{0.95}$$

$$=\left| \frac{1}{\frac{1}{\delta_H-\delta_L}+e^{Y(t)+dY(t)_{0.95}}} - \frac{1}{\frac{1}{\delta_H-\delta_L}+e^{Y(t)}} \right|$$

(5)

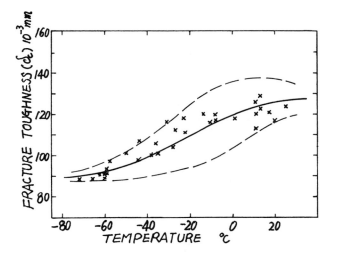

Fig.1 Regression curve
of $\delta_c(t)$ for 16Mn steel

It was singnificant that all the additional testing points were located within the confidence regions. Therefore, the regression coefficient, to could defined as the characteristic transition temperature of COD, and 2b marked the half width of the transition region. We are sure that it is correct to compare the fracture behavior of steel on the basis of "$\delta_c(t)-t$" curve, especialy on the basis of t_o and 2b values.

It must be pointed out that, for the four pressure vessel steels, the widths of ranges of transition which amounted to about 52.8^oC to 76^oC were close to that of about 66.7^oC which was obtained by Pellini thirty years ago. Based on his experiment results, he concluded that, in most plates of medium strength steels used in ships and naval vessels, containing small weld

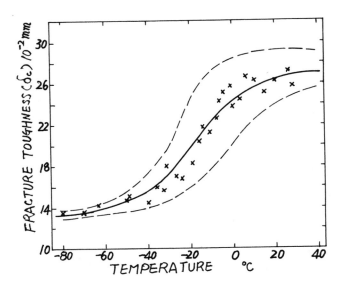

Fig.2 Regression curve
of $\delta_c(t)$ for 20G steel

Concerning of C_v is an energy index and δ_c is a dimension index, we measured both C_v(kgf-m) and the lateral expansion MLE(mm) of V-notched specimens in the same temperature range for those kinds of steel, and determined the characteristic transition temperatures of $C_v(t)$ and MLE(t) curves in term of t_{o1} and t_{o2}. The correlation method of temperature shift between $\delta_c(t)$ and $C_v(t+\Delta t_1)$ and between $\delta_c(t)$ and MLE($t+\Delta t_2$) is used, we obtain the following two formulas:

$$\delta_c(t) = a_1 + b_1 C_v(t+\Delta t_1) \qquad (6)$$
$$\delta_c(t) = a_2 + b_2 MLE(t+\Delta t_2) \qquad (7)$$

where

$$\Delta t_1 = t_{o1} - t_o$$
$$\Delta t_2 = t_{o2} - t_o$$

a_1, b_2, a_2, b_2 are regression coefficients.

The correlation coefficients for formula (6) are about 0.72-0.79, and 0.88-0.96 for formula (7). Therefore it is clear that calculating $\delta_c(t)$ from MLE is better than from C_v.

cracks, the ductile-brittle transition of fracture failure appears in range of temperatures from NDT to NDT+66.7°C. However, our experimented results on temperature behavior of fracture toughness showed that there exists ductile-brittle transition temperature range of fracture failure for most medium and low sthength steels.

RELATION BETWEEN $\delta_c(t)$
AND $C_v(t)$ OR MLE(t)

Due to difficulty of measuring $\delta_c(t)$, when steels to be selected, Charpy impact energy C_v of V-notched specimen are used usually as the toughness index of steel. In some code an empirical formula between $C_v(t)$ and $\delta_c(t)$ was suggested, in JWES-2805 it is:

$$\delta_c(t) = 0.01 C_v(T+112-\sigma_y-5\sqrt{t}\ )$$

where T is working temperature, t represents plate width.

Conclusion

1. The nonlinear function (1) may be selected as statistically defined reference curve of COD for most of pressure vessel steels. Two regression coefficients t_o and b indicate characteristic behavior of ductile-brittle transition in fracture failure of these steels.

2. Although the temperature is changed, transition tendency of $\delta_c(t)$, $C_v(t)$ and MLE(t) is similar and the relation between $\delta_c(t)$ and $C_v(t)$, especially between $\delta_c(t)$ and MLE(t) may be established by the correlation method of temperature shift.

INELASTIC ANALYSIS OF A HOT SPOT ON A HEAVY VESSEL WALL

Kenneth L. Baumert
Air Products & Chemicals, Inc.
Allentown, Pennsylvania, USA

David A. Secrist
ICRC, Subsidiary of Air Products
Ipswich, Massachusetts, USA

ABSTRACT

Heavy wall, 0.2-0.3 m (8-12 inches), reactor vessels are common in the refining process industries. Many of these vessels are used in process operations involving hydrogen at high pressures and temperatures. When processing conditions involve three-phase flow, maintaining a uniform temperature distribution in the vessel becomes a concern. Maldistributed flow can result in a hot spot on the vessel wall.

This hot spot condition was simulated by computer analysis on a reactor vessel with a 0.25 m (10-inch) thick wall. The results indicate that stresses are relieved after a short period of time (1000-3000 hours) because of metal creep. The bulge on the ID is small and probably within the original vessel tolerance. Very high residual stresses are imparted to the steel after cooldown. This could result in a fatigue failure because of normal startups and shutdowns. Finally, if the vessel is fabricated from heat treated steel, the elevated temperature can temper the alloy to below minimum design properties.

HEAVY WALL REACTOR VESSELS are common to refining and chemical industries. Accordingly, many of these vessels are used in processes which involve hydrogen at high pressures and temperatures. When internal reactor conditions also involve three-phase flow, the maintenance of an itnernal uniform temperature distribution becomes a concern. Maldistributed flow can create a thermal upset condition on the vessel wall, locally overheat the metal and cause a loss of strength with time.

Several failures of heavy vessels, 0.2-0.3 m (8-12 inches) wall, have occurred during the past 15 years. Some suspect that a concentrated "hot spot" near or on the vessel wall caused at least one of those ruptures. It is not yet well understood what mechanism precipiates the event.

Despite postmortem investigations, several technical questions remain unanswered with respect to vessel durability and performance during the hot spot lifetime.

This paper describes an engineering assessment of structural integrity and additional safe operating time once a vessel has been subjected to such an upset condition.

ASSUMPTIONS FOR ANALYSIS - Chosen on basis of operating experiences, the problem parameters set the physical and thermal limits of a local upset. The hot spot was assumed large enough to damage the vessel and yet not so large as to be detected by normal plant instrumentation. It is presumed to occur during normal operation and to be independent of any other process condition.

A 0.6 m by 0.6 m (24" by 24") hot spot was assumed to create a constant and uniform, 593°C (1100°F), temperature over the inner wall surface. Internal reactor process conditions were taken to be 450°C (840°F) and 13.8 MPa (2000 psig), which remained constant throughout the upset scenario. The startup transient is presumed to be much shorter than either the long-term damage time from either Larson-Miller parameter loss of tensile properties, or from time until onset of tertiary creep.

The vessel base material (ASTM A-387, grade 22, class 2, 2.25 Cr-1.0 Mo steel) was assumed homogeneous and isotropic without imperfections which affect thermal properties or mechanical behavior. The strength of class 2 material is obtained by quenching heat treatment. Tests performed on heavy section steel plates have exhibited some strength variance across the thickness due to non-uniform cooling during heat treatment. Hence, the greatest wall thickness attainable is limited by the physics of heat removal from the interior of the slab.

Vessel external insulation was taken as a typical calcium silicate material mechanically bonded to the wall. Perfect thermal contact between the metal wall and the installed insulation was assumed. No interfacial temperature drops or other effects which would alter the external convective heat transfer coefficient were included. A typical thickness of 0.13 m (5.25") was selected.

FINITE ELEMENT MODELING WITH ANSYS[1] - The well-known ANSYS computer program was employed for the solution of this engineering problem. Its method of solution is a matrix inversion technique based on a finite element idealization. Only three-dimensional isoparametric 8-node solid elements were used for both thermal (STIF70) and mechanical (STIF45) calculations.

The total solution was completed in two phases. A heat transfer analysis was first computed to establish the proper thermal profiles within the wall. The temperature distribution was verified by comparison to closed-form solutions. Then the program was restarted, the temperatures inserted, and the mechanical analysis begun. The non-linear ANSYS formulation of plasticity and creep were selected.

The vessel sector evaluated was 0.9 m (36") in longitudinal height by 37.4 angular degrees along the circumferential direction, 1.16 m (46.4") arc length. Its internal radius was 1.66 m (66") and wall thickness 0.26 m (10.25"). The dimensions were calculated to provide sufficient attenuation lengths for the changes in stresses and thermal profiles to be satisfactorily reduced to normal values at the boundaries. Computed stresses at the boundaries were compared to and agreed with thick cylinder analytical results.

The circular arc of the vessel has been approximated by a series of right trapezoidal prisms. 540 solid (brick) elements, described by 800 nodes, were sufficient to minimize curvature effects on both the stress and thermal solutions. The radial and angular mesh sizes were not uniform, but were variably spaced to resolve the large thermal and stress gradients expected to occur near the hot spot location. No transition or curved side elements were used. A pictorial description of the section model will be found in Figures 1 through 4, with brief temperature and stress results in Figure 5 through 7.

RESULTS AND DISCUSSION - Figures 8 and 9 provide the location for elements 1, 4, 33, 36, 505 and 508. Table 1 presents the principal stresses for normal, 450°C (840°F), operating conditions as well as for conditions with the 593°C (1100°F) hot spot applied (at 0 hours) as well as 10, 30, 100, 300, 1000 and 3000 hours later. The effects of creep deformation are obvious on the stresses. The circumferential (Y) and axial (Z) stressee change radically, whereas the radial (X) stress is essentially unaffected.

Over a 3000-hour time span, element 1 displays the most drastic change, with the hoop stress decreasing in magnitude from -55.8 MPa (-8.1 ksi) to -9 MPa (-1.3 ksi). The axial stress decreases from -137 MPa (-19.9 ksi) to -26.2 MPa (-3.9 ksi). Element 4, in the outer portion of the wall, reflects an increase in hoop stress from 39.3 MPa (5.7 ksi) to 51.7 MPa (7.5 ksi). Elements 33 and 36 show little variance over time, as both decrease in hoop stress while slightly increasing in axial stress. Elements 505 and 508 are unaffected by the hot spot for all practical purposes.

The evaluation of the stress state upon vessel cooldown is interesting. Table 2 shows the residual principal stress components after vessel cooldown. Cold residual stress state conditions are shown after 1000 hours and 3000 hours of creep deformation. Hot vessel stresses for normal operating conditions and upon the initial application of the 1100°F hot spot discontinuity are shown for comparison.

Element 1 experiences large reversals of the hoop and axial stresses: normal operation is 91.7 MPa (13.3 ksi); during the hot spot it is -55.8 MPa (-8.1 ksi); and after 3000 hours with cooldown it is 66.2 MPa (9.6 ksi). Similarly, for the axial stress: normal, 39.3 MPa (5.7 ksi); after hot spot, -137 MPa (-19.9 ksi); after 3000 hours plus cooldown, 155 MPa (22.5 ksi). Element 4 does not show such large magnitude changes as element 1. Elements 34 and 36 are also affected by the hot spot; the hoop stresses increase by a factor of about 2. However, there are no large residual tensile stresses after cooldown. Elements 505 and 508 are essentially unaffected by either the hot spot or creep deformations.

Tables 3 and 4 represent hot radial displacements of discrete points located on the vessel wall. The displacement of lines 1-1, 1'-1', and 2-2 (shown on Figure 8) show smooth increases of displacement from the hot areas to cooler ones. Line 2'-2' displacement shows no effects. Normal operation causes a radial outward expansion of the vessel and a slight thickening of the wall due to pressure loading and thermal expansion. The displacement values suggest that the vessel would bulge in the vicinity of the hot spot, but that the bulge is quite small relative to the initial thermal expansion. Results for line 1'-1' at (R,Z) location, 1.81 m, 0.5 m (76.25", 2.00"), suggest a bulge of 0.0136 m-0.012 m (0.540"-0.478"), 0.0016 m (0.062" or 62 mils). Measuring this small increase during vessel operation will be quite difficult. Relative to the initial thickening of the wall, the hot spot and subsequent material creep thickens the wall near the hot spot and thins it elsewhere.

The creep strain components for selected elements have been examined. The location of the elements on the inside face of the wall is shown in Figure 9. The values of strains are tabulated as Tables 5 and 6. Given the geometry of this vessel and the restraint provided by the large masses of relatively cold metal surrounding the hot spot area, the radial component of creep strain was judged the most likely to exhibit growth. Elements 1 and 37 showed the largest radial tensile strain of 0.11 to 0.12%. This is considerably below the acceptable value of 1%.

288

The other strain components of these elements are an order of magnitude smaller.

All the elements which showed the most creep strain activity (1, 5, 9, 37, 41,45 and 81) also show a compressive creep strain in the axial direction after cooldown. It is important to realize that the normal material yield strain is in the 0.2% range. Thus, creep strains are nominally one-half of yield.

CONCLUSIONS - 1. The 593°C (1100°F) hot spot is not likely to damage the vessel during the 3000 hour period before the material itself degrades to minimum design class 2 properties. In this case, temperature alone will cause the degradation. After 3000 hours, ASTM A387 Gr. 2, Cl. 2 material is no longer acceptable relative to ASME[2] code standards.

2. The vessel deformations due to a local hot spot will not be measurable from the outside when compared to expansions during startup or cooldown. The deflections are easily hidden by the insulation and are not large enough to tear the exterior protective covering. Additional local deformations caused by creep are negligible and undetectable realtive to the displacements encountered during normal process operations.

3. The maximum local radial deformation caused by the hot spot is on the order of 0.0016 m (0.062"). It would require precise instrumentation to measure small distortions caused by creep.

4. Large residual tensile stresses are frozen into the metal after cooldown if significant creep has taken place. The tensile stresses in element 1 can be larger than ASME allowable design stresses. The principal stress has also altered direction. These conditions are most important in metal fatigue. Damage is caused by repeated stress reversals which would accompany startup and shutdown operations.

ACKNOWLEDGEMENT

This work was performed under DOE contract DE-AC05-78OR03054.

REFERENCES

1. G. J. DeSalvo and J. A. Swanson, "ANSYS-Engineering Analysis System Users Manual", Rev. 4.1, Swanson Analysis Systems, Inc., March 1983.

2. ASME Boiler and Pressure Code, Section VIII, Div. 2, 1980.

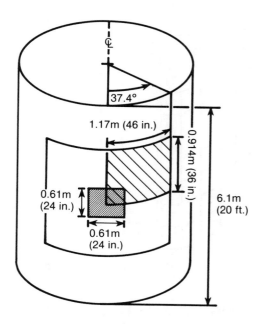

FIGURE 1. VESSEL SECTION MODEL

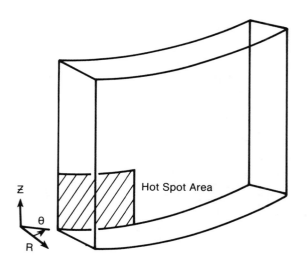

FIGURE 2. QUARTER – SYMMETRY SLAB

Z
R
33
34
35
36
36
29
32
25
28
21
24
17
20
13
16
9
12
5
8
1
2
3
4
4
Element Formed By Constant
Axial Increment

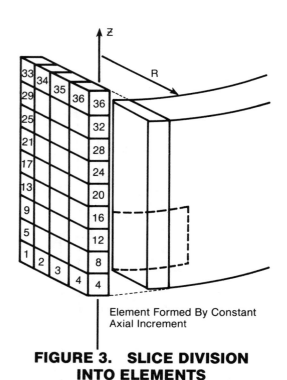

FIGURE 3. SLICE DIVISION INTO ELEMENTS

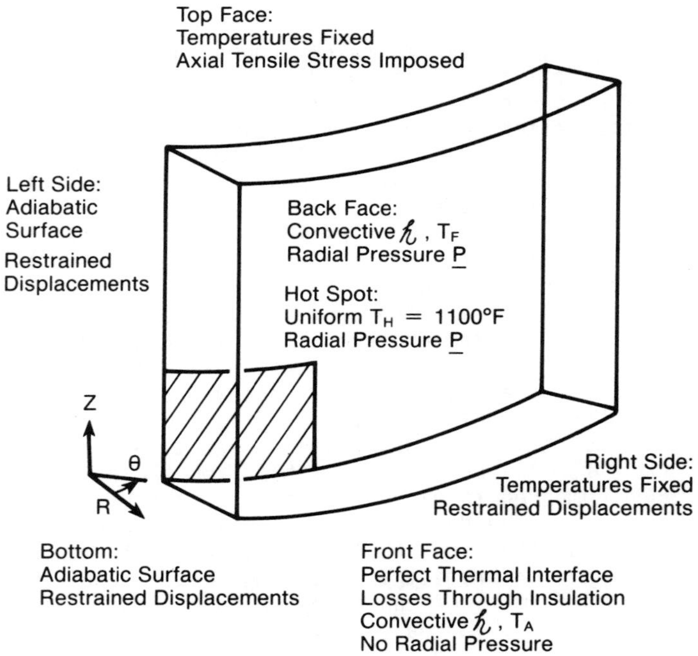

Top Face:
Temperatures Fixed
Axial Tensile Stress Imposed
Left Side:
Adiabatic
Surface
Restrained
Displacements
Back Face:
Convective h, T_F
Radial Pressure P
Hot Spot:
Uniform T_H = 1100°F
Radial Pressure P
Z
θ
R
Right Side:
Temperatures Fixed
Restrained Displacements
Bottom:
Adiabatic Surface
Restrained Displacements
Front Face:
Perfect Thermal Interface
Losses Through Insulation
Convective h, T_A
No Radial Pressure

FIGURE 4. APPLIED BOUNDARY CONDITIONS

704 (1300)
676 (1250)
649 (1200)
621 (1150)
593 (1100)
565 (1050)
537 (1000)
510 (950)
482 (900)
454 (850)
427 (800)
Temperature, °C (°F)
704 (1300) H.S.
649 (1200) H.S.
593 (1100) H.S.
537 (1000) H.S.
Normal
Z
R
θ
0
5
10
15
20
25
30
35
Theta Location, in Angular Degrees

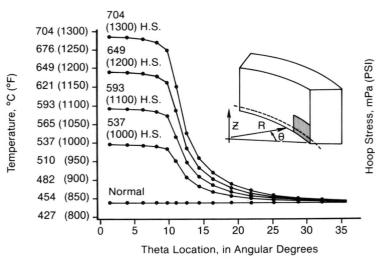

FIGURE 5. TEMPERATURES VS. THETA LOCATION

138 (20000)
103 (15000)
69 (10000)
34.5 (5000)
0 (0)
− 34.5 (− 5000)
− 69 (− 10000)
− 103 (15000)
− 138 (− 20000)
− 172 (− 25000)
Hoop Stress, mPa (PSI)
Normal
537 (1000) H.S.
593 (1100) H.S.
649 (1200) H.S.
0
5
10
15
20
25
30
35
Theta Location, in Angular Degrees

FIGURE 6. HOOP STRESSES VS. THETA LOCATION

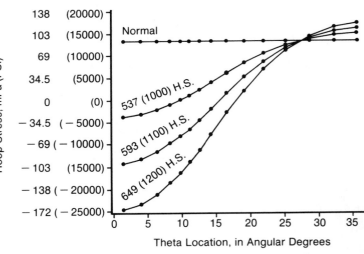

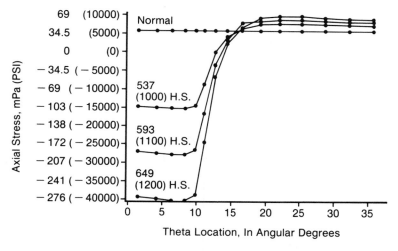

FIGURE 7. AXIAL STRESSES VS. THETA LOCATION

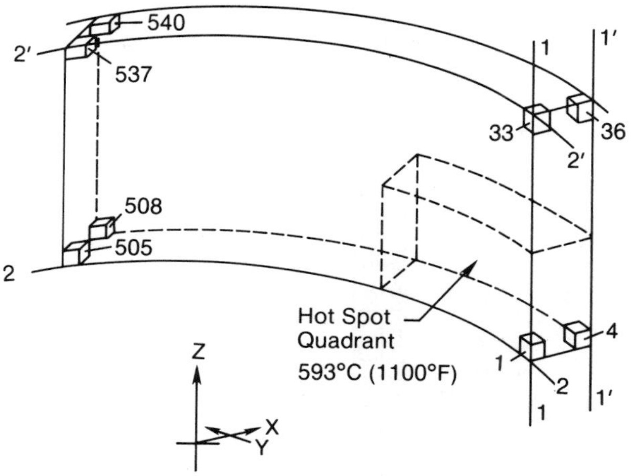

FIGURE 8. LOCATION OF ELEMENTS AND DISPLACEMENT LINES

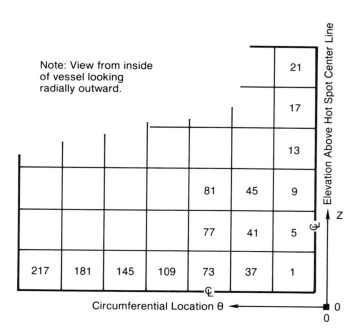

FIGURE 9. ELEMENT IDENTIFICATION AND LOCATIONS FOR USE WITH TABLES 5 AND 6

TABLE 1

Centroidal Stresses for Various Elements

Element 1

Time Before and After Hot Spot, Hr	Stress, MPa (psi) σx	Stress, MPa (psi) σy	Stress, MPa (psi) σz
<0 (Before Hot Spot)	-12.5 (-1815)	91.6 (13293)	39.3 (5697)
0	-14.6 (-2116)	-55.8 (-8093)	-137 (-19919)
10	-16.3 (-2368)	-43.8 (-6350)	-82.9 (-12025)
30	-15.8 (-2300)	-34.7 (-5041)	-64.8 (-9402)
100	-15.0 (-2181)	-24.6 (-3572)	-47.1 (-6831)
300	-14.5 (-2100)	-18.4 (-2663)	-34.8 (-5046)
1000	-14.1 (-2048)	-14.2 (-2060)	-27.4 (-3974)
3000	-13.9 (-2021)	-9.0 (-1306)	-26.1 (-3781)

Element 4

Time Before and After Hot Spot, Hr	Stress, MPa (psi) σx	Stress, MPa (psi) σy	Stress, MPa (psi) σz
<0 (Before Hot Spot)	-2.21 (-324)	85.7 (12432)	42.4 (6156)
0	-3.21 (-467)	39.6 (5740)	16.9 (2449)
10	-3.90 (-567)	46.7 (6779)	15.6 (2257)
30	-3.85 (-558)	50.1 (7269)	14.6 (2115)
100	-3.67 (-532)	52.9 (7680)	14.2 (2059)
300	-3.57 (-518)	54.0 (7839)	14.6 (2112)
1000	-3.15 (-457)	53.9 (7813)	16.2 (2348)
3000	-1.78 (-259)	52.0 (7544)	19.7 (2864)

Element 33

Time Before and After Hot Spot, Hr	σx	σy	σz
<0 (Before Hot Spot)	-12.5 (-1810)	94.3 (13676)	41.0 (5949)
0	-11.1 (-1617)	201 (29149)	40.1 (5819)
10	-10.7 (-1556)	193 (28023)	40.7 (5904)
30	-10.7 (-1547)	187 (27078)	41.0 (5949)
100	-11.0 (-1591)	178 (25840)	41.2 (5978)
300	-11.2 (-1633)	171 (24867)	41.3 (5996)
1000	-11.4 (-1648)	166 (24111)	41.5 (6018)
3000	-11.1 (-1611)	161 (23398)	41.7 (6054)

Element 36

Time Before and After Hot Spot, Hr	σx	σy	σz
<0 (Before Hot Spot)	-2.27 (-329)	87.2 (12650)	41.4 (6004)
0	-1.56 (-227)	137 (19949)	40.0 (5804)
10	-1.43 (-207)	133 (19363)	39.9 (5791)
30	-1.34 (-194)	131 (19067)	39.9 (5785)
100	-1.33 (-193)	129 (18680)	39.9 (5782)
300	-1.39 (-202)	126 (18280)	39.9 (5782)
1000	-1.43 (-207)	124 (17977)	39.9 (5779)
3000	-1.37 (-199)	124 (17959)	39.8 (5769)

Element 505

Time Before and After Hot Spot, Hr	σx	σy	σz
<0 (Before Hot Spot)	-12.5 (-1821)	91.7 (13302)	39.2 (5697)
0	-12.3 (-1785)	109 (15805)	52.6 (7630)
10	-12.3 (-1781)	109 (15755)	49.9 (7238)
30	-12.2 (-1778)	108 (15660)	48.3 (7004)
100	-12.2 (-1777)	107 (15524)	46.3 (6710)
300	-12.3 (-1781)	106 (15399)	44.5 (6457)
1000	-12.3 (-1785)	105 (15284)	43.2 (6260)
3000	-12.3 (-1784)	105 (15195)	41.9 (6084)

Element 508

Time Before and After Hot Spot, Hr	σx	σy	σz
<0 (Before Hot Spot)	-2.27 (-329)	85.8 (12441)	42.4 (6158)
0	-2.07 (-301)	97.4 (14128)	91.8 (13312)
10	-1.99 (-289)	97.8 (14186)	92.8 (13456)
30	-2.03 (-295)	97.9 (14194)	93.1 (13511)
100	-2.01 (-291)	97.6 (14152)	93.4 (13542)
300	-2.01 (-292)	97.2 (14094)	93.3 (13527)
1000	-2.03 (-295)	97.8 (14032)	93.4 (13552)
3000	-2.02 (-293)	95.9 (13917)	93.8 (13600)

TABLE 2

Residual Centroidal Stresses After Creep and Cooldown

Total Time of Hot Spot, Hr	Element 1 Stress, MPa (psi)			Element 4 Stress, MPa (psi)		
	σ_x	σ_y	σ_z	σ_x	σ_y	σ_z
Normal Operating Conditions, T=450°C (840°F), t<0	-12.51 (-1815)	91.6 (13290)	39.3 (5697)	-2.2 (-324)	85.7 (12432)	42.4 (6156)
Hot Spot Spike T=593°C (1100°F), t=0	-14.6 (-2116)	-55.8 (-8094)	-137 (-19919)	-3.2 (-467)	39.6 (5740)	16.9 (2449)
1000	0.917 (133)	58.6 (8497)	154 (22316)	-0.517 (-75)	21.2 (3072)	-1.42 (-206)
2000	1.0 (145)	62.6 (9079)	155 (22552)	0.013 (2)	20.3 (2948)	1.4 (200)
3000	1.06 (154)	66.2 (9601)	155 (22501)	0.38 (55)	19.6 (2847)	3.6 (516)

Total Time of Hot Spot, Hr	Element 33 Stress, MPa (psi)			Element 36 Stress, MPa (psi)		
	σ_x	σ_y	σ_z	σ_x	σ_y	σ_z
Normal Operating Conditions, T=450°C (840°F), t<0	-12.5 (-1810)	94.3 (13676)	41 (5949)	-2.3 (-329)	87.2 (12650)	41.4 (6004)
Hot Spot Spike T=593°C (1100°F), t=0	-11.1 (-1617)	201 (29149)	40.1 (5819)	-1.56 (-277)	137 (19949)	40 (5804)
1000	-0.51 (-75)	-45.8 (-6640)	-1.76 (-256)	0.11 (17)	-18.4 (-2674)	-0.26 (-38)
.2000	-0.57 (-83)	-49.4 (-7170)	-1.90 (-275)	0.13 (19)	-18.8 (-2722)	-0.32 (-47)
3000	-0.61 (-89)	-52 (-7536)	-1.97 (-286)	0.14 (20)	-18.4 (-2671)	-0.38 (-55)

Total Time of Hot Spot, Hr	Element 505 Stress, MPa (psi)			Element 508 Stress, MPa (psi)		
	σ_x	σ_y	σ_z	σ_x	σ_y	σ_z
Normal Operating Conditions, T=450°C (840°F), t<0	-12.5 (-1821)	91.7 (13302)	39.3 (5697)	-2.3 (-329)	85.8 (12441)	42.4 (6158)
Hot Spot Spike T=593°C (1100°F), t=0	-12.3 (-1785)	109 (15805)	52.6 (7630)	-2.1 (-301)	97.4 (14128)	91.8 (13312)
1000	-0.05 (-8)	-4.3 (-630)	-12.6 (-1834)	-0.05 (-1)	-0.84 (-122)	1.9 (278)
3000	-0.05 (-9)	-5.1 (-737)	-14.3 (-2067)	-0.05 (-9)	-2.0 (-289)	2.6 (382)

TABLE 3

Hot Displacements for Line 1-1 θ = 0 R = 1.68 m (66 in)

Radial Displacements, m (in)

Element Number/Time at Temperature	Normal	0 Hours	10	30	100	300	1000	3000
0	0.011 (0.420)	0.011 (0.447)	0.011 (0.448)	0.011 (0.449)	0.011 (0.452)	0.012 (0.455)	0.012 (0.457)	0.012 (0.461)
4	0.011 (0.420)	0.011 (0.446)	0.011 (0.448)	0.011 (0.449)	0.011 (0.451)	0.011 (0.454)	0.012 (0.457)	0.012 (0.460)
8	0.011 (0.420)	0.011 (0.445)	0.011 (0.446)	0.011 (0.447)	0.011 (0.450)	0.011 (0.452)	0.011 (0.454)	0.012 (0.457)
12	0.011 (0.420)	0.011 (0.444)	0.011 (0.445)	0.011 (0.446)	0.011 (0.448)	0.011 (0.450)	0.011 (0.453)	0.012 (0.455)
16	0.011 (0.421)	0.011 (0.446)	0.011 (0.447)	0.011 (0.447)	0.011 (0.449)	0.011 (0.451)	0.011 (0.452)	0.012 (0.455)
20	0.011 (0.421)	0.011 (0.445)	0.011 (0.445)	0.011 (0.445)	0.011 (0.446)	0.011 (0.448)	0.011 (0.449)	0.011 (0.451)
24	0.011 (0.421)	0.011 (0.441)	0.011 (0.441)	0.011 (0.441)	0.011 (0.442)	0.011 (0.443)	0.011 (0.444)	0.011 (0.446)
28	0.011 (0.421)	0.011 (0.436)	0.011 (0.436)	0.011 (0.436)	0.011 (0.436)	0.011 (0.437)	0.011 (0.438)	0.011 (0.439)
32	0.011 (0.421)	0.011 (0.430)	0.011 (0.430)	0.011 (0.430)	0.011 (0.430)	0.011 (0.431)	0.011 (0.431)	0.011 (0.432)
36	0.011 (0.421)	0.011 (0.424)	0.011 (0.424)	0.011 (0.424)	0.011 (0.424)	0.011 (0.424)	0.011 (0.424)	0.011 (0.425)

Hot Displacements for Line 1'-1' θ = 0 R = 1.94 m (76.25 in)

Element Number/Time at Temperature	Normal	0 Hours	10	30	100	300	1000	3000
0	0.012 (0.478)	0.013 (0.525)	0.013 (0.528)	0.013 (0.529)	0.013 (0.532)	0.014 (0.535)	0.014 (0.537)	0.014 (0.540)
4	0.012 (0.478)	0.013 (0.525)	0.013 (0.527)	0.013 (0.528)	0.013 (0.531)	0.013 (0.533)	0.014 (0.536)	0.014 (0.538)
8	0.012 (0.478)	0.013 (0.522)	0.013 (0.524)	0.013 (0.525)	0.013 (0.527)	0.013 (0.529)	0.013 (0.531)	0.014 (0.534)
12	0.012 (0.478)	0.013 (0.518)	0.013 (0.519)	0.013 (0.520)	0.013 (0.522)	0.013 (0.523)	0.013 (0.525)	0.013 (0.527)
16	0.012 (0.478)	0.013 (0.513)	0.013 (0.513)	0.013 (0.514)	0.013 (0.515)	0.013 (0.516)	0.013 (0.518)	0.013 (0.520)
20	0.012 (0.478)	0.013 (0.507)	0.013 (0.507)	0.013 (0.507)	0.013 (0.508)	0.013 (0.509)	0.013 (0.510)	0.013 (0.512)
24	0.012 (0.478)	0.013 (0.510)	0.013 (0.501)	0.013 (0.501)	0.013 (0.501)	0.013 (0.502)	0.013 (0.503)	0.013 (0.504)
28	0.012 (0.479)	0.012 (0.494)	0.012 (0.494)	0.012 (0.494)	0.012 (0.494)	0.012 (0.495)	0.012 (0.495)	0.012 (0.496)
32	0.012 (0.479)	0.012 (0.487)	0.012 (0.487)	0.012 (0.487)	0.012 (0.487)	0.012 (0.487)	0.012 (0.488)	0.012 (0.488)
36	0.012 (0.479)	0.012 (0.480)	0.012 (0.480)	0.012 (0.480)	0.012 (0.480)	0.012 (0.480)	0.012 (0.480)	0.012 (0.480)

TABLE 4

Hot Displacements for Line 2-2   Z = 0   R = 1.68 m (66 in)
Radial Displacements, m (in)

/Time at Temperature	Normal	0 Hours	10	30	100	300	1000	3000
0	0.011 (0.420)	0.011 (0.447)	0.011 (0.448)	0.011 (0.449)	0.011 (0.452)	0.012 (0.455)	0.012 (0.457)	0.012 (0.461)
2.8	0.011 (0.420)	0.011 (0.447)	0.011 (0.448)	0.011 (0.450)	0.011 (0.452)	0.012 (0.455)	0.012 (0.458)	0.012 (0.461)
5.2	0.011 (0.420)	0.011 (0.447)	0.011 (0.449)	0.011 (0.450)	0.011 (0.453)	0.012 (0.455)	0.012 (0.458)	0.012 (0.461)
7.4	0.011 (0.420)	0.011 (0.448)	0.011 (0.450)	0.011 (0.451)	0.011 (0.453)	0.012 (0.456)	0.012 (0.458)	0.012 (0.461)
9.1	0.011 (0.420)	0.011 (0.450)	0.011 (0.451)	0.011 (0.452)	0.012 (0.455)	0.012 (0.457)	0.012 (0.459)	0.012 (0.462)
10.4	0.011 (0.420)	0.011 (0.451)	0.011 (0.453)	0.011 (0.454)	0.012 (0.457)	0.012 (0.459)	0.012 (0.461)	0.012 (0.464)
11.7	0.011 (0.420)	0.011 (0.454)	0.012 (0.456)	0.012 (0.457)	0.012 (0.460)	0.012 (0.462)	0.012 (0.464)	0.012 (0.467)
13.4	0.011 (0.420)	0.012 (0.457)	0.012 (0.459)	0.012 (0.460)	0.012 (0.463)	0.012 (0.465)	0.012 (0.467)	0.012 (0.470)
15.2	0.011 (0.420)	0.012 (0.459)	0.012 (0.461)	0.012 (0.462)	0.012 (0.465)	0.012 (0.467)	0.012 (0.469)	0.012 (0.471)
17.8	0.011 (0.420)	0.012 (0.461)	0.012 (0.463)	0.012 (0.464)	0.012 (0.466)	0.012 (0.469)	0.012 (0.471)	0.012 (0.473)
20.4	0.011 (0.420)	0.012 (0.462)	0.012 (0.464)	0.012 (0.465)	0.012 (0.467)	0.012 (0.469)	0.012 (0.472)	0.012 (0.474)
23.4	0.011 (0.420)	0.012 (0.463)	0.012 (0.464)	0.012 (0.466)	0.012 (0.468)	0.012 (0.470)	0.012 (0.472)	0.012 (0.474)
26.9	0.011 (0.420)	0.012 (0.463)	0.012 (0.464)	0.012 (0.467)	0.012 (0.468)	0.012 (0.470)	0.012 (0.472)	0.012 (0.474)
30.4	0.011 (0.420)	0.012 (0.463)	0.012 (0.464)	0.012 (0.466)	0.012 (0.468)	0.012 (0.470)	0.012 (0.472)	0.012 (0.474)
33.9	0.011 (0.420)	0.012 (0.463)	0.012 (0.464)	0.012 (0.465)	0.012 (0.468)	0.012 (0.470)	0.012 (0.472)	0.012 (0.474)
37.4	0.011 (0.420)	0.012 (0.463)	0.012 (0.464)	0.012 (0.465)	0.012 (0.468)	0.012 (0.470)	0.012 (0.472)	0.012 (0.474)

TABLE 5

Secondary Creep Strains

Tabled by Component After 3000 Hours

All locations are element centroids
0.019 (.75) m (in) deep inside vessel wall.
See Figure 9 for element identification.

Element	Strain, m/m (in/in)		
No.	Radial	Hoop	Axial
1	0.00118	0.00032	-0.0015
5	0.00098	0.00024	-0.0012
9	0.0005	0.00013	-0.00063
37	0.0011	0.00037	-0.0015
41	0.00092	0.0003	-0.0012
45	0.00048	0.00019	-0.0007
73	0.00097	0.00044	-0.0014
77	0.00083	0.00039	-0.0012
81	0.00044	0.00029	-0.00073
13	-0.00006	0.00035	-0.00029
17	-0.00029	0.00059	-0.00031
21	-0.00033	0.00055	-0.00022
109	0.00084	0.00051	-0.0014
145	0.00068	0.00055	-0.0012
181	0.00031	0.00032	-0.0006
217	0.00002	0.00003	-0.00005

TABLE 6

Secondary Creep Strains

Tabled by Component After 3000 Hours

Locations are element centroids at
0.019 (0.75), 0.63 (2.5), 0.124 (4.875), and
0.210 (8.25) m (in) for elements 1, 2, 3, and 4
respectively (through-wall).

Through-Wall Elements Near Hot Spot Centerline

$\bar{\theta} = 1.4°$ Z = 0.05 (2) m (in)

Vessel Radius, m (in)	Element No.	Strain, m/m (in/in)		
		Radial	Hoop	Axial
1.69 (66.75)	1	0.00118	0.00032	-0.0015
1.74 (68.5)	2	0.00067	0.00043	-0.0011
1.80 (70.875)	3	0.0001	0.00051	-0.00061
1.89 (74.25)	4	-0.00051	0.00061	-0.0001

Locations are element centroids at
0.019 (0.75), 0.63 (2.5), 0.124 (4.875), and
0.210 (8.25) m (in) for elements 537, 538, 539, and 540
respectively (through-wall).

Through-Wall Elements Far from Hot Spot

$\bar{\theta} = 35.65°$ Z = 0.864 (34) m (in)

Vessel Radius, m (in)	Element No.	Strain, m/m (in/in)		
		Radial	Hoop	Axial
1.69 (66.75)	537	-0.000006	0	0.000006
1.74 (68.5)	538	-0.000007	0.000001	0.000005
1.80 (70.875)	539	-0.000007	0.000002	0.000004
1.89 (74.25)	540	-0.000007	0.000004	0.000002

ACCURACY AND PRECISION OF MECHANICAL TEST DATA GENERATED USING COMPUTERIZED TESTING SYSTEMS

Mitchell R. Jones
Instron Corporation
Canton, Massachusetts, USA

ABSTRACT

Several sources and types of errors are easily overlooked when using computerized mechanical testing systems to measure material properties such as modulus, offset yield, plastic strain, or fracture toughness. Accuracy and precision errors can accumulate within the computerized test system and cause artificially high property variances, misinterpretation of results, or erroneous output. These issues may be illustrated by tracing the flow of information from the test specimen to the computer output. Transducers, signal conditioners, digital converters, and computer processors are reviewed to establish the importance of accuracy, resolution, and speed. Data acquisition (including signal conditioning and conversion) and analysis are studied to determine the significance of various errors. Algorithms for averaging of incoming data and for steepest slope determination are discussed. Statistical analysis of properties derived from least squares line regressions establishes the statistical component of the uncertainty in the measured property. It also sets the minimum amount of data needed in a database for accurate reconstruction of the specimen's load-strain curve. Examples of the effects of these factors in typical metals tension, low cycle fatigue, elastic fracture toughness (KIC), and elastic-plastic fracture toughness (JIC) tests are given where appropriate. While these error sources are insidious, careful error analysis reduces most errors to an acceptable level.

TO ASSURE ACCURATE RESULTS, many factors must be examined when measuring material properties with computerized testing systems. Various sources of uncertainty are illustrated by following the flow of information from the test specimen to the interpretation of the computer output (see Fig. 1). The effects of analog and digital electronics, software, and statistics on the accuracy of material property measurements should be understood. Under the best circumstances, the only significant errors in the final results are directly attributable to the accuracy of the analog load and strain (or displacement) measurement systems. However, less than ideal conditions can arise in computerized systems from several sources: excessive analog signal noise, insufficient analog to digital converter resolution, data offset due to sequential sampling of transducer channels, incorrect slope determination, or insufficient quantity of data. Thus, configuring a computerized materials testing system consists of choosing hardware, software, and data storage capabilities such that overall accuracy is maintained without unnecessary cost and complexity.

Although beyond the scope of this work, attention to assumptions of uniform stress, environment, and specimen structure are necessary for accurate results, and an acceptable level of accuracy in the results must be established.

TRANSDUCERS AND SIGNAL CONDITIONERS

When an equipment manufacturer publishes an accuracy specification for a load, strain, or displacement measurement system, this may or may not include the combined accuracy of the transducer, signal conditioner, and readout device (e.g., analog recorder). The first step, therefore, when establishing the accuracy of data generated using computerized instrumentation is to subtract, if necessary, the accuracy of the readout device from the specified overall measurement system accuracy.

Transducers and their conditioners have many specifications which may effect overall system accuracy. Under most circumstances, linearity, repeatability, extensometer gage length accuracy, and signal-to-noise ratio are primary considerations.

NOISE - Unlike the other measurement system specifications, noise is influenced by the laboratory environment. Minimizing it is a complex task because there are many sources in a materials testing system. Noise observed at the output of the transducer signal conditioner is primarily due to three phenomena. External to the testing system are electromagnetic interference and differences in electrical potential at the grounding point of the transducer body and the conditioner chassis. Inherent in the system is thermal noise [1].

In practice, noise from external sources is usually minimized by a combination of the following (see Fig. 2): making the signal travelling from the transducer to the conditioner as strong as possible; shielding the cable that carries the signal by surrounding all conductors with a braided conducting jacket; twisting the conductors; grounding the shield at one end of the cable; and grounding the body of the transducer to the conditioner chassis. Strengthening the signal is accomplished by maximizing the transducer excitation (balanced against the undesirable effects of excessive excitation) and preamplifying the signal at the transducer. Grounding the cable shield at one end minimizes the penetration of electric fields into the area between the twisted wires carrying the transducer signal [2]. Twisting the wires reduces interelectrode capacitance and minimizes the area where electromagnetic interference can penetrate. Equalizing the ground potential of the transducer body and of the conditioner chassis is accomplished by providing an electrical path between them although perfect equality is difficult to achieve.

Thermal noise is caused by random electron motion in the strain gages and in the signal conditioner amplifiers. It is an inherent attribute of most types of conditioners and strain gage transducers. It can't be eliminated and sets a lower bound on the conditioner output noise level.

OTHER SPECIFICATIONS - In addition to the transducer and conditioner specifications already mentioned, hysteresis, drift, temperature stability, frequency response (band width), and physical transducer resonance may become important in some tests. Resolution, sensitivity, zero return, and span stability are generally not significant but should not be ignored in demanding applications.

SIGNAL CONVERSION

Conversion of the signal from analog to digital form is an important step. Two things determine the minimum number of bits in the analog to digital (A/D) converter. First, random noise in the analog signal should be resolvable in the converted digital signal.

Second, the accuracy of the analog signal should not be substantially diminished by the A/D resolution. The factors determining conversion rate should also be considered, along with the possibility of data offset.

RESOLUTION - An upper bound on the number of bits in the A/D converter is set by establishing the smallest practical noise level among the channels to be converted. An A/D converter with the correct number of bits will report noise in the last bit. Thus, even though the last bit will toggle constantly due to noise, changes in the signal of the same magnitudes as the noise can be detected by appropriate computer software averaging. For example, if the signal contains no less than X volts of peak to peak noise then the number of volts per bit should equal X:

$$\frac{\text{Voltage Range}}{2^n} = \text{Volts per bit} \qquad (1)$$

$$n = \ln (\text{Voltage Range/Noise})/\ln 2 \qquad (2)$$

where n is the number of bits.

If fewer bits are used, then the ability of the system to resolve minute changes in load or strain is compromised. For example, noise can be resolved in a system where the voltage range is -10 to +10 volts and noise is $\pm$.005V if the A/D converter has $\ln(20V/.005V)/\ln2$ or 12 bits.

ACCURACY - The A/D converter's resolution should be small compared to the accuracy of the signal being converted. Specifically, the number of volts per bit should be less than the measurement system's accuracy expressed in volts. In this way no substantial uncertainty is added to the signal during A/D conversion.

Furthermore, a computerized system should be an improvement over analog readouts. Thus, the A/D converter should resolve at least as well as a skilled individual reads an analog plot with a ruler.

CONVERSION RATE - The application of the testing equipment determines the desired conversion rate. Conversion rate is the number of load and strain data pairs or load, strain, and displacement triplets per unit time. It is fixed by the ratio of the desired number of data points per test (or cycle) and the duration of the test (or cycle). The section on "statistical analysis of least squares lines" quantitatively explains how many data points per test are really necessary. The overall conversion rate, i.e., the rate of data influx available to the computer application software, is generally determined by the A/D converter and the speed of the computer processor in conjunction with the data data acquisition software. With current technology, affordable and sufficiently fast A/D converters make the computer processor and data acquisition software most often the limiting factors.

DATA OFFSET - This term refers to the error in load, strain, or displacement which can occur when these channels are sampled sequentially. For example, a finite amount of time elapses between samples of load and strain so it is possible that the reported strain is not the same as the value which occurred at the moment of load sampling. Data offset has no effect when measuring slope (for measurment of modulus or crack length) because the data is linear and all points are offset the same amount.

If load is sampled before strain, and the rate of strain increase (or decrease) is constant, then the effect of offset is to shift all data points by the same amount. Under these conditions, therefore, data offset is not an issue when measuring offset yield stress, plastic strain, KIC, or energy absorbed by the specimen. This statement also applies if load is changing at a constant rate and strain is sampled before load. However, measurement of total strain at offset yield stress or maximum cycle stress during low cycle fatigue requires awareness of offset effects. The magnitude of the offset error is determined by the test duration (or cycle duration) and the total time required for the A/D converter to switch channels and make a single conversion. Note that this time interval is typically much smaller than the overall conversion rate including computer software.

The magnitude of the data offset can be seen in the following example. The desired measurement is total strain at offset yield stress of steel. If the time to reach yield is 1 second and the A/D channel switch and sample time is 50E–06 seconds, then the error in strain is the rate of straining times the A/D switch and sample time, or ((.003 m/m)/1 sec) (50 E–06 sec) which is .15E–06 m/m. The relative error is (.15 E–06 m/m)/(.003 m/m) which equals .005%. In this example, the offset error is significantly small.

COMPUTERS

PROCESSORS – Precision and speed are the significant computer processor parameters. Precision refers to the number of bits used in computer memory to represent a number. Because some numbers cannot be represented exactly in a finite number of binary digits, these numbers will cause precision errors when manipulated repeatedly. Errors caused by insufficient precision are manifested as inaccurate results and may be difficult to detect. Fortunately, materials testing doesn't require calculations where the same number is manipulated over and over.

As mentioned earlier, the speed of the processor and data acquisition software usually limit the rate of data input. As faster computers become affordable, it will be possible to measure properties at higher rates of deformation.

SOFTWARE – Two types of algorithms may affect the accuracy of the computer output. These are raw data averaging and automatic steepest slope measurement.

Raw data averaging refers to application of a moving average to load and strain/ displacement data before it is analyzed. If the software allows the user to select the number of points to be used in the moving average, the user is given access to a flexible software "filter" which tends to reduce data scatter. The software author should provide a default number of points to be averaged which suits the software application. However, it is possible that this default will smooth out desired specimen response characteristics.

Algorithms which automatically find the steepest portion of a specimen response curve have the advantage of requiring no user input and perform this task more consistently than humans can. With these algorithms, modulus, offset yield, or KIC determination can be automated to a higher degree. However, their usefulness is debatable because different algorithms may produce different results for the same curve. Reproducible, accurate results can only be obtained by verifying the algorithm's reliability under demanding conditions such as gradually decreasing slope. A plot of time vs. the difference between the specimen response curve and the steepest-slope line quickly reveals the accuracy of the algorithm.

STATISTICAL DATA ANALYSIS

Statistical analysis of digitized stress-strain, load-crack opening displacement (COD), and J-crack advance curves can be revealing. Such an analysis quantifies the statistical component of the uncertainty (as opposed to the accuracy and precision components) in measured properties derived from the slope or y–intercept of a least squares line, such as modulus, offset yield, plastic strain, crack length, KIC, and JIC. Statistical analysis also quantitatively establishes whether a sufficient number of data pairs were used to generate the regression line or if the data were too scattered, such that the minimum amount of data necessary for accurate results are collected. This becomes especially important if the raw data are to be archived or data based.

Statistical uncertainty in calculated slope and y-intercept data arises because the quantity of data is finite and the quality of data is imperfect. Consequently, calculated slope and intercept values represent an estimate of the "true" values. The true values would be those for an analysis of a very large number of data pairs. The uncertainty associated with the estimate is quantified by calculating "confidence limits" for a specified confidence level (e.g., 95%). The true value lies within the confidence limits with the specified confidence. Thus, confidence limits represent the statistical uncertainty and are calculated as follows [3]:

$$\text{for slope } m,$$
$$\text{slope confidence limits} = m \pm ts_m \qquad (3)$$
$$\text{and for intercept } b,$$
$$\text{intercept confidence limits} = b \pm ts_b \qquad (4)$$

where t is the "t statistic" for the desired level of confidence (the value of the "t statistic" is obtained from appropriate statistical tables), and s_m is the estimated slope variability:

$$s_m = s_e / \sqrt{SS_x} \qquad (5)$$

$$\text{with } s_e = \sqrt{S/(n-2)} \qquad (6)$$

$$S = SS_y - m^2(SS_x) \qquad (7)$$

$$SS_y = \Sigma y_i^2 - n\overline{y}^2 \qquad (8)$$

$$\text{and } \quad SS_x = \Sigma x_i^2 - n\overline{x}^2 \qquad (9)$$

with $\overline{x}, \overline{y}$ = averages.

s_b is the estimated intercept variability:

$$s_b = (s_e(\sqrt{\Sigma x_i^2})) / \sqrt{n(SS_x)} \qquad (10)$$

Calculation of slope confidence limits directly produces the statistical uncertainty of modulus measurment. If the crack length of a crack propagation specimen is measured by the compliance technique, plugging the upper and lower compliance confidence limits into the crack length-specimen compliance equation produces the crack length confidence limits [4].

To calculate the statistical uncertainty of offset yield stress or KIC, in cases where KIC is the intersection of the load-COD curve and the 5% compliance offset line, one simply assumes that the origin of the load vs. deflection curve is the true origin. After taking the strain or slope offset, a deviation in the slope of the offset line equal to the statistical uncertainty of the regression slope is taken. The minimum and maximum stress or load can now be determined empirically, defining the statistical uncertainty in offset yield stress or KIC.

When measuring plastic strain, a procedure similar to that for offset yield will reveal the statistical uncertainty of the measurement (see Fig. 3). The point on the load-strain curve that corresponds to the total strain for which the plastic strain component is desired must be located. From this point, the modulus minimum and maximum slopes (as defined by the modulus confidence limits) are used to extrapolate back to zero load. The confidence limits for plastic strain are the intersections of the modulus limits and the strain axis.

Analysis of the statistical uncertainty of JIC, whether determined by the single specimen or multiple specimen techniques, can be slightly more complex. The statistical uncertainty in JIC is conservatively estimated by finding the y-intercept confidence limits of the J-crack advance regression line. However, JIC represents the intercept of the regression line and the theoretical blunting line, not the y-axis. A better estimate of the statistical uncertainty in JIC is calculated by first performing a coordinate transformation on the qualified J-crack advance data. The points must be rotated about the origin so that the blunting line is coincident with the y-axis. The confidence of the y-intercept (which is the blunting line intercept) can now be calculated with Eq. (4).

By calculating statistical confidence limits, the question of sufficient quantity and quality of data needed to represent the specimen response curve is answered. If the statistical uncertainty causes the overall uncertainty of the result to rise to an undesirable level, then remedial action is needed. Statistical uncertainty can usually be reduced by gathering more data or reducing the scatter. More data can be gathered by increasing the rate of data acquisition or increasing the test duration. Data scatter is diminished by applying a moving average to incoming data, electronically filtering the analog signal, reducing electrical noise (if possible), or choosing more appropriate transducers such that the signal-to-noise ratio is improved.

SUMMARY

Under optimum conditions, the accuracy of data generated using computerized testing systems can be calculated by applying a conventional error analysis with the load and strain/displacement measurement system accuracy specifications. However, additional uncertainty may arise from a number of easily unnoticed sources. These are insufficient analog signal-to-noise ratio, inadequate A/D converter resolution, data offset due to sequential sampling of transducer channels, incorrect rate of data sampling, inappropriate software algorithms for smoothing of the specimen response curve or automatic steepest slope determination, and insufficient data for high statistical confidence. Application of statistical analysis to properties derived from least-squares regression lines quantifies the statistical component of uncertainty, while conventional error analysis yields the accuracy and precision components. The quantity of data needed to accurately represent a specimen load-strain or load-COD curve in a database is determined with statistical analysis. With careful error analysis, properties of acceptable accuracy can be measured.

REFERENCES

1. Unpublished research, D.H. Howling, "Noise in Servohydraulic Control Systems and the Effects and Merits of Load Ranging," Instron Corporation Applications Laboratory Report #113 (1981), Instron Corporation, 100 Royall Street, Canton, Massachusetts 02021

2. H.W. Oh, Noise Reduction Techniques in Electronic Systems, p. 198, John Wiley and Sons, New York, New York (1976)

3. F. Mostellar, et al, Beginning Statistics with Data Analysis, P. 360, Addison-Wesley Publishing Company, Inc., Reading, Massachusetts (1983)

4. D.A. Jablonski, "Computerized Single-Specimen J-R Curve Determination for Compact Tension and Three-Point Bend Specimens," in Automated Test Methods for Fracture and Fatigue Crack Growth, ASTM STP 877, W.H. Cullen, R.W. Landgraf, L.R. Kaisand, and J.H. Underwood, Eds., American Society for Testing and Materials, Philadelphia, Pennsylvania (1985)

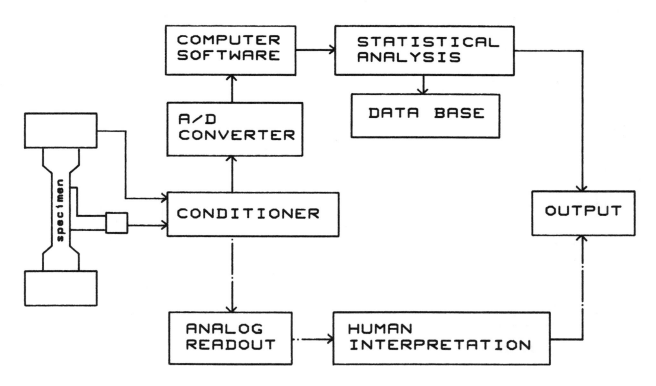

Fig. 1. Possible Paths from Specimen to Output

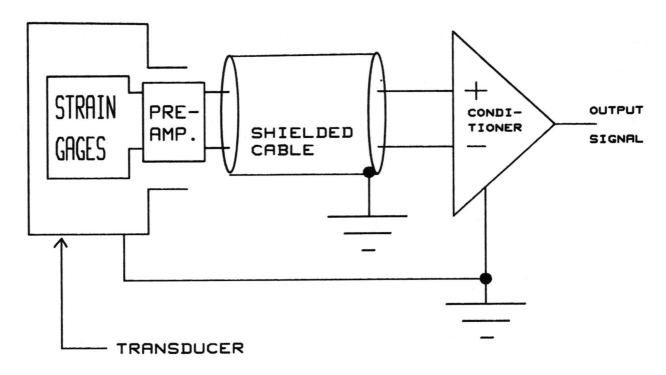

Fig 2.   Schematic of an Analog Measurement System

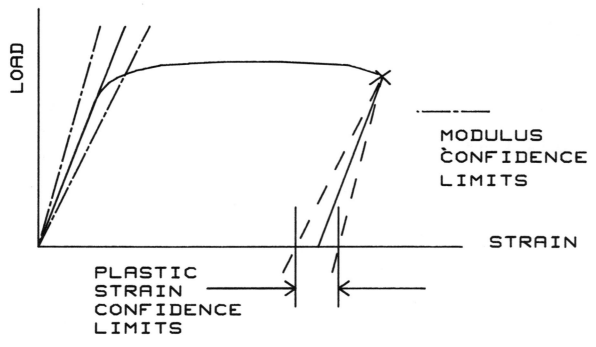

Fig. 3.   Statistical Uncertainty of Plastic Strain

PROBABILISTIC SAFETY ANALYSIS
OF A PRESSURE VESSEL

J. Watanabe
R&D Division
The Japan Steel Works, Ltd.
Tokyo, Japan

T. Iwadate
Research Laboratory
The Japan Steel Works, Ltd.
Muroran, Japan

Reactor vessels for high temperature-pressure hydrogen units may be deteriorated during long years service. A safety factor K_{IC}/K_I needs to be maintained for a flawed reactor. Both K_{IC} and K_I have probablistic nature, and reliability of obtained K_{IC}/K_I is important for management to decide repair or replace. A practical example reactor made of 2 1/4Cr-1Mo steel was investigated, and it was found that currently available ultrasonic testing technique has sufficient accuracy.

INTRODUCTION

Prediction of the remaining life of plant equipment is one of today's major tasks for engineers in various industries[1,2]. They collect relevant data such as fabrication records, operational history, past inspection records, material damage data and so forth, and the data are studied employing available most advanced techniques. However, the state of the art in this area has still unsolved problems, and extensive efforts are being paid.

Managements sometimes face a very critical and difficult problem on the repair or replace decision, because it requires not only technical but also economical and legal considerations. Engineers must prepare technical data which help management decide repair or replace, but the data are not always 100 percent accurate. If the management is taught accuracy of the technical data, it will help him balance technical, economical and legal factors.

In this paper, a probabilistic safety analysis of a high temperature-pressure hydrogen service reactor will be presented. Thick wall reactors, commonly made of 2 1/4Cr-1Mo steel, is the heart of hydro-processing units in today's modern petroleum refinery[3]. Since evolution of the modern hydrocracking process about 30 years ago, many heavy wall reactors are being used, and they are getting older. As is common in many welded constructions, the reactors have some discontinuities, and 2 1/4Cr-1Mo steel may lose its toughness during service due to temper embrittlement[3]. Past experiences showed some failures in the reactors (Table 1)[4]. Safety of the reactor has critical importance, and it requires periodical examination of the integrity of the reactors. When a reactor explodes, relevant engineers and managements will be involved in technical, economical and legal issues.

SAFETY FACTOR K_{IC}/K_I

An unstable fracture of a reactor from a crack is most dangerous one, and in order to

---
*J. WATANABE is Director, R & D Division, The Japan Steel Works, Ltd., Tokyo, Japan.

*T. IWADATE is Senior Engineer, Research Laboratory, The Japan Steel Works, Ltd., Muroran, Japan.

Table 1   Diagnosis of Cr-Mo pressure vessel failures and remedies[4].

Symptom	Cause	Remedy
Fissure Decarburization	Hydrogen attack	Selection of proper grade steel
Crack at Type 347 stainless steel weld corner	Hydrogen embrittlement of Type 347 stainless steel weld	Optimization of weld configuration, chemical composition & welding sequence
Disbonding of stainless steel weld overlay	Hydrogen-assisted cracking	Selection of proper welding parameters and flux
Brittle fracture	Temper embrittlement of 2 1/4Cr-1Mo steel	Control of chemical composition
Crack at attachment weld joint	Creep cracking of 1 1/4Cr-1/2Mo HAZ	Selection of proper grade steel, Smooth weld configuration

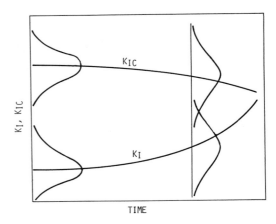

Fig. 1-Schematic showing probabilistic nature of K_{IC} and K_I.

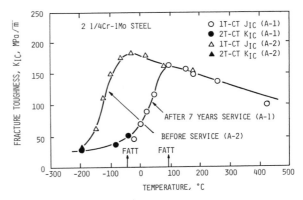

Fig. 2-Fracture toughness versus temperature relationship of a 2 1/4Cr-1Mo steel[5].

prevent it some level of safety factor K_{IC}/K_I (K_{IC} : fracture toughness, K_I : stress intensity factor at crack tip) must be maintained. For a stable fracture, some level of safety factor σ_B/σ or σ_{YS}/σ (σ_B : ultimate tensile strength, σ_{YS} : yield strength, σ: working stress) is kept.

As demonstrated by Table 1, presence of cracks or crack-like defects are postulated in a reactor after many years service. Consequently not only safety factors σ_B/σ and σ_{YS}/σ but also K_{IC}/K_I must be examined. Fracture toughness K_{IC} has probabilistic nature, and it may deteriorate during service. Cracks may grow during service with some probabilistic nature, and the stress intensity factor K_I at crack tip will also be probabilistic as shown in Fig. 1.

RELIABILITY OF K_{IC}/K_I ASSESSMENT

For a practical example case reliability of the K_{IC}/K_I assessment, especially the effect of nondestructive testing will be examined.

1. Steel

A 2 1/4Cr-1Mo steel with J-factor 310, yield strength 421MPa, Charpy-V upper shelf energy 147Joule, service hours 80,000 hours is used. 2 1/4Cr-1Mo steels may be temper embrittled during survice at approximately 400°C (Fig. 2), and potential to the

embrittlement is manifested by J-factor = (Si + Mn)(P + Sn) x 10^4 [4]. A steel with J-factor of 310 has relatively high tendency to the embrittlement.

A method to predict fracture toughness of 2 1/4Cr-1Mo steels with various confidence limits has been proposed by T. Iwadate et al[5], and the fracture toughness versus temperature relationship shown in Fig. 3 was obtained for the steel investigated.

2. Defect and K_I

Fig. 4 shows an example of a crack in the reactor. Inside of 2 1/4Cr-1Mo steel shell is weld overlaied by stainless steel, and on its surface internal attachments are welded on. Cracks may initiate from the attachment

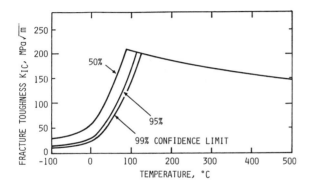

Fig. 3-Predicted fracture toughness K versus temperature relationship after 80,000 hours service[5].

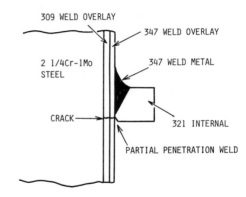

Fig. 4-Schematic drawing of prior existing crack.

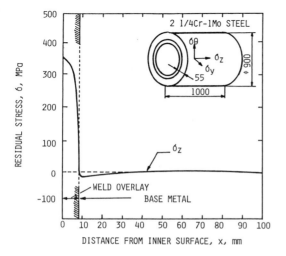

Fig. 5-Residual stress distribution measured by Sacks' method[5].

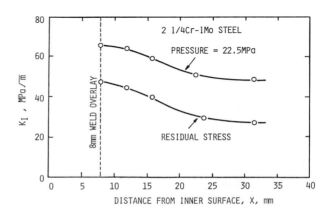

Fig. 6-Stress intensity factors due to the residual stress and internal pressure[5].

corners and grow into 2 1/4Cr-1Mo base metal during service. Residual stress exists in base metal/weld overlay neighbourhood, and a measured result is shown in Fig. 5[5]. For through cracks with various depths, the stress intensity factor K_I was obtained for the residual stress alone and the residual stress and an internal pressure of 22.5MPa (Fig. 6)[5].

3. Reliability of ultrasonic testing and its effect on K_{IC}/K_I assessment

Ultrasonic testing from the outside surface of the reactor shell is used for detection and sizing of the defect shown in Fig. 4. The

ultrasonic testing technique currently available has a detection capability as shown in Fig. 7 and Eqs. 1 and 2. Sizing capability is demonstrated by Fig. 8 and Eq. 3.

Detection

$$a < 7mm \quad D(a) = 0 \quad \ldots\ldots\ldots\ldots\ldots \quad (1)$$

$$a \geq 7mm \quad D(a) = (1 - 7/a) \quad \ldots\ldots\ldots \quad (2)$$

Sizing

$$f(A/a) = \frac{1}{\sqrt{2\pi}\,\sigma}\exp\left[-\frac{1}{2}\left(\frac{A-a}{\sigma}\right)^2\right]\ldots \quad (3)$$

$$\sigma = 4.5mm$$

Assume

Metal temperature   40°C

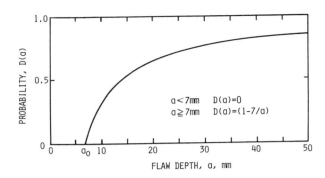

Fig. 7-Detection capability of ultrasonic
testing.

Fracture toughness K_{IC}	Fig. 3
Crack depth	20 mm
K_I distribution	Fig. 6.

Effect of ultrasonic test capability on K_{IC}/K_I
was obtained, where K_I value was calculated
using various detection and sizing capabilities shown in Figs. 7 and 8 (Fig. 9).

The results showed relatively small effect
of the ultrasonic sizing capability, although
the better the sizing accuracy the higher the
K_{IC}/K_I. In this particular case, the safety
factor K_{IC}/K_I is very low. If the standard
deviation σ is 4.5mm, which is currently
available by a skilled ultrasonic operator, it
is impossible to maintain a safety factor 1.4
with 99% confidence limit. K_{IC}/K_I of 1.4
corresponds to a safety factor of 2 stress-
wise, and it may be considered as the minimum
required safety factor for safe operation.

If it is judged that a confidence limit of
50% is enough, the reactor is still safe.
However, if a confidence limit of as high as
99% is required, the reactor must be repaired
or replaced. The other solution is found to
increase the ultrasonic sizing capability, but
it is not immediately available.

It must be emphasized that the sizing
capability discussed above is only applicable
to the defects which are known to exist.
In other words, probability of detection of

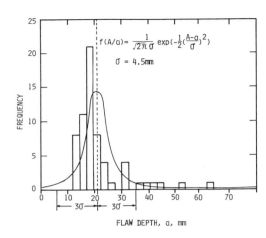

Fig. 8-Sizing capability of ultrasonic
testing.

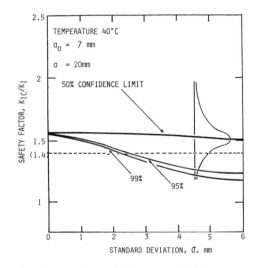

Fig. 9-Ultrasonic sizing capability versus
K_{IC}/K_I.

wholely unknown defects is not included in
this argument. Inspectors are advised in
advance that cracks may exist in the area
shown in Fig. 4. Such advanced information
can be obtained through past experiences, and
when past experiences are not available,
inspection programs must be very careful and
elaborate.

SUMMARY

Heavy wall reactor vessels made of 2 1/4Cr-
1Mo steel have been used for many years in
high temperature-pressure hydrogen units.

During service, the reactor steel may be temper embrittled, and crack may initiate and grow into base metal.

Safety of a flawed reactor is assessed by K_{IC}/K_I. For a reactor made of 2 1/4Cr-1Mo steel with relatively high potential to temper embrittlement, K_{IC} versus metal temperature relationship was obtained for various confidence limits. The stress intensity factor K_I was calculated considering internal pressure and residual stress in the neighborhood of weld overlay. K_{IC}/K_I of different confidence limits was obtained for various ultrasonic testing reliability, and the results showed relatively small effect of ultrasonic sizing capability. In this case, advanced knowledge of crack existence is assumed, otherwise careful and elaborate inspection programs are required.

REFERENCES

1. C.E. Jaske : ASME PVP-vol.98-1, 1985, P.3.
2. British Standard Institution : PD 6510, 1983 "A Review of the Present State of the Art of Assessing Remanent Life of Pressure Vessels and Pressurized Systems Designed for High Temperature Service", 1983.
3. W.E. Erwin, J.G. Kerr : WRC Bulletin 275, 1982.
4. J. Watanabe : Proc. Fifth Internatinal Conf. on Pressure Vessel Technology, ASME, 1985, vol.III, P.51.
5. T. Iwadate, J. Watanabe and Y. Tanaka : Trans. ASME, J. of Pressure Vessel Technology, Aug. 1985, vol.107, P.230.

ESTIMATION OF THE PROBABILITY OF FLAW DETECTION

Fujia Lin, Yushan Huang
Xi'an China

ABSTRACT

A statistical method for determining the flaw detection probability has been developed. By using the formula presented in this paper, the lower confidence limit of the probability of flaw detection with the given confidence level for a sample of arbitrary size can be calculated simply from the common F-distribution. The formula presented is exact as well as simple. As an example, the flaw detection probability data with 95% confidence is obtained from the results of inspecting corner flaws at holes in 50 specimens.

THE RELIABILITY PREDICTION and the damage tolerance analysis of aircraft structures based on the principles of fracture mechanics require the information of the detective probability of potential flaw in a component. The flaw detection probability is influenced by many factors. If the conditions of nondestructive testing inspection process, the material of specimens and the human factors are fixed the flaw detection probability is the function of the flaw size. In general, the larger the flaw, the higher the detective probability.

The method to estimate the flaw detection probability by experiment has been discussed by some papers [1-3]. However, in those papers the formulae to calculate the lower confidence limit of the flaw detection probability are rather complicated and approximate. The formula presented in this paper is more exact and simpler.

FLAW DETECTION PROBABILIT

Suppose there are n cracks with size a in a sample. Let sn be the detected number out of the n cracks by a given inspection procedure. Then the point estimate of the crack detection probability, p, is $\hat{p}=\dfrac{s_n}{n}$. Because $\hat{p}$ is a random variable, it is not exactly equal to the true detective probability, p. It may be larger or smaller than p. Therfore, the lower confidence limit p_L for the detective probability p must be considered to be a reliable estimation of the detective probability.

Because there are only two possible results of the inspection for each crack in an experiment, the crack is either detected or undetected, the random variable sn is a binomial distribution. For a given confidence level $1-\alpha$, the p_L is determined by the equation

$$P\{S \geqslant S_n\} = \sum_{i=S_n}^{n} C_n^i p_L^i (1-p_L)^{n-i} = \alpha \qquad (1)$$

If n is equal to or smaller than 25, the table

of binomial cumulative distribution can be used to find the lower confidence limit p_L. When n is greater than 25, the p_L may be approximatly ditermined by Poisson distribution or normal distribution [1-3]. For example, if n=45, s_n=43, 1-α=0.95, when the approximate method is used, the approximate value of p_L is 0.894. On the other hand, the exact value must be 0.867. The difference between the approximate value and the exact can not be ignored. In fact, the exact value of p_L can be calculated by the formula

$$p_L = \frac{f_2}{f_2 + f_1 x} \qquad (2)$$

where $f_1 = 2(n - s_n + 1)$
$f_2 = 2s_n$
x can be obtained from the F-distribution table corresponding to the degrees of freedom f_1 and f_2 for given α. That is

$$P\{F > x\} = \alpha \qquad (3)$$

PROOF of formula (2). Let

$$I(y, \gamma, \theta) = \frac{\Gamma(\gamma + \theta)}{\Gamma(\gamma)\Gamma(\theta)} \int_0^y t^{\gamma-1}(1-t)^{\theta-1} dt \qquad (4)$$

is the Beta distribution function with parameters γ and θ.

The probability density function of F-distribution corresponting to the degrees of freedom f_1 and f_2 is

$$p(x, f_1, f_2) = \frac{\Gamma\left(\frac{f_1+f_2}{2}\right)}{\Gamma\left(\frac{f_1}{2}\right)\Gamma\left(\frac{f_2}{2}\right)} f_1^{\frac{f_1}{2}} f_2^{\frac{f_2}{2}}$$

$$\cdot \frac{x^{\frac{f_1}{2}-1}}{(f_2 + f_1 x)^{\frac{f_1+f_2}{2}}} \quad (x > 0) \qquad (5)$$

Introducing the variable

$$Y = \frac{f_1 F}{f_2 + f_1 F} \qquad (6)$$

and using the law to transform the distributions, we obtain from (5)

$$P\{Y \leq y\} = I(y, \frac{f_1}{2}, \frac{f_2}{2}) \qquad (7)$$

That is the Beta distribution corresponding to the parameters $\gamma = \frac{f_1}{2}$, $\theta = \frac{f_2}{2}$.
Then the distribution function of F can be easily expressed by Beta distribution function, i.e

$$P\{F > x\} = I(\frac{f_2}{f_2 + f_1 x}, \frac{f_2}{2}, \frac{f_1}{2}) \qquad (8)$$

On the other hand, by means of integration by parts we obtain

$$\sum_{i=S_n}^{n} C_n^i p_L^i (1 - p_L)^{n-i} = n C_{n-1}^{S_n-1} \int_0^{p_L} t^{S_n-1}(1-t)^{n-S_n} dt$$

$$= I(p_L, S_n, n - s_n + 1) \qquad (9)$$

Comparing formula (8) with (9), we obtain

$f_1 = 2(n - s_n + 1)$
$f_2 = 2s_n$
$p_L = \frac{f_2}{f_2 + f_1 x}$

EXAMPLE

A set of 50 machined plates (approximately 30mm x 300mm x 5mm) with a hole with diameter 6mm is prepared. These specimens are made of 45[#] steel. About half of the specimens contain fatigue crack ranging in size from 0.4mm to 2mm. All of the cracks are corner cracks at the hole. The magnetic particle inspections have been conducted repectedly and independently for all of the specimens. The experimental data and the results of statistical analysis are shown in table 1.

Table 1—Statistical Analysis of Inspection Data
$(1-\alpha = 0.95)$

crack size (mm)	n	s_n	$\hat{p}$	f_1	f_2	x	p_L
0.41–0.60	513	158	0.308	712	316	1.18	0.273
0.61–0.80	264	152	0.576	226	304	1.23	0.522
0.81–1.00	111	95	0.856	34	190	1.49	0.789
1.01–1.20	157	141	0.898	34	282	1.47	0.849
1.21–1.40	191	175	0.916	34	350	1.46	0.876
1.41–1.60	225	213	0.947	26	426	1.52	0.915
1.61–1.80	176	169	0.960	16	338	1.68	0.926
1.81–2.00	65	65	1.000	2	130	3.07	0.953

REFERENCES

1. P. F. Packman, H. S. Pearson, J. S. Ouens and G. Young "Definition of Fatigue Cracks through Nondestructive Testing", J. Materials, Vol.4, No 3, (1969)

2. P. F. Packman, J. K. Malpani, F. M. Wells, "Probability of Flaw Detection for Use in Fracture Control Plans", Strength and Structure of Solid Materials, Tokyo (1976)

3. J. K. Malpuni, "Reliability of Flaw Detection by Nondestructive Inspection and Its Application to Fracture Mechanics Design and Life Analysis", D. Phil. Thesis, University of Vanderbilt (1976)

4. A. Hald, "Statistical Theory with Engineering Application", New York (1955)

ULTRASONIC TESTING IMPROVEMENTS WITH AN ADVANCED ULTRASONIC INSPECTION SYSTEM

K.F. Schmidt, J.R. Quinn
NES/Dynacon
Concord, California, USA

The Ultrasonic Data Recording and Processing System (UDRPS) has been in field use for eighteen months. A number of applications in various industries have been successfully developed. The improvements in testing reliability, repeatability and economic consequence obtainable with such an advanced technique in particular applications are discussed.

The NES/Dynacon Ultrasonic Data Recording and Processing System (UDRPS) has seen wide spread application in several industries in the last one and a half years. In the nuclear power industry, applications have been made to PWR & BWR reactor vessels, PWR noz zles, BWR pipe welds, PWR steam generator shells and steam line piping. Other utility applications have been made in turbine disc bores and keyways, disc rims in the blade attachment area and to boresonic inspection. Applications in high volume casting inspections have been developed. Several uses in airframe composite structures have beenmade.

The following is a brief review of several of these applications. A short treatise on the principals of operation may prove helpful at this point.

The UDRPS theory of operation is based on a rejection of amplitude discrimina tion as a reliable means of detection of indications. UDRPS utilizes a signal to noise ratio detection critia combined with a pattern recognition approach. Specifically, the pattern recognized in the amplitude-time domain is the target motion line exhibited by all valid reflectors within the field of view of a moving transducer. The UDRPS algorithm performs an integration of the signal plus noise received from each target motion line and a contin uously updated local average noise estimate. The ratio of the total signal plus noise to local noise is then used to discriminate targets on a threshold basis. The threshold is set by direct observation of the part to be inspected or to calibration blocks were are alleged to be identical. This algorithm is known as automatic target detection or ATD. Target location is derived by deconvolving the transducer beam spread from the measured target motion line; generating a centroidoftheline.

The features of the UDRPS design which have contributed to its rapid and seemingly widespread application appear tobe:

- Real time ATD and centroiding providing an overall volumetric index to all indications within the inspected volume on an amplitude independent basis.

- Recording of all video or RF ultrasonic data over the inspected volume without gates or thresholds.

- High speed scanning with multiple transducers permitting rapid inspection of large volumes of materials.

- Continous scanning without interruption for data record manipulation.

The list of applications is too long to be completely discussed here; some of the more interesting follow.

Nuclear Reactor Vessel Inspection

Two specific objectives have been addressed with UDRPS in PWR vessel inspection. These are the detection and through wall sizing of underclad or near surface cracks and cracks buried deep within the full penetration weld.

An extensive evaluation of UDRPS capability in this application has been performed by the EPRI NDE Center in Charlotte, N.C. The evaluation was performed between April 1983 and July 1985.

For underclad crack detection, six specially constructed plates with various clad weld overlays were inspected in double blind tests. The plates are a realistic simulation of PWR vessel walls. The flaws were a combination of fatigue cracks and saw cuts. As the detection process is dominated by clad noise, the saw cuts are, in this application, a technically justifiable and economic substitutefor fatiguecracks.

Eighty-nine underclad defects (3 to 25 mm deep) were presented to UDRPS. All data were collected in a predetermined, raster scan pattern using either 60 or 70 degree dual L-wave transducers. All data analysis was made only from the digitally recorded data andimages.

All defects were detected by this process. Binary statistics indicated that a probability of detection (POD) of 95% was demonstrated with a 95% confidence level by these results. Sizing results were dependent upon the transducer used. The 70 degree dual L-wave probe is well known to be optimal for detec tion but inaccurate in sizing. A 45 degree dual L-wave probe is far better for sizing. The root mean square error for the former was 0.16 inches while in the latter case the error was 0.12 inches.

In the evaluation of buried flaws, three heavy section blocks containing 34 defects were inspected. The clad was a combination of automatically and manually applied material. Each block was inspected by a combination of transducers. The best performance was obtained with 1 MHz, 45 degree, shear wave and 2 MHz, 45 degree,L-wave transducers. All flaws were detected although not necessarily with the same transducer.

Binary statistics indicated a POD of 89% at a 95% confidence level. This is the best result obtainable from this limited sample set. A sizing error of 0.14 inches was obtained with a 45 degree 1 MHz S-wavetransducer.

Several other manual inspection teams using various procedures also tested onthese sameblocks.

All teams performed at substantially lower POD's and sizing accuracies. The measured performance of the UDRPS approach appears to duplicate in field ready state the best published research laboratory techniques. A typical example of the UDRPS data from this evaluation may be informative. In Figure 1 a heavy section test block is illustrated. The block contains eight intentionally included fatigue cracks and one near surface, long slag line. The intended defects are well known as to location and size. In the figure, eight spatially segregated groups of targets are shown. The long slag line is also evident. The eight valid indications are immediately evident.

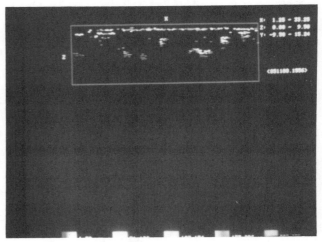

Figure 1

Valid target groups demonstrate detection of eight intended fatigue cracks, underclad slag line and calibration holes (left edge). DAC correction was not used.

Boiling Water Reactor Pipe Weld Inspection

UDRPS has been applied to pipe inspec tion over the last two years. The system has been integrated with two types of pipe scanners. A number of reactor piping inspections have been performed. The problem of austenitic pipe scanning is one of geometric resolution and crack-geometry discrim ination. Detection of circumferentia lly oriented cracks is not difficult. Resolution is a problem due to the close proximity of weld root and pos sible counter bores. Crack-geometry discrimination is a problem due to the possibility of a crack very close to a root and possible cracks on top of counterbores.

In the initial round of NRC recog- nized qualification tests, performed in 1984, six of seven candidates past the exam. In January 1985, four candidates quali fied to size stress corrosion cracks in BWR piping. The recent requalification tests have proven much more difficult. The inital group of UDRPS operators all failed. A second round of candidates have now proven successful.

An example of pipe data is shown in Figure 2. The top half of the figure contains three perpendicular views of all the valid target locations. In the lower half of the image, the locations of all valid traveling targets with a signal to noise ratio above a threshold and a spatial cor- relation value above another thres- hold value are shown. This image is a feature map in the parlance cur- rently in vogue. The weld root and crack are clearly evident.

Steam Turbine Inspection

Utility power generation turbines tend to be large, expensive machines with very long lead replacement times. Inspection of the rotor shaft bore, pressed-on disk bore and anti- rotation keyway lock, and the blade attachment structure are key to the avoidance of in-service failures and

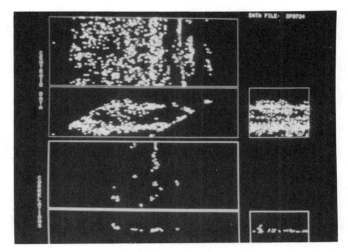

Figure 2

Pipe Image, Amplitude map of valid targets above in three perpendicular views. Amplitude independent correlated feature map below with the weld root as a vertical, diffuse line and cracks are short targets to the left and right (below).

potentially long forced outages. Inspections by normal practice do not provide an actual record of the unprocessed data; rather, what is provided are conclus ions based on a technican's on the spot analysis of any significant part of the data set. The deligence of such on the spot data review may be less than desired by the owner of the turbine.

UDRPS in combination with a variety of custom built manipulators affords a complete archival record of the exact condition of the critical volumes of the inspectable turbine structure. All indications are accu- rately detected and recorded.

The potential benefit is to establish a historical record of the exact condi tion of the turbine and thereby make possible a detailed tracking of the state of the structure throughout its service life.

Carbon Graphite Inspection

Airframes are now in manufacture which contain a substantial percent- age of composite material. Complete composite airframes are now under development. Composite inspections usually involve large areas but thin sections. The capabilities of UDRPS to do reactor vessel inspections are easily transfer red to this problem.

A typical composite test sample is shown in Figure 3A. The sample is about 0.5 inches thick and is composed of carbon graphite. The sample con tains mica discs layed at various levels in the material as test objects, simulating delaminations. Unintended flaws are also present.

The sample was scanned at 5 MHz on a fine grid. A reflector plate was placed beneath the sample. Since all data is digitized on a range of just before the top surface of the sample down to the reflector plate, any depth of C-scan plane may be reconstructed from the recorded data base. Thus, at a given depth, flaws at that depth appears as bright reflectors, while those at a shallower depths cast shadows. The reflector plate image is a shadowgram of the discontinuities within the sample (Figure 3B).

Alunimum Die Casting Inspection

An aluminmun diecast part manufactured in thousands per day has been demonstrated as leading to premature compo nent field failures at an undesireable rate. These failures appear to result from fatigue crack growth between subsurface pores in regions of high stress. A conservative accept-reject criteria would be to reject all parts having a single isolated pore within the critical subsurface volume. A more economic and fracture mechanically correct criteria would be to reject upon volumetric, areal or linear density specifications. Since pore shape and orientation affects the amplitude returned from a pore, a real-time, amplitude independent target detection criteria is well suited to thisapplication.

A series of 0.013 inch diameter side drilled holes were drilled into a part at 0.10, 0.15 and 0.20 inches below the as cast surface. For economic reasons, the inspection should occur on the part as cast, before any surface machining is invested in the component. The surface through which the sound must enter has 0.20 inch deep linear serations in a spline shaft-like pattern.

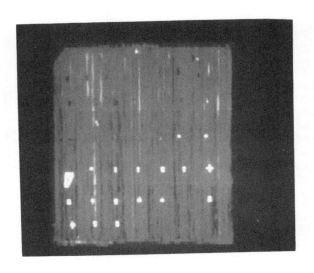

Figure 3

A: Pulse echo image of composite plate delaminations in material at this level are white, shadows (black) are from delaminations at a higher level.

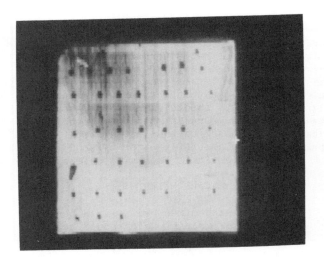

Figure 3

B: Reflector plate image of composite plate.

The image shown in Figure 4 was taken
with 5 MHz, 45° shear waves. The six
side drilled holes were not intended
to be of equal length. A natural
flaw is visible at top, right center.
The lower half of the figure was
obtained from a calculation of
spatial correlation of the targets
without regard to amplitude. the
coding represents the degree of
correlation (red, low; blue- violet,
high). A threshold placed on the
degree of correlation display imposes
the appropriate porosity density
criteria.

The technique has the potential of
greatly decreasing the field failure
rates of components while avoiding
rejection of serviceable components.

The cases cited above begin to illustrate the potential of an advanced
ultrasonic inspection technology. In
each case, substantial cost benefit
to the affected industry can be
demonstrated.   Significant
performance advantages have also been
demonstrated in some applications.

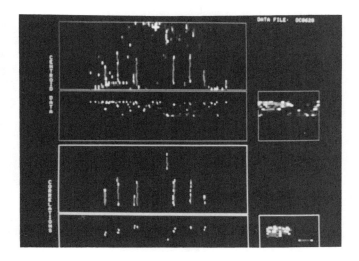

Figure 4

Valid target image of cast
aluminum part, upper half.
Amplitude independent correlation
feature map of all extended
of all extended targets.

THE RELIABILITY ANALYSIS
OF AIRCRAFT STRUCTURES

Depei Zhu
Northwestern Polytechnical University
Xian, China

Abstract

This paper presents a set of fundamental equations for reliability analysis of aircraft structures. For a single-critical-point structure, it considers the following factors: the static strength of structure, initial crack, the initiation and propagation and unstability of fatigue crack, the residual strength of structure, the statistical distribution of load, the periods of overhaul, accident damage, the communication of damage among fleets etc. Based on this mathematical model, the influence of the various factors to reliability can be analyzed quantitively, and the various criteria for fatigue design of aircraft structure can be evaluated from the aspect of reliability.

THE ESSENTIAL PURPOSE OF AEROSTRUCTURAL DESIGN is to ensure the reliability of structures. This is achieved usually using the various design criteria, e.g. static criterion; safe-life, fail-safe, or damage tolerance in fatigue design etc. The aeronautical history has shown that the previous criteria are neither sufficient nor necessary for aircraft reliability. A series of accidents due to structural failure has prompted the development of the concept of reliability analysis for aircraft structures. In later 1950's a reliability criterion was suggested[1], then in early 1970's the computative methods were presented[2,3]. There are a number of papers (see Reference 4) to show the importance of reliability analysis.

The reliability analysis of aircraft structures is a special case of the system reliability problem. It is easy to obtain the relation between structural reliability and componental reliability if the components are statistically independent[5,6]. Considering the correlation among the components and the redistribution of stresses due to the local damage, the problem becomes very complicated.

Even for single-critical-point structure the mathematical model for reliability analysis is far more complicated than the common statistical decay models. The instantaneous risk rate depends upon current crack length and the statistical distribution of external load. Moreover, the distribution of crack length is a result of previous load history and crack detection, and the probability of detecting cracks is dicided by inspection procedure and is affected by the communication of any accidents.

This paper aims to consider the above factors as many as possible in presenting the fundamental equations of reliability analysis for single-critical-point structure. In computation it is necessary to know the statistical distribution of each factor and to simplify them. It may be pointed out that some situations discussed in References 2,3, 4,7 may be summarised into simplified cases of the mathematical models of this paper. The formulas are given in more detail in Reference 8.

NOMENCLATURE

All in nomenclature are dimensionless.

$d(n_s)$	Probability rate of detecting cracks at n_s
$F(\ )$	Distribution function
$H(n_s)$	Probability of failure up to n_s
l	The value of L
L	Crack length
n	Life
n_i	Life at i-th inspection (overhaul)
n_s	Current life
N	Number of load applications
$\bar{N}$	Reference life
$p(\ )$	Probability density function (PDF)
$p_D(l)$	Probability rate of detecting crack length l
P	Probability
P_{Di}	Probability of detecting cracks at i-th inspection
$\bar{P}_{Di}$	$= 1 - P_{Di}$
$P_D(l)$	Probability of detecting crack length l
$r(n_s)$	Risk of failure at n_s
$r^*(n_s)$	Probability rate of failure at n_s
$R(n_s)$	Survival probability up to n_s
R_s	Residual strength (and its value)

s	The value of S
S	External load
S_m	The external load in stable level flight
x	The value of ξ
y	The value of η
z	The value of ζ
α	The percent of elements containing initial crack in parent population
α_K	The percent of unexpected crack at life n_K
δ	The symbol denoting ignorable term
ζ	Life of crack initiation
η	Initial crack length
μ_0	Nominal static strength
ξ	Residual strength parameter
$\varphi(l)$	The function of nominal residual strength to crack length l

PREPARATIVE FORMULAS

Denote the propagation function of fatigue crack as

$$L = L(n, \eta, \zeta) \tag{1}$$

where

$$\eta \sim F_\eta(y), \quad (y > 0) \tag{2}$$

$$\zeta \sim F_\zeta(z), \quad (z > 0, \ y = 0) \tag{3}$$

Dealing with the new structures without initial crack, we have

$$F_\eta(0) = P\{\eta \leqslant 0\} = 1 - \alpha \tag{4}$$

and denote

$$l(n, 0, z) = g(n, z) \tag{5}$$

$$g(n, z) = 0 \quad (n \leqslant z) \tag{6}$$

For the structures with initial crack,

$$l(0, y, z) = y \tag{7}$$

Suppose the initial crack propagats as

$$l(n, y, z) = g(n + n_0, z_0)$$
$$= h(n, y) \quad (y > 0) \tag{8}$$

where z_0 is the nominal life of crack initiation and n_0 satisfies

$$y = g(n_0, z_0) \tag{9}$$

Denote the residual strength as

$$R_8 = R_8(\xi, L) \tag{10}$$

where

$$\xi \sim F_\xi(x), \quad (x > 0) \tag{11}$$

and simply suppose

$$R_8(x, l) = x\varphi(l)\mu_0 \tag{12}$$

$$\varphi(0) = 1 \tag{13}$$

Denote the exceeding probability of external load as

$$F_8(s) = P\{S > s\} \tag{14}$$

and necessarily

$$F_8(s_m) = 1 \tag{15}$$

Suppose the PDF are $p_1(x)$, $p_{23}(y, z)$ respectively. $p_{23}(y, z)$ is of Dirac mode due to its discrete change at $y = 0$ and $z = z_0$, and its integral is in Lebesgue's sense, If we divide the parent po-

pulation into two subsets with and without initial crack, the conditional PDF are

$$p_2(y) = \frac{\dfrac{dF_\eta(y)}{dy}}{\alpha}, \quad (y > 0, \ z = z_0) \tag{16}$$

$$p_8(z) = \frac{\dfrac{dF_\zeta(z)}{dz}}{1 - \alpha}, \quad (z > 0, \ y = 0) \tag{17}$$

respectively, they are continuous functions, and the integrals are Riemann's.

THE CASE WITHOUT INSPECTION

All aircraft keep in service before failure happening. The survival probability of an element after $\triangle n \times \widetilde{N}$ load applications is

$$(1 - F_8(R_8))^{\triangle N} = (1 - F_8(R_8))^{\triangle n \times \widetilde{N}} \tag{18}$$

When $\triangle n \to 0$ it can be written formally as

$$e^{-F_8(R_8)\triangle n \times \widetilde{N}} \tag{19}$$

For finite $\triangle n$, we should use

$$F_8^*(R_8) = ln\frac{1}{1 - F_8(R_8)} \tag{20}$$

instead of $F_8(R_8)$. When $R_8 \to S_m$, it is concluded from eqs. (15), (20) that

$$e^{-F_8^*(R_8)} \to 0 \tag{21}$$

which denotes failure is a certain event. However, the error of life in using eq. (19) instead of eq. (20) is always ignorable.

The survival probability or probability in service is

$$R(n_8) = \iiint_0^\infty e^{-\int_0^{n_8} F(n, x, y, z)dn} p_1(x)p_{23}(y, z)dxdydz \tag{22}$$

where $F(n, \xi, \eta, \zeta)$ is obtained from $F_8(R_8)$ and eqs. (10), (1) and can be written as

$$F_0(n, x, z) = F(n, x, o, z) \quad (y = 0) \tag{23}$$

$$F_a(n, x, y) = F(n, x, y, z_0) \quad (y > 0) \tag{24}$$

For structures without crack, from eqs. (12), (13)

$$F_0(n, x, z) = F_8(x\mu_0) \times \widetilde{N} \quad (n \leqslant z) \tag{25}$$

On the other hand, for a very long crack, from eq. (21)

$$e^{-\int_0^{n_8} F(n, x, y, z)dn} = 0 \quad (R_8 \leqslant S_m) \tag{26}$$

From eqs. (10), (6), (8)

$$R_8(x, g(n, z)) \leqslant S_m \quad (y = 0) \tag{27}$$

$$R_8(x, h(n, y)) \leqslant S_m \quad (y > 0) \tag{28}$$

322

They result in

$$n_0(x,z) \leqslant n \qquad (y=0) \qquad (29)$$

$$n_a(x,y) \leqslant n \qquad (y>0) \qquad (30)$$

so that eq. (22) turns into

$$R(n_s) = (1-\alpha) \iint_{n_0(x,z)=n_s} e^{-\int_0^{n_s} F_0(n,x,z)dn} p_1(x)p_3(z)dxdz$$

$$+ \alpha \iint_{n_a(x,y)=n_s} e^{-\int_0^{n_s} F_a(n,x,y)dn} p_1(x)p_2(y)dxdy \qquad (31)$$

or

$$R(n_s) = (1-\alpha) \iint_0 e^{-\int_0^{n_s} F_0(n,x,z)dn} p_1(x)p_3(z)dxdz$$

$$+ \alpha \iint_0 e^{-\int_0^{n_s} F_a(n,x,y)dn} p_1(x)p_2(y)dxdy - \delta R(n_s) \qquad (32)$$

Other probabilities are

$$H(n_s) = 1 - R(n_s) \qquad (33)$$

$$r^*(n_s) = \frac{dH(n_s)}{dn_s} = -\frac{dR(n_s)}{dn_s} \qquad (34)$$

$$r(n_s) = r^*(n_s)/R(n_s) \qquad (35)$$

ROUTINE INSPECTION

We mathematically turn the frequent inspections into continuous inspection. Suppose the interval of inspections is $\triangle n_1$, the condition for transform is

$$e^{-p_D(l)\triangle n_1} = 1 - P_D(l) \qquad (36)$$

or

$$p_D(l) = \frac{1}{\triangle n_1} \ln \frac{1}{1-P_D(l)} \qquad (37)$$

where

$$p_D(l) = p_D(g(n,z)) \qquad (y=0) \qquad (38)$$

$$p_D(l) = p_D(n(n,y)) \qquad (y>0) \qquad (39)$$

For the case with both continuous inspection and intervallic inspections (overhauls),

$$R(n_s) = (1-\alpha) \iint_{n_0(x,z)=n_s} e^{-\int_0^{n_s}(F_0(n,x,z)+p_D(g(n,z)))dn} \times$$

$$\times \prod_{i=1}^{K} \bar{P}_{Di}(g(n_i,z))p_1(x)p_3(z)dxdz$$

$$+ \alpha \iint_{n_a(x,y)=n_s} e^{-\int_0^{n_s}(F_a(n,x,y)+p_D(h(n,y)))dn} \times$$

$$\times \prod_{i=1}^{K} \bar{P}_{Di}(h(n_i,y))p_1(x)p_2(y)dxdy$$

$$(n_K < n_s < n_{K+1}) \qquad (40)$$

$$r^*(n_s) = (1-\alpha) \iint_{n_0(x,z)=n_s} K_0(n_s,x,z) \times$$

$$\times \prod_{i=1}^{K} \bar{P}_{Di}(g(n_i,z))p_1(x)p_3(z)dxdz$$

$$+ \alpha \iint_{n_a(x,y)=n_s} K_a(n_s,x,y) \times$$

$$\times \prod_{i=1}^{K} \bar{P}_{Di}(h(n_i,y))p_1(x)p_2(y)dxdy + \delta r^*(n_s)$$

$$(n_K < n_s < n_{K+1}) \qquad (41)$$

where

$$K_0(n_s,x,z) = F_0(n_s,x,z)e^{-\int_0^{n_s}(F_0(n,x,z)+p_D(g((n,z)))dn} \qquad (42)$$

$$K_a(n_s,x,y) = F_a(n_s,x,y)e^{-\int_0^{n_s}(F_a(n,x,y)+p_D(h(n,y)))dn} \qquad (43)$$

$$H(n_s) = \sum_{j=1}^{K-1} H_j + \triangle H_K(n_s) + \delta H(n_s) \qquad (n_K < n_s < n_{K+1}) \qquad (44)$$

where

$$H_j = \int_{n_j}^{n_{j+1}} r^*(n_s)dn_s \qquad (45)$$

$$\triangle H_K(n_s) = \int_{n_K}^{n_s} r^*(n_s)dn_s \qquad (46)$$

UNEXPECTED DAMAGE IN SERVICE

Suppose the probability of unexpected damage at n_u is α_u, when $n_s > n_u$ the probability in service is

$$(1-\alpha_u)R(n_s) + \alpha_u R(n_u) \iint_{n_a(x,y)=n_s-n_u} K(x,y)dxdy \qquad (47)$$

where $K(x,y)$ is

$$e^{-\int_0^{n_s-n_u}(F_a(n-n_u,x,y)+p_D(h(n-n_u,y)))dn} \times$$

$$\times \prod_{i=j}^{K} \bar{P}_{Di}(h(n_j,y))p_1(x)p_2(y)$$

$$(n_s > n_u, n_K < n_s < n_{K+1}, n_{j-1} < n_u < n_j) \qquad (48)$$

$R(n_s)$ is obtained from eq. (40); $p_2(y)$ differs from the original one.

COMMUNICATION OF DAMAGE

The case after any damage is denoted by subscript 2 to be distinguished from the original case denoted by subscript 1, the probability rate of failure is

323

$$[R_1(n_S)]^{m-1}r_1^*(n_S) + (m-1)\int_0^{n_S}[R_1(n)]^{m-2}[r_1^*(n)+d_1(n)] \times$$

$$\times[(1-\alpha)\iint_{n_0(x,z)=n_S} \bar{F}_{02}(n_S,x,z)e^{-\int_0^n(F_{01}(n,x,z)+p_{D1}(g(n,z)))dn} \times e^{-\int_n^{n_S}(F_{02}(n,x,z)+p_{D2}(g(n,z)))dn} \times$$

$$\times\prod_{i=1}^{K_n}\bar{P}_{D1i}(g(n_{1i},z)) \times \prod_{j=1}^{O_n}\bar{P}_{02j}(g(n_{2j},z))p_1(x)p_3(z)dxdz +$$

$$+\alpha\iint_{n_a(x,y)=n_S} \bar{F}_{a2}(n_S,x,y)e^{-\int_0^n(F_{a1}(n,x,y)+p_{D1}(h(n,y)))dn} \times e^{-\int_n^{n_S}F_{a2}(n,x,y)+p_{D2}(h(n,y)))dn} \times$$

$$\times\prod_{i=1}^{K_n}\bar{P}_{D1i}(h(n_{1i},y)) \times \prod_{j=1}^{O_n}\bar{P}_{D2j}(h(n_{2j},y))p_1(x)p_2(y)dxdy]dn$$

$$(n_{1Kn}<n<n_{1(Kn+1)}, \quad n_{2j}=n+\triangle n_{2j}, \quad \triangle n_{20n}<n_S-n_u<\triangle n_{2(On+1)}) \tag{49}$$

ACKNOWLEDGE

Sincere thanks go to my director Professor Yushan Huang and to Lecturer Fujia Lin for reading the manuscript and for helpful suggestions.

REFERENCES

1. Ferrari, R. M., Milligan, I. S., Rice, M. R., and Weston, M. R., in Proceedings, ICAF-AGARD Symposium on Full Scale Fatigue Testing of Aircraft Structures, Amsterdam, June 1959, Pergamon Press, London, 1961, pp. 413-426.
2. Eggwertz, S., Investigation of Fatigue Life and Residual Strength of Wing Panel Reliability, ASTM STP 511, 1972, pp. 75-105.
3. Payne, A. O., A Reliability Approach to the Fatigue of Structures, ASTM STP 511, 1972, pp. 106-155.
4. Zhengfei Tian, Fujia Lin and Yushan Huang, The Reliability Analysis of Wing Beam under the Influence of Several Factors (in Chinese), NPU SHJ 8501, 1985.
5. Eggwertz, S. and Lindsjo, G., Analysis of the Probability of Collapse of a Fail-Safe Aircraft Structure Consisting of Parallel Elements, FFA Rep. 102, 1965.
6. Heller, R. A. and Donat, R. C., Fatigue Failure of a Redundant Structure, ASTM STP 404, 1966, pp. 55-66.
7. Fujia Lin and Yushan Huang, The Reliability Analysis of the Main Members in Wings Used in a Pair (in Chinese), ACTA Aeronautica et Astronautica Sinica, Vol.4 No.3, 1983.
8. Depei Zhu, The Mathematical Model for Reliability Analysis of Aircraft Structures, NPU SHJ 8551, 1985.

HOW MUCH SAFETY? WHO DECIDES?

C.O. Smith
Rose-Hulman Institute of Technology
Terre Haute, Indiana, USA

ABSTRACT

All hardware and equipment has some probability of causing injury to the user. Control of, and responsibility for, this probability lies with the designer. Two major questions are critical: (1) What is the magnitude of probability of injury and who determines this? (2) What is an acceptable magnitude of probability of injury and who determines this? The designer must devote much attention to answering these questions since he is ultimately responsible. This paper discusses assessment of (1) danger, and (2) the level of acceptability. Suggestions are made to assist designers.

PRODUCTS LIABILITY LITIGATION [1] in the United States (which also applies to all manufacuters selling in the United States) places great significance on the question of safety of a given product, whether it is (1) a consumer item potentially affecting large numbers of people or (2) an item intended for industrial use such as a power press potentially affecting relatively few people. Any manufactured item, no matter how properly and carefully designed and manufactured, has some probability of causing injury to the user. This probability may range from very large to very small. Englightened self-interest, apart from any legal considerations, would suggest reducing the probability of injury to an acceptable level. What, however, is an acceptable level? How does a designer deal with the problem?

HAZARD, RISK AND DANGER

The words "hazard," "risk," and "danger" are often used interchangeably but they are clearly different concepts in personal injury litigation (Philo and Rine [2]).

HAZARD - A hazard is a condition, or changing set of circumstances, which presents an injury potential. Hazards created by products may be:
    jaws of a power press,
    a helmet with insufficient energy-
        absorbing system,
    a vehicle with inadequate cornering
        capacity,
    a fork lift truck without overhead
        guard or load back rest,
    an unguarded power saw without an
        antikickback device,
    a toxic chemical.
or any one of innumerable others.

RISK - Risk is exposure to injury (or loss). In evaluating personal injury potential, risk is the probability of injury. Risk is affected by proximity, exposure, attention arresters, noise, light, experience and intelligence of the user, etc. While the point seems obvious, too many are willing to categorize risk without knowing the amount of exposure. An elevator with a ten person capacity in a busy office building, in comparison with an elevator with a five person capacity in a quiet apartment house presents, not twice, but hundreds of times the risk.

DANGER - Danger is the unreasonable or unacceptable combination of hazard and risk. [It is commonly held that any risk of serious injury or death is unreasonable or unacceptable if reasonable accident prevention methods could eliminate it. A 100% risk of injury can be reasonable and acceptable if the injury is minimal and the risk is recognized by the user.]

UNREASONABLY DANGEROUS

The American Law Institute [3] says "unreasonably dangerous" means that: "The article sold must be dangerous to an extent beyond that which would be contemplated by the ordinary consumer who purchases it, with the ordinary knowledge common to the community as to its characteristics. Good whiskey is not unreasonably dangerous merely because it will make some people drunk, and is especially dangerous to alcoholics; but bad whiskey, containing a dangerous amount of fusel oil, is unreasonably dangerous."

The American Law Institute [3] also says, relative to unavoidably unsafe products, that:

"There are some products which, in the present state of human knowledge, are quite incapable of being made safe for their intended and ordinary use ... Such a product, properly prepared, and accompanied by proper directions and warnings, is not defective, nor is it unreasonably dangerous."

The "unreasonably" in "unreasonably dangerous" is the word that permits courts to weigh risks (in the sense of exposure to injury or loss) against benefits. Snyder [4] has said: "...not all products that might pose a danger to the consumer will be considered legally defective, and in determining whether a product has a condition not contemplated by the consumer which presents an unreasonable danger to him, a court must look to many factors. Among these are:

(1) The seriousness and likelihood of the danger created
(2) The obviousness of the danger
(3) The social value of the product-frequently expressed as its usefulness and social desirability
(4) Whether the danger can be eliminated without severely decreasing the social value of the product or increasing its cost
(5) The existence of safety standards in the industry and whether the product conforms to such standards."

IMPLIED QUESTIONS FOR DESIGNERS

The above definitions may be quite acceptable for legal purposes. For designers, who desire as much quantification as possible, these definitions are little more than guidelines and require designers to answer two questions:
(1) (a) What is the danger (combination of hazard and risk)?
    (b) Who determines this?
(2) (a) Is the danger acceptable (or reasonable)?
    (b) Who determines this?
Designers have the responsibility for answering these questions before their products are sold and used by customers. This responsibility must be taken very seriously.

ASSESSING DANGER

As Lowrance [5] suggests, measuring danger, i.e., the combination of hazard and risk (probability and severity of injury), is

an empirical, scientific activity.

It follows that designers are better qualified by education and experience than most people to measure danger. Presumably designers will use organized approaches to cope with the complexity. A design-audit checklist, suggested by Corley [6], can be very helpful.

One obvious place for assessing danger is the design review process. Design review is a most valuable aid for the designer, but is not a substitute for adequate design and engineering. A formal design review board, or panel, specific to the individual product, should be composed of: (1) a variety of design engineers (mechanical, electrical, human factors, reliability, safety, packaging, etc.); (2) representatives from management, sales, insurance, legal, and finance departments; (3) a products liability attorney; (4) representatives from vendors and users, if possible. These individuals should not only be expert but practical. They should be able to evaluate and constructively criticize so the designer can make further analyses and investigations, as appropriate. Finding proper answers to the right questions can substantially reduce danger and potential products liability litigation or eliminate them.

When the design work is completed, including the work of a design review panel, how does one know if it has been effective? Using a product safety program evaluation checklist, such as suggested by Bartels [7], may provide an answer.

ASSESSING ACCEPTABLE DANGER

As Lowrance [5] suggests, judging the acceptability of danger is a normative, political activity.

Assessing danger is not a simple matter. Assessing the acceptability of danger is far more complex. As Lowrance [5] says: "By employing the word 'acceptable' it emphasizes that safety decisions are relativistic and judgmental. It immediately elicits the crucial questions, 'Acceptable in whose view?' and 'Acceptable in what terms?' and 'Acceptable for whom?' Further, it avoids all implication that safety is an intrinsic, absolute, measurable property of things." These questions are crucial. They are also extremely difficult to answer.

FOCUS OF ASSESSMENT TASK - In assessing acceptable danger, one major task is determining the distribution of danger, benefits, and costs. This is both a political issue and an empirical matter. It involves questions such as:
(1) Who will actually be paying the costs?
(2) Will those who benefit be those who pay?
(3) Will those endangered be those who benefit?
Answers to these questions may be based on

quantifiable data but often must be based on estimates or surveys. A related major task is to determine the equity of distribution of danger, benefits, and costs. This is a question of fairness and social justice for which answers are a matter of personal and societal value judgment.

CHANGING CONDITIONS - Changing conditions alter the level of danger one may consider acceptable. The danger of dashing across the street against a red light with vehicles moving on the green light is unacceptable to most people on a pleasant, sunny day. Some of the same people will dash across the street under the same traffic conditions when it is raining hard. The danger in the latter situation is obviously greater than in the former but the individual perceives it as acceptable at that point.

Both danger and its acceptability can change with time. For example, Ackerman [8] points out that the explosion of a steam boiler in Hartford, Conn., on 2 March 1854 (killing 21 and seriously injuring 50 others) started a small group thinking that a remedy could be found if the cause were known. The War Between the States (1861-1865) caused abandonment of this effort. Interest revived with a boiler explosion on the Mississippi River steamer "Sultana" in April 1865, with the loss of more than 1200 lives. It wasn't until 1911, however, that the American Society of Mechanical Engineers (ASME) started an effort leading to formulation of the Boiler and Pressure Vessel Code which has been so widely adopted. In the early days of steam boilers people apparently regarded such explosions as "Acts of God" or due to "natural causes" and accepted the danger (albeit, perhaps reluctantly). Over the years, there have been many fatalities and injuries due to problems with steam boilers. Since the beginning of use of the atom as a power source (despite various problems, e.g., the well-publicized Three Mile Island situation), there have been very few fatalities, injuries, or exposure to radiation associated with nuclear powered plants. Nonetheless, it appears that large segments of the U.S. public consider nuclear-fueled plants unacceptable, yet accept fossil-fueled plants (despite potential problems such as environmental pollution and acid rain).

Another example of attitudes changing with time is the rise of "consumerism" in the United States which has led to agencies such as the Consumer Product Safety Commission and the Environmental Protection Agency.

CULTURAL BIAS - Acceptability of danger may differ on a cultural or national basis. For example, in Japan, one can obtain beer and sake from coin-operated dispensers at any time. In the United States, buyers of alcoholic beverages must be a minimum age with sales restricted to specific hours in licensed stores. In the United States, one regularly reads newspaper accounts of serious injury and/or fatalities from automobile accidents in which a driver had been drinking. Laws exist in the United States forbidding driving while intoxicated. A common view is that these are neither stringent nor vigorously enforced. Similar accidents are rare in Japan where every licensed driver is regarded as a professional driver and punishment for driving while intoxicated is swift and severe. One can only infer that (1) the acceptability of obtaining and using alcohol and (2) the acceptability of driving after using it differ greatly between the United States and Japan.

Differences in cultural or regional attitudes among many groups within a country such as the United States make assessing an acceptable level of danger a very difficult task.

ASSESSOR OF ACCEPTABILITY

Who determines the acceptable level of danger? In terms of ability to judge acceptability, designers and/or engineers are no better qualified than any other group of people and, in general, are less qualified than many others. It is often alleged that engineers (because of their inherent characteristics, education and experience) are less sensitive to societal influences of their work and products than others. Like most stereotypes, there is some truth in this view. But if not the designer, then who?

Some have proposed that competition will establish the level of acceptable danger of a given product. This appears much too simplistic. Obviously cost (direct financial burden) influences acceptablility. Cost, however, should include all that must be paid in terms of manpower, material resources, social options, individual freedom, and other goods, as well as direct financial burden. There is also substantial evidence that some segments of industry (all manufacturers producing a given type of competing products) are delinquent, in that use of their products results in serious injury or death although ample technology exists to provide reasonable redesign and/or accident prevention measures at relatively low cost, which could reduce, if not completely eliminate, the danger.

One approach is to establish a danger review panel, specific to the product. This panel should be composed of people with a variety of capabilities. Certainly engineers, especially human factors engineers, should be on the panel. As with the design review panel, management, legal counsel, etc., should be represented. But there should also be a number of people, such as sociologists, psychologists, and consumers/users, who are not company employees. The latter group has no vested interest in the product and would provide a better view of public acceptability of danger. The danger review panel would be charged with assessing the acceptability of danger and making consensus recommendations (substantial agreement, i.e., much more than a simple majority but not necessarily unanimous) to top management which then decides on an appropriate course of action.

One might also conduct an "opinion" poll, somewhat like a market survey, which would obtain substantial input from the general public especially if the product is a consumer item. In conducting such a survey, however, there are two major considerations: (1) obtaining an adequate sample truly representative of users and (2) phrasing questions very carefully, with extensive "cross check" questions, so there is minimum bias in the responses.

The foregoing comments represent suggestions. I have no illusions that they provide complete answers. A specific assessment method or routine must be developed for each individual product. In that context, there is no single assessment model that will serve all situations as indicated by Gray [9]: "Many technology assessment theorists suggest that there is a uniform, formal, assessment model that can be used as a standard for actual assessment projects. In fact, technology assessment has so many different variations that assessment theory can be expected to routinely conflict with assessment practice."

Reference to Lowrance [5] and to Schwing & Albers [10] can be helpful in developing further sensitivity to assessing an acceptable level of danger.

POTENTIALLY ACCEPTABLE LEVEL OF DANGER

It has been suggested that the average United States resident appears willing to accept danger of fatalities of about 1 in 10^6 per hour for air or automobile travel and thus this might be considered a reasonable level for other products that involve public safety. This may be a level (although certainly debatable) which is rational for the public as a whole, but it may not be perceived as such by a bereaved family.

In a somewhat different aspect, statistics indicate the danger of dying from an automobile accident in the United States is about 1 in 4000 per year. (Obviously, the danger of being injured is greater.) One can infer that people in the United States consider this an acceptable level of danger in view of the fact that little effort is expended in tyring to decrease the accident rate. At the same time, the National Highway Traffic Safety Administration indicates that about 50% of fatal traffic accidents in the United States involve drunken drivers. If alcohol related fatalities could be eliminated, the danger would decrease to about 1 in 8000. Two groups, MADD (Mothers Against Drunk Driving) and SADD (Students Against Driving Drunk), are unwilling to accept alcohol related accidents and are exerting efforts to sharply reduce such accidents.

One aspect of a potentially acceptable level of danger is the manner in which it is stated. Engineers would undoubtedly prefer to state the level in terms of probability. The general public, however, might well prefer it otherwise, or even unstated. For example, the general public must be aware of the fatalities from automobile accidents in the United States. It is entirely possible that if the automobile manufacturers were to point out that there is about one chance in 4000 of any individual being killed, and a much greater chance of being injured, in an automobile in any one year, the attitude of the public might be different. It must be further recognized, however, that while it is possible to reduce the level of danger to an extremely small number, danger can not be completely eliminated, no matter how much effort is expended.

In my opinion, there is no one level of acceptable danger. Each situation must be judged independently. Some guidance is provided by a California Supreme Court decision [11] (cited by Peters [12]) which said that a product is defective in design if there is "excessive preventable danger." This exists if a danger-benefit analysis shows that dangers outweigh benefits when measured against five criteria: (1) the hazard, or severity of consequences in the event of failure; (2) risk, or probability that such a hazard will occur; (3) technical feasibility of safer alternative designs; (4) economic feasibility of safer alternative designs; and (5) possible adverse consequences (hazard) to user and product resulting from alternative designs. In general, United States courts are saying that if a product can be made safer at a reasonable cost without introducing other dangers, then such action must be taken. If there is substantial potential of injury and no technically feasible safer alternative design can be found, the danger may still be judged acceptable if potential benefits outweigh potential damages. For example, kitchen knives are hazardous and the risk is high. So far, no one has devised a means of making such a knife safer nor has anyone devised another way of accomplish-ing the same purpose. The implicit judgement is that benefits outweigh potential damages.

SUMMARY

The task of assessing danger is difficult. The task of assessing acceptable danger is far more difficult, but not impossible. Technical people are generally capable of, and qualified for, measuring hazard, risk, and danger. Technical people, however, are no better qualified than the general public to make value judgements of the level of danger acceptable to the general public. Nonetheless, the assessment and a decision must be made.

The designer (1) must assess the danger and determine its acceptable level, (2) needs assistance from the general public in determining an acceptable level of danger, (3) has the ultimate responsibility for whatever level of acceptability is chosen and built into the product.

If the designer has performed his task thoroughly and carefully and can prove it through records, he may still be found liable for damages if sued; but statistics indicate average damages awarded to the plaintiff will be significantly smaller than if there is no documented assessment effort.

REFERENCES

1. Smith, C. O.: "Products Liability: Are You Vulnerable?" Prentice-Hall, Inc., Englewood Cliffs, NJ, 1981.

2. Philo, H. M. and Rine, N. J.: "The Danger Never Was Obvious." Journal of Products Liability, Vol. 1, No. 1, 1977, pp. 12-19.

3. American Law Institute: Restatement of the Law, Second, Torts, 2d, Vol. 2, American Law Institute Publishers, St. Paul, MN., 1965.

4. Snyder, D. J.: "Perspectives of the Legal System.: ASME Design Engineering Conference, Chicago, IL, May 1977, ASME Paper 77-DE-22.

5. Lowrance, W. W.: "Of Acceptable Risk." William Kaufman, Inc., Los Altos, CA, 1976.

6. Corley, G. W.: "Design Assurance Practices and Procedures--A Product Liability Prevention Tool." Journal of Products Liability, Vol. 2, No. 1, 1978, pp. 1-16.

7. Bartels, H. G.: "How to Evaluate a Product Safety Effort." Hazard Prevention, Nov/Dec 1977, pp. 4-6.

8. Ackerman, A. J.: "America's Abundant Electricity Due to the ASME Boiler Code: It All Began With an Appalling Disaster." ASME Paper 78-WA/TS-2.

9. Gray, Lewis: "Doing a Technology Assessment? Beware of Assessment Theory." ASME Paper 83-WA/TS-7.

10. Schwing, R. A. and W. A. Albers, Coeditors, "Societal Risk Assessment," Plenum Press, New York, NY, 1980.

11. --Barker vs Lull Engineering Co. 20C. 3d 413.

12. Peters, G.A.: "New Product Safety Legal Requirements." Hazard Prevention, Sep/Oct 1978. pp 21-23.

APPENDIX

PRODUCTS LIABILITY IN THE UNITED STATES

Products liability is a legal term which encompasses the action whereby an injured party (plaintiff) seeks to recover damages for personal injury or property loss from a producer and/or seller (defendant) when the plaintiff alleges that a defective product caused the injury or loss. The term includes commercial loss by a consumer or business operation due to alleged failure or inadequate performance of a product.

If a product liability suit is entered against a company, the plaintiff's attorney and his technical experts will attempt to convince the jury that the company did not exercise reasonable care in one or more of the following areas, and because the company did not, an innocent party was injured.

(1) The product was defectively designed.
(2) The materials used in the product were unsuitable or defective.
(3) The product was defectively manufactured.
(4) The product lacked sufficient quality control.
(5) The product was not tested for performance.
(6) There was a failure to foresee the consequences of ordinary wear and tear and improper maintenance on the part of the user.
(7) There was a failure to foresee unintended, as well as intended, uses.
(8) The product had insufficient or inadequate safety devices.
(9) The product had safety devices which malfunctioned.
(10) The product failed to comply with existing codes and safety standards.
(11) Instructions were inadequate or lacking for the proper and safe use of the product.
(12) The product was not accompanied by proper warnings.
(13) The manufacturer did not notify all owners of the product that new safety devices were available, or that a design modification to improve the safety of the product had been made and was available.

The situation is schematically summarized in Fig. 1.

Although products liability has received great attention in the past two decades with development of a substantial body of law, primarily from decisions of various courts, it is not new. The first law code known to be in writing was established by Hammurabi, King of Babylon, about 4000 years ago. One item in that code says: "If a builder build a house for a man and do not make its construction firm and the house which he has built collapse and cause the death of the owner of the house--that builder shall be put to death." That clearly is a more drastic penalty than any now being rendered in the United States!

Who may be a plaintiff? Essentially any consumer, user, or bystander may seek to recover for injuries or damages caused by a defective and unreasonably dangerous product.

Who may be a defendant? Everyone (whether a corporation, business organization, or individual) who had some degree of responsibilitiy for a given product, from its inception as an idea or concept to its purchase and use, is potentially vulnerable as a defendant in products liability litigation.

There are three legal bases on which a products liability lawsuit can be based: negligence, breach of warranty, or strict liability. All three are predicated on the fault system, i.e., a person whose conduct causes injury to another is required to fully and fairly compensate the injured party.

The basic method of imposing liability on a defendant requires the plaintiff to prove the defendant acted in a negligent manner. Under the negligence theory in products liability, the plaintiff must essentially establish proof of specific negligence, a rather difficult task.

A user of a product may, as a result of express oral or written statements, or implication, reasonably rely on the manufacturer's express or implied assurances (including advertising material): (1) as to the quality, condition and merchantability of goods, and (2) that the manufactured goods are safe for their intended purpose and use. If the user relies on these assurances and is injured, he can sue on the basis of breach of warranty.

Both negligence and breach of warranty theories require proof of some fault on the part of the defendant, i.e., focus is on the action of an individual. Strict liability, however, focuses on the product itself and is based on Section 402A of the Restatement of Torts, 2d, i.e.:

Special Liability of Seller of Product for Physical Harm to User or Consumer

(1) One who sells any product in a defective condition unreasonably dangerous to the user of consumer or to his property is subject to liability for physical harm thereby caused to the ultimate user or consumer, or to his property, if
   (a) the seller is engaged in the business of selling such a product, and
   (b) it is expected to and does reach the user or consumer without substantial change in the condition in which it is sold.

(2) The rule stated in Subsection (1) applies although
   (a) the seller has exercised all possible care in the preparation and sale of his product, and

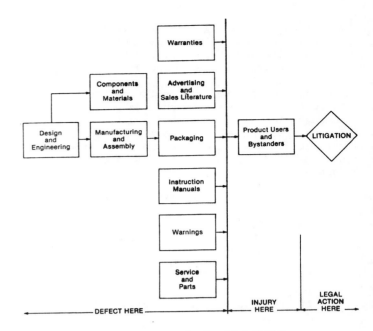

Fig. 1 — Essence of products liability.

   (b) the user or consumer has not bought the product from or entered into any contractual relation with the seller.

The principle arguments justifying strict liability of the manufacturer are: (1) such liability tends to promote safety by stimulating the manufacturer to replace what he has considered as adequate methods of production, inspection, and testing with improved methods; (2) the consumer lacks the skill and means to investigate the soundness of a product and to establish the element of fault; (3) the manufacturer is normally in a position to spread the harm caused by a defective product by increasing prices enough to cover the cost of increased insurance or of judgments; (4) in effect, strict liability was previously imposed indirectly by a series of warranty actions beginning with one by the consumer against the retailer who then sued his supplier and so on through a chain which ultimately reached the manufacturer. This waste is eliminated when the consumer is permitted to sue the manufacturer directly.

Manufacturing defects are defined in terms of specification conformance, user satisfaction and deviation from the norm. A failure to conform with stated specifications is a rather obvious manufacturing defect and not a new definition. The aspect of user satisfaction is perhaps less well recognized, but it has been a legal definition for some time that a manufacturing defect exists when there is such a departure from some quality characteristic that the product or service does not satisfy user requirements. The Barker case [11] added the third definition, i.e., deviation from the norm. Under a test using this definition, a jury can be instructed that a manufacturing

330

defect occurs when: (1) a product comes off an assembly line in a substandard condition, (2) when the product differs from the manufacturer's intended result, or (3) when the product differs from other ostensibly identical units of the same product line. Focus and emphasis are clearly on a "maverick" (non conforming) product.

Design defects are defined in terms of ordinary consumer expectations and excessive preventable danger. The product may be considered defective if it failed to perform as safely as an ordinary consumer would expect. This is interpreted in the context of intended use (or uses) in a reasonably foreseeable manner where "foreseeable" has the same meaning as "predicted" in failure-modes-and-effects, fault-tree, or hazard analyses. It appears that many "ordinary" consumers would have no concept of how safe a product should, or could, be without expectations created by statements in sales material, inferences from mass media, general assumptions regarding mdern technology, and faith in corporate enterprise.

A product is also defective in design if there is excessive preventable danger. This exists when a danger-benefit analysis shows that dangers outweigh benefits when measured in terms of five criteria: (1) gravity of the danger posed by the design (i.e., severity of the conquences in the event of a failure); (2) probability (including frequency and exposure of the failure mode) that such a danger will occur; (3) technical feasibility of a safer alternative design, including possible remedies or corrective action; (4) economic feasibility of possible alternative designs or corrective actions; and (5) possible adverse consequences to the product and consumer which would result from alternative designs. Additional relevant factors may be included, but design adequacy is evaluated in terms of a balance between benefits from the product and risks of danger. Quantification is not required but may be desirable to increase credibility. This places a burden of proof on the defendant. Once the plaintiff proves the product is a proximate cause of injury, the defendant must prove the benefits outweighed the dangers.

MAINTENANCE RELATED FAILURES

C.O. Smith
Rose-Hulman Institute of Technology
Terre Haute, Indiana

ABSTRACT

Many operational failures which result in personal injury and/or property damage are attributed to defective material. Examination, however, often indicates the real cause is related to maintenance. Maintenance-related failures can be grouped by basic causes: (1) Failure to perform maintenance, or inattention to detail by the maintenance worker; (2) Design defects which make good maintenance very difficult and/or unsafe; (3) Failure by the worker to recognize the consequences of certain "short cut" or "easier" maintenance methods; (4) Combinations of the above. This paper illustrates these points by discussing specific examples. Suggestions are made for reducing the problems of maintenance-related failures.

IT IS WELL KNOWN that products of various kinds fail, resulting in damages, both personal injury and commercial losses. In many of these situations, the injured party initiates litigation against the manufacturer or supplier alleging manufacturing and/or design defects. In some situations, the allegations are well-founded. One aspect of product failure which, although highly significant, receives far less attention is the detrimental effect on life and performance of products resulting from inadequate maintenance and repair practices.

As efforts increase to conserve energy, conserve material, and substitute for unavailable materials, there will be increased use of maintenance, repair and rejuvenation practices. In time, this will create still more problems and failures if maintenance and repair are not done properly.

Product failures in maintenance or repair situations commonly relate to: (1) Failure to perform maintenance or performing maintenance negligently; (2) Lack of safety or difficulty in performing maintenance due to inadequate design; (3) Use of "short cut," "routine," presumably easier, maintenance methods in which the worker does not recognize or understand the potential consequences due to ignorance (or limited understanding) of fundamentals and/or basic phenomena; (4) Combinations of the above. Let us consider some specific examples as illustrations.

STAINLESS STEEL EXTINGUISHER

An employee, while involved in cleaning a jeep (fitted out and used as a patrol vehicle and small fire truck), was in the process of lifting out a fire extinguisher and putting it on the ground when it "exploded." The barrel (or shell) "took off" and landed about 49m (160 ft) away. The man was struck on the right side of his face and head, receiving extensive injuries, and suffering lacerations on the fingers of his right hand. [1]

The fire extinguisher was about six years old. It had been in the patrol jeep that entire period, except when the jeep was being cleaned. There was no evidence that this specific extinguisher had been inspected since purchase although it was normal practice to have regular inspection of all extinguishers on the premises.

The fire extinguisher was a dry chemical powder type made of stainless steel containing 9.1 kg (20 lb) of powder. A trigger in the handle normally discharged the contents of a carbon dioxide cartridge with the powder then being blown out through a hose and nozzle. The dish-shaped bottom of the extinguisher and the powder contents were found at the scene of the accident. The CO_2 cartridge had been discharged in a normal manner and had not been exploded from within. There was no evidence of blockage of the hose or powder discharge nozzle.

Fig. 1 shows a typical section of the failure edge of the dished end. The darker regions of this fracture surface were colored with variations and mixtures of reddish-brown and bluish-gray, in contrast with some bright and shiny areas. This indicates partial failure of the metal prior to the accident.

Laboratory examination identified the material as austenitic stainless steel which failed due to grain boundary corrosion as shown in Fig. 2. The jeep was used to carry calcium chloride to be spread on icy steps. The extinguisher sat in a "well" in which water occasionally collected. A combination of any chloride with moisture in a relatively confined space would almost certainly lead to grain boundary corrosion of an austenitic stainless steel "sensitized" by heat treatment during its processing history.

Fig. 1 - Typical portion of failure edge of stainless steel fire extinguisher. Contrast between light and dark areas is obvious and indicates partial failure prior to final failure.

Fig. 2 - Radial cross-section with failure edge at left side. A partial network is apparent at grain boundaries.
Electrolytic etch in oxalic acid   70X

This failure occurred under a relatively rare set of circumstances, a situation which might be difficult to anticipate or foresee. The mechanism of grain boundary corrosion in stainless steels, however, has been known for

decades and one can argue that the manufacturer should have given an additional heat treatment to remove this possibility. More pertinent in this context, however, is failure of the owner to perform routine, periodic, inspection and pressure tests of this specific extinguisher. A pressure test in a cage, as prescribed by the National Fire Protection Association, would undoubtedly have detected the developing weakness due to corrosion and thus minimized the probability of injury.

LIQUID PROPANE (LP) GAS BOTTLE

A large number of liquid propane or butane (LP) gas bottles are used on construction projects and in recreational applications. The design of most of these is similar to that shown in Fig. 3. Such a bottle is designed as a pressure vessel with a hood or guard welded to the top. This hood protects the pipe entering the bottle and provides a handle for transporting the bottle. [2]

Fig. 3 - Typical LP gas bottle

A bottle of this type had been filled early on a cold morning and carried to a job site to be used as a gas supply for a welding torch. The welder was delayed about four hours in starting. Soon after he started welding, the bottle exploded and engulfed him and a fellow worker in flames, causing fatal burns to both.

Inspection of the failed bottle showed severe corrosion in the area indicated by an arrow in Fig. 4. Corrosion pits penetrating to about 60% of the wall thickness were also observed. It seems obvious that proper maintenance procedures would include periodic cleaning and painting to minimize deterioration and pitting. If pitting had been observed, there is no question that the bottle should have been taken out of service. In this situation, no information was available relative to inspection and maintenance practices of the construction company. It should be noted that pits are not always

Fig. 4 - LP gas bottle with arrow showing region of severe pitting (corrosion).

observable in the best of circumstances. In this situation, it would be very difficult to observe pits, or to clean, in the restricted region of the weld.

Although poor maintenance may have contributed to this failure, it seems clear that the design is a major problem. Welds tend to be more susceptible to corrosion than base material. These welds were located in a region in which water collects leading to acceleration of corrosion. A redesign which did not allow water to collect would be a distinct improvement.

It is noted that a warning is attached to the bottle shown in Figs. 3 and 4. This warning, however, makes general comments relative to the gas contents and the handling thereof. It says absolutely nothing about maintenance procedures or potential problems (e.g., corrosion) with the bottle itself. In litigation, a plaintiff's attorney would argue that this is, in itself, a design defect.

SAND MIXER

A foundry core-sand mixer had a fixed base with a horizontal arm (trough) which could be rotated about a vertical axis. Ingredients were fed in at the fixed end of the trough and were mixed as they were moved to the swinging end of the trough. Mixing was accomplished by an auger which rotated in the trough. Control was exercised through a switch box (Fig. 5) on the swinging end of the trough. [3]

A worker was cleaning the auger and trough with an air chisel when a fellow workman (not one who normally used the mixer) accidentally hit the toggle switch which caused the auger to rotate. The worker's left hand was caught between the auger and trough and seriously injured.

The length of the radius from the vertical center line of the machine (through the fixed base) to the toggle switch is about 2-1/4 m (7.3 ft). The base of the toggle switch is 1.6

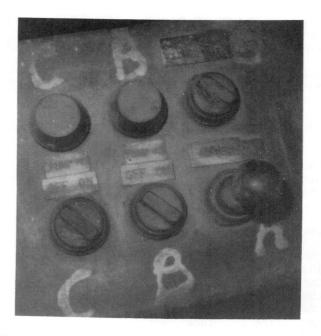

Fig. 5 - Switch box (at "free end" of sand mixing trough) which controlled the amounts of sand, binder, and catalyst. The toggle switch (lower right) controlled rotation of the auger.

m (62 in) from the floor. The switch box containing the toggle switch and other switches is at a lower elevation than the trough, with the toggle switch being about 0.56 m (22 in) from the end of the trough. The worker was standing on a portable platform about 0.61 m (24 in) high. The top of the toggle switch is 76 mm (3 in) above the top of the box. The other five switches on the box are all about 25 mm (1 in) above the top of the box. The toggle is easily moved. Three of the other five switches require a rotation. The other two are push buttons.

A maintenance manual was provided with the mixer. It was clear that the manufacturer recognized the need for cleaning auger and trough. No preferred cleaning procedure was indicated although the manual did say: "The auger section being readily removable permits removal for cleaning by soaking or blasting, if desired." There was no statement of any kind in the manual relating to personnel safety while cleaning (or operating) the mixer.

A panel door on the fixed base of the mixer had a yellow sign (64 mm x 70 mm; 2-1/2 in x 2-3/4 in) which said: "Caution: Power to be turned off before operating control panel door or working on machine." This warning is ambiguous. The worker interpreted it to apply when working inside the base. The manufacturer presumably intended that it apply at any time the mixer was being repaired or cleaned.

The trough had a red sign (50 mm x 100 mm; 2 in x 4in) on the exterior which said: "Caution: Do not open trough while machine is in operation." This clearly implies such action is dangerous. One can infer that having the trough open while cleaning the auger is also dangerous. At the same time, the trough must be open to clean the auger, if it is in position in the trough, and the toggle switch must be operated to make the auger turn to expose new surface to be cleaned.

The arrangement of switches on the control box is such that the toggle switch clearly stands out from the others, in physical size, in accessibility, and in ease of operation. While this is appropriate in operating the mixer to fill core boxes without inadvertently changing the mixture of sand, binder and catalyst, it does make the toggle switch significantly more "vulnerable" to inadvertent operation.

A design defect can be shown. Rotation of the auger in the trough generated pinch points in which a worker could be seriously injured. The probability of exposure was high, despite the warning on the side of the trough, as the trough could be opened with the auger continuing to rotate. Putting a limit switch on the trough cover was a technically feasible alternative. If the cover were lifted, the mixer would shut off, the auger would cease rotating, and no pinch points would be generated. While a limit switch would add to the cost of the mixer, this would be economically feasible since it would be a small fraction of the total cost. There would be no adverse effects of installing a limit switch. Putting a wire cage over the switch box and toggle switch was a technically feasible alternative which would permit ready access to the toggle switch when operation was intended but would minimize accidental operation. The added cost of the cage would be economically feasible since it would be a small fraction of the total cost. There would be no adverse effects· from this cage.

TWIST DRILL FAILURE

The common twist drill is readily available in a multitude of places and sold by the millions each month in the United States. Although most twist drills are well made, they do break for a variety of reasons, in part because the vast usage encompasses such a range of situations (undesirable to optimum) and users (inept to expert). [4]

A 19.0 (3/4 in) stud had broken in the vertical wall of a metalworking machine known as an upsetter. A parallel vertical wall left a limited amount of space in which men could work. A pilot hole had been drilled with a 4.76 mm (3/16 in) drill. The drill was held in a Jacobs chuck in a portable drill press which, in turn, was held to the work piece by an electromagnet. After the pilot hole was finished, the drill press was removed, the pilot drill was replaced by a 15.97 mm (5/8 in) drill, and the press was repositioned. One man was doing the drilling while another man was pumping oil into the hole. When the 15.97 mm

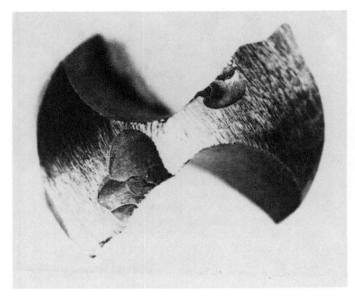

Fig. 6 - Tip from broken drill showing end view. Approximately 4.8X

(5/8 in) drill was about 12.7 mm (1/2 in) to 19.0 mm (3/4 in) into the pilot hole, there was a "bang." The high speed drill shattered, causing a chip to lodge in the right eye of the oiler. Nine pieces were recovered, including the drill tip shown in Fig. 6.

The cause of failure lies in a combination of actual use conditions and a defectively sharpened drill tip. The conditions of use, including a smaller pilot hole, a portable drill press, relocation of the drill between drilling operations, and a questionable supply of coolant, placed an abnormal demand on the drill. Examination of Fig. 6 indicates one cutting lip is about 7.25 mm (0.286 in) long while the other is about 8.02 mm (0.316 in) long. The shorter lip contacts the work before the longer lip. The shorter lip, therefore, bears all of the initial drilling stresses. The larger of the two chipped areas along the cutting edges in Fig. 6 is on the shorter lip. The chisel edge (the short edge dividing the two cutting lips) was about 0.27 mm (0.018 in) off center. The broken point also had improper clearance angles.

It seems obvious that the point of the broken drill was not the original point put on by the manufacturer but came from regrinding. It is highly likely this sharpening was done on a grinding wheel with the drill held manually. Such operation is obviously not well-controlled. Proper control is possible using drill-sharpening jigs that have been available at reasonable cost for many years. Recognizing that manual holding of drills is a rather common practice, one can only infer that most workmen do not recognize the potential danger of this practice.

CHIPPING SLEDGEHAMMER

Radavich [5] details a situation in which 9 kg (20 lb) sledgehammers were used to remove wedges holding dies in forging presses. One worker was struck in the throat by a chip from a fellow worker's sledgehammer. Lung damage resulted and suit was entered against the maker of the hammer and the supplier of the steel alleging defective material.

Visual examination of the hammer found a number of craters on the edges of the striking faces where chips had ejected. The original chamfer of the striking faces was nearly gone and the hammer was about 6 mm (1/4 in) shorter than a comparable new hammer. Samples from the failed hammer were mounted and examined. Samples which contained old craters (missing chips) showed a white layer on the surface of the crater while other areas of the edge of the hammer face showed conical or moonshaped white layers. Higher magnification of the white layers showed cracks in the layers.

The mounted samples were subjected to hardness tests. The core material gave Rockwell C values of 42-46 while the white areas gave values of about 65. This high hardness is typical of untempered martensite. Fig. 7 shows hardness indentations which indicate the large difference in hardness between the core and the white layer.

Repeated use of hammers on hard materials promotes spreading or mushrooming of the edge of the hammer face. This deformed material must be removed, normally by grinding or using a torch. The manufacturer had a published policy of taking in used sledgehammers, reprocessing them, and returning them. Although warnings were issued about the consequences of uncontrolled redressing of faces, it requires less effort to redress a hammer in the plant than to ship it out and wait for its return.

In the worker's plant, it was routine practice to remove the wooden handles and hold the hammer head to a grinding wheel until the edges were redressed. When the hammer became too hot to hold, it was thrown into a five gallon pail of water. No matter how correct the initial condition of the hammer, the heat of grinding was sufficient to produce a temperature greater than 1400°F (sufficient to austenitize the AISI 1060 steel). Quenching in water produced martensite, a hard, unstable constituent highly susceptible to cracking and chipping when the hammer impacted any other hard surface. This microstructural change occurred in only a small, but highly critical, portion of the hammer. This appears to be a situation in which the maintenance men were totally unaware of the potential consequences of their regular maintenance practice.

SUMMARY OF EXAMPLES

It is obvious in these examples that no single cause is responsible for failure and resulting

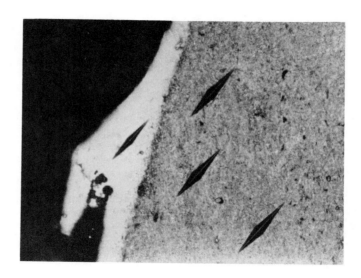

Fig. 7 - Microhardness impressions in section of chipped sledge hammer showing difference in hardness between core metal and white layer. 280X

injury. At the same time, however, at least one major aspect can be found in each case. With the fire extinguisher, neglecting routine inspection was a highly significant factor in the failure.

Delinquent design was a major feature in failure of the LP bottle and injury caused with the sand mixer. No doubt both manufacturers believed a good design had been produced for sale. It also seems clear that they failed to foresee potential problems. Foreseeing the actual circumstance of use of the fire extinguisher might well be considered unreasonable but there was nothing unique or unusual about the circumstances of use of the LP bottle or the sand mixer. The U.S. courts are saying that a manufacturer is responsible for designing a product which can be safely used in foreseeable, but unintended, situations as well as intended situations. Both manufacturers recognized some potential problems: the LP bottle maker attached a warning relative to the use of the contents; the maker of the sand mixer provided warnings which appear to be ambiguous. Does a manufacturer have a duty to warn? Definitely yes, when the danger from a product exceeds that anticipated or understood by the ordinary consumer or user. The test of "ordinary user" and his expectation is the key. Admittedly, there is no easy or simple definition of "ordinary user" or what he expects. It should also be noted that a warning is effective only when: (1) The message is received (seen); (2) The message is understood; and (3) The user acts in accordance with the message. Obviously, a manufacturer has no control over the third aspect but he does, however, have substantial control over the other two.

In the cases of the twist drill and the sledgehammer, it seems clear that a major element was ignorance, or unawareness, of the workers of the potential consequences of their common, usual, maintenance practices.

SUGGESTIONS FOR REDUCING THE PROBLEMS

It is relatively easy to set forth or propose general "solutions" but it may be extremely difficult to translate them into effective action. In situations in which routine maintenance procedures are neglected, an easily stated, obvious solution is to say that such procedures must be used and that good administrative control will ensure their use. This implies greatly increased effort and emphasis on control—which does not come easily or readily while "getting the job done" in the everyday, busy working world.

Responsibility for design defects (deficiencies or shortcomings) can be rather clearly focused since the designer is ultimately responsible. It is easy to say he should do his task more carefully, thoughtfully, and thoroughly but it is difficult for him to achieve this. In addition, even the most highly trained, competent, professionals are sometimes guilty of mistakes or oversights. Talbot [6] provides further insight and suggestions. A very valuable aid for the designer is use of formal design review procedures in which special emphasis is placed on all foreseeable uses or applications (intended and unintended) and how safety can be provided in these uses. This is seldom an easy task but certainly is one of enlightened self-interest.

For the problem of ignorance of fundamentals or basic phenomena, educating repair and maintenance people is in order. The best method might be through company-sponsored programs—presented by experts who are capable of "getting across" to the workers. This is not a simple task. This eduction might also be done "on the job" by having knowledgeable workers devote a portion of their working time to guiding other workers, either formally or informally, in proper procedures. In this process, it is highly desirable that the workers think such education is their idea. Perhaps participation and clear support by trade unions might be heplful in this respect. Hansing [7], Mann [8], and Wireman [9] offer some suggestions in the matter of educating shop personnel.

Many repair and maintenance operations are too small to provide such programs and are highly concerned with day-to-day operations. Perhaps special sections in trade magazines could illustrate the problems with specific case histories. Greater effort can be made for improved administrative control. Whatever the method, however, there must be education, greater knowledge and understanding, and more pride in doing a good job.

The point is that design deficiencies and/or repair and maintenance practices in a great variety of situations can lead to reduced performance, reduced product life, and/or failure in products. Any of these may lead to dangerous and unsafe situations which can result in injuries and fatalities. Corrective action is imperative.

CONCLUSION

Maintenance-related failures, as shown by the examples cited, appear to fit the situation (paraphrased somewhat) described by Allaway's statement: "I find that progress is hardly ever limited by the need for new knowledge but nearly always by the failure to make effective use of existing knowledge." [10]

The message is clear! We must try harder, we must check, double check, even triple check!

REFERENCES

[1] Smith, C. O., "Failure in Two Fire Extinguishers," Journal of Engineering Materials and Technology, October 1975, pp 367-370.

[2] Talbot, T. F., "Design, Maintenance and Failures," ASME Paper 84-DE-1.

[3] Smith, C. O., "Design Deficiencies? Devastation!" ASME Paper 81-WA/DE-7.

[4] Smith, C. O., "Investigation of Twist Drill Failure," SAMPE Quarterly, October 1979, pp 47-51.

[5] Radavich, J. F., "Chip Away," Proceedings, Annual Reliability and Maintainability Symposium, 1978, pp 370-373 (IEEE).

[6] Talbot, T. F., "Reducing Failure With Better Design and Clearer Maintenance Instructions," National Design Engineering Show & ASME Conference, McCormick Place, Chicago, IL, 11-14 March 1985, Session 6.4.

[7] Hansing, R. F., "The Critical Skilled Worker Shortage-How to Approach The Problem," National Plant Engineering & Maintenance Conference, McCormick Place, Chicago, IL, 26-29 March 1984, Session P 1.12

[8] Mann, L., "Do Formalized Preventive Maintenance Programs Lead To Over Maintenance?", National Plant Engineering & Maintenance Conference, McCormick Place, Chicago, IL, 26-29 March 1984, Session P 2.5.

[9] Wiseman, T., "How To Train Unskilled Maintenance Personnel 'High Tech' Equipment," National Plant Engineering & Maintenance Conference, McCormick

Place, Chicago, IL, 26-29 March 1984,
Session P 2.10b.

[10] Allaway, P. A., "Opening Address," Third
National Reliability Conference,
Institute of Quality Assurance, United
Kingdom, April 1981.

COMPARISON BETWEEN EXPERIMENTAL AND ANALYTICAL (EMPIRICAL) RESULTS FOR SURFACE FLAWS

W. G. Reuter
EG&G Idaho, Inc.
Idaho Falls, Idaho, USA

ABSTRACT

The primary concern of this research is the ability to predict the behavior (initiation of crack growth, crack growth, and instability) of structures containing defects. But, due to present limited capabilities the emphasis is on crack initiation. This paper presents comparisons between experimental results obtained from surface-flawed specimens with predictions based on K_{Ic} and J_{Ic} plus a few existing analytical solutions. The conditions evaluated range from linear elastic to fully plastic. Because the conditions for a J-controlled field were exceeded in some instances, an empirical correlation is also provided.

WHEN ESTIMATING THE INTEGRITY of a structure, the most accurate approach is to conduct tests of actual components containing natural defects loaded in a manner to simulate the structure. When this is not practical the next best technique is to use an appropriately designed and tested specimen, such as the surface-cracked specimen, to provide assurance of structural adequacy.

The most conservative approach is to use crack initiation as the basis for estimating structural integrity. For brittle materials, those that satisfy linear-elastic fracture mechanics (LEFM) criteria, crack initiation is generally synonymous with unstable crack growth (catastrophic failure). The generally accepted approach for predicting conservative crack initiation conditions for LEFM is to use plane strain fracture toughness (K_{Ic}) and analytical procedures such as those described in Reference 1. As the material becomes more ductile, crack initiation is no longer synonymous with catastrophic failure since crack initiation is generally followed by stable crack growth. Test techniques provided in Reference 2 may be used to experimentally determine J_{Ic} (J integral for crack initiation), but there are no generally accepted methods for predicting crack initiation of a general surface crack.

Research has been, and is continuing to be, conducted at the Idaho National Engineering Laboratory (INEL) on the behavior of surface cracks. The research has been primarily with elastic-plastic and fully plastic conditions, but some tests have been conducted under LEFM conditions.

This paper compares experimental results obtained from surface-flawed specimens with predictions based on K_{Ic} for LEFM and on J_{Ic} or an empirical approach for elastic-plastic and plastic conditions.

TEST PROCEDURES AND RESULTS

The three materials tested were Type 304 stainless steel (SS), ASTM A710, and Ti-15-3. The latter two materials were used for elastic conditions where the ferritic steel was tested on the lower shelf. The A710, when tested in the lower region of the ductile-brittle transition zone, typified elastic-plastic behavior. The A710, when tested in the upper region of the ductile-brittle transition zone, and the 304 SS typified the fully plastic behavior.

Standard fracture toughness tests were used to measure K_{Ic} and J_{Ic} and tests of specimens containing surface flaws provided experimental results for comparison with predictions. All of the specimens were cracked in the T-S orientation. Because of limited plate thickness, larger compact specimens were also fabricated for the A710 in the T-L orientation. The description of the tests and results follow for each material.

Acoustic emission techniques were used to detect the load associated with initiation of crack growth. Metallographic examination of some of the specimens verified that initiation of crack growth had occurred at or prior to the crack initiation loads as detected by acoustic emission.

ANNEALED TYPE 304 SS - Measurements of J_{Ic} were attempted but the specimen size did not satisfy conditions for a J-controlled field; therefore, estimates of J_{Ic} were made (see Table 1).

Thirty surface-flawed specimens were fabricated from three 13-mm thick plates of three different heats. The gage section of the specimens was 6.4-mm thick by 177.8-mm wide by 355.6-mm long. Surface defects were electric-discharge machined (EDM) in the center of the gage length. Elliptical, rectangular, and triangular-shaped EDM defects were used to evaluate the sensitivity of test results to defect configuration. The specimens were all tested monotonically under tensile loading. The results are summarized in Table 2.

ASTM A710 - Measurements of K_{Ic} and/or J_{Ic} were made using techniques described in ASTM E399[3] and ASTM E813,[2] respectively. The test results are summarized in Table 1.

Surface-flawed specimens were fabricated from the 31.8-mm thick plate where the specimen mid-thickness corresponded to the center of the plate thickness. The gage section of the specimens was either 6.35 mm thick by 101.6 or 152 mm wide by 223 or 310 mm long or

12.7 x 101.6 x 223 mm. Surface defects were fatigue precracked in the center of the gage length and the specimens were tested monotonically under tensile loading. The results are summarized in Table 3.

Ti-15-3 - Measurements of K_{Ic} were made at 297 K using three-point bend specimens and procedures described in Reference 3. The test results are summarized in Table 1.

Surface-cracked specimens were fabricated from the transverse orientation of the plate. The gage section of the specimens were 5.08-mm thick by either 50.8 or 63.5-mm wide by 127-mm long. Surface cracks were fatigue precracked in the center of the gage length and the specimens were tested under tensile or bending loads. The results are summarized in Table 4.

ANALYSIS AND DISCUSSION OF RESULTS

A comparison of predictions and experimental results for LEFM, elastic-plastic, and fully plastic conditions are presented below.

LEFM - The defect configuration and crack initiation stresses in Tables 3 (at 200 K) and 4 are used with Reference 1 solutions to calculate K_{crit} (maximum applied stress intensity factor at either the free surface or at maximum crack depth). These values of K_{crit} are plotted as a ratio of K_{crit}/K_{Ic} versus cF^2 (a parameter used to quantify the severity of the surface crack[4-7]) in Figure 1. The solid symbols representing tensile test results show

Table 1 - Summary of Mechanical Properties

Material	Test Temperature (K)	Yield Strength (MPa)	Flow Stress (MPa)	J_{Ic} (kJ/m^2)	K_{Ic} (MPa·m$^{1/2}$)
304-1	297	253	458	estimated to be 700	--
304-2	297	260	432	estimated to be 700	--
304-3	297	254	438	estimated to be 700	--
A710	200	500	591	10.2	45.8
A710	255	467	548	18.6[1]	61.0
A710	297	452	524	39.6[1]	89.1
Ti-15-3	297	1452	1486	--	40.8

[1]Transition region.

Table 2 - Surface Crack Dimensions and Experimental Stresses for Type 304 SS

Specimen Number	Crack Depth, a (mm)	Crack Length, 2c (mm)	Configuration[1]	Thickness, t (mm)	Width (mm)	Crack Initiation Stress, σ_{init} (MPa)
2	1.829	5.842	el.	6.274	177.9	254.5
4	1.524	20.320	tri.	6.229	177.8	306.6
5	4.826	50.622	tri.	6.248	176.5	210.3
6	4.674	11.430	tri.	6.223	177.4	252.0
7	4.953	10.744	el.	6.248	177.5	276.2
8	1.651	18.110	el.	6.426	178.4	271.8
9	4.775	49.886	el.	6.147	177.4	255.2
13	5.080	50.165	el.	6.299	175.9	291.7
14	3.302	36.068	el.	6.502	178.9	282.4
15	1.778	18.047	tri.	6.248	176.3	275.2
16	3.429	8.255	tri.	6.274	177.8	249.4
17	3.429	7.518	el.	6.198	178.0	292.6
19	2.946	7.620	tri.	6.350	177.8	246.4
20	3.353	3.962	tri.	6.223	177.8	296.9
24	3.150	37.465	tri.	6.223	177.8	251.5
25	1.600	4.318	el.	6.147	177.3	244.3
27	3.404	51.102	rect.	6.401	175.3	292.8
29	1.575	17.780	rect.	6.172	177.8	243.4
30	4.724	10.465	rect.	6.223	177.8	241.4
32	3.277	7.620	rect.	6.299	177.8	258.4
34	1.524	19.431	tri.	6.121	177.8	301.8
35	1.524	3.353	rect.	6.147	177.6	336.4
36	1.448	4.318	el.	6.121	177.8	260.9

(1) el.--elliptical, tri.--triangular, and rect.--rectangular.

Table 3 - Surface Crack Dimensions and Experimental Stresses for ASTM A710

Specimen Identification	Depth, a (mm)	Length, 2c (mm)	Thickness, t (mm)	Width, W (mm)	Crack Initiation, σ_{exp} (MPa)
Tested at 200 K					
B-2	2.29	7.37	6.35	101.7	590.0
7	4.06	26.98	6.45	101.4	362.9
13	4.50	41.22	6.38	101.4	275.3
25	4.22	6.60	6.27	100.8	579.6
27	4.29	9.09	6.45	101.3	408.5
B-27	5.00	17.02	12.75	102.4	464.7
B-32	7.19	32.18	12.85	102.0	409.2
B-41	7.62	18.36	12.83	101.9	447.5

Table 3 - (Continued)

Specimen Identification	Depth, a (mm)	Length, 2c (mm)	Thickness, t (mm)	Width, W (mm)	Crack Initiation, σ_{exp} (MPa)
Tested at 255 K					
14	4.62	39.50	6.35	101.3	316.7
22	2.97	7.11	6.50	101.4	505.1
30	4.42	8.89	6.32	101.0	529.9
B-1	1.90	6.73	6.36	101.7	623.1
B-14	1.27	2.54	6.43	101.8	621.7
B-28	4.19	13.31	12.70	102.2	610.0
B-31	5.77	26.80	12.80	102.1	
B-36	1.65	3.66	12.57	101.7	549.9
B-39	5.21	11.81	12.78	101.7	469.2
B-44	7.11	17.78	12.72	101.9	400.2
Tested at 297 K					
3	2.31	7.75	6.40	101.7	449.2
9	4.34	26.16	6.50	101.5	481.6
10	2.54	25.91	6.35	101.3	489.9
15	4.57	40.00	6.50	101.5	362.4
18	1.14	2.41	6.50	101.5	502.3
24	2.54	6.22	6.55	101.4	529.2
32	4.32	9.52	6.53	101.7	533.4
B-4	4.27	26.67	6.60	101.8	462.3
B-11	4.47	40.97	6.43	101.8	358.0
B-17	2.74	6.48	6.35	101.6	459.2
B-18	3.34	7.24	6.30	101.1	529.6
B-34	1.78	3.56	12.80	102.3	520.4
B-42	7.19	18.16	12.75	101.8	507.8
B-45	5.92	25.65	12.75	101.6	454.7
C-1	1.98	6.81	6.32	152.4	507.9
C-9	4.24	40.51	6.50	152.4	583.8
C-10	3.68	25.60	6.38	151.7	506.3
C-21	4.27	40.13	6.30	152.5	441.2
C-22	4.45	8.89	6.35	151.9	576.9
C-30	4.50	9.53	6.35	152.5	496.4

that Reference 1 solutions provide generally conservative predictions since for only two tests is $K_{crit}/K_{Ic} < 1.0$. For bending loads (open symbols) the use of Reference 1 solutions resulted in substantial conservatism which appears to be a function of cF^2.

Similar comparisons made from specimens fabricated from silicon carbide and tested in bending showed K_{crit}/K_{Ic} ranging from 0.96 to 1.34 but $cF^2 < 0.9$ mm. (These results are presented elsewhere in this conference.) These data provide verification of applicability of Reference 1 solutions for bending loads as long as $cF^2 < 1.0$ mm.

At the present time the reasons for the conservatism due to using Reference 1 solutions

for bending loads is unknown, but research is presently underway using moire interferometry to experimentally determine K_{crit} at the free surface for comparison with calculations.

ELASTIC-PLASTIC AND PLASTIC CONDITIONS - The distinction between the elastic-plastic and plastic conditions is that yielding is contained in the former and not in the latter. Therefore, satisfaction of a J-controlled field,[8] i.e.,

$$t, b > 25 \, J_{Ic}/\bar{\sigma} \quad \text{for bend tests} \quad (1)$$

$$t, b > 200 \, J_{Ic}/\bar{\sigma} \quad \text{for center-cracked tensile specimens} \quad (2)$$

where $\bar{\sigma}$ is the flow stress,

Table 4 - Fracture Toughness Results of Ti-15-3 Tested at 297 K

Specimen Number	Thickness (mm)	Width (mm)	Crack Depth (mm)	Crack Length (mm)	P_{max} (kN) [1]
1	5.11	50.8	0.66	50.8	216.4
2	5.11	50.8	0.64	50.8	207.7
3	5.11	50.8	0.51	50.8	251.4
4	5.08	50.8	0.84	50.8	(16.5)
5	5.00	50.7	0.61	50.7	202.4
					(18.1)
6	5.18	50.8	4.47	28.1	74.3
7	5.11	50.8	4.52	28.17	76.5
8	5.11	50.8	4.19	26.59	(12.5)
9	5.13	50.8	4.44	27.81	81.0
10	5.05	63.5	4.78	34.47	78.5
11	5.08	63.5	4.80	35.51	75.8
12	5.08	63.5	4.83	34.80	(13.7)
14	5.06	63.5	4.85	33.93	88.7
15	5.11	63.5	3.58	8.20	175.7 [2]
	--	--	--	--	(27.4)
16	5.10	63.5	3.89	8.33	192.2

(1) Bend test with 50.8 mm span ( ), otherwise tensile test.

(2) Specimen failed at pin hole under tensile load, then tested in bending.

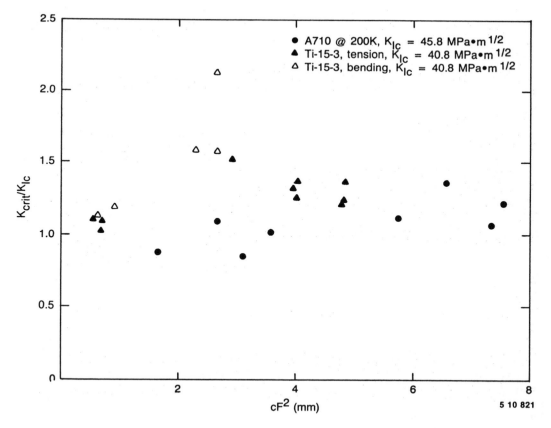

Figure 1. Ratio of K_{crit} to K_{Ic}, for elastic conditions, versus cF^2.

identified those conditions that are within the elastic-plastic region. At the present time, the conditions required for a surface flaw are not established and, as will be discussed later, the applicable value of J_{Ic} is questionable.

For the A710 surface-cracked specimens tested at 255 K, crack initiation occurred by cleavage for all eleven specimens with several exhibiting substantial plastic zone sizes. For surface-cracked specimens tested at 297 K crack initiation and crack penetration of the wall thickness occurred by dimple rupture for all 31 specimens. (Those specimens not included in Table 3 were tested without using acoustic emission.) For these specimens dimple rupture was the final failure mechanism for 70% of the tests.

The predicted values of J_{exp} for crack initiation were calculated based on an equation developed for center-cracked plates.[9] Paris, in Reference 10, suggested that the addition of several specific terms, including a geometry correction term, could modify the equation for use with cylinders containing a surface crack. The modified equation is:

$$J = \sigma_{ys} \ \epsilon_{ys} \ a \ \left| F\left(\frac{\sigma}{\sigma_{ys}}\right)\right| \ \left[\frac{M^2}{Q}\right] \qquad (3)$$

where

    J is J integral

    σ_{ys} is yield strength

    ϵ_{ys} is yield strain = σ_{ys}/E

    E is Young's modulus

    a is depth of a surface crack

    $\left|F\left(\frac{\sigma}{\sigma_{ys}}\right)\right|$ is stress correction term where

    σ and σ_{ys} are true stresses

    M is correction factor for front and back face

    Q is defect geometry correction factor.

Because of space limitations, the reader is referred to Reference 11 for a more detailed explanation of Eq. (3). The approach used in Reference 11 was to substitute J_{Ic} and predict the stress corresponding to crack initiation. The ratio of predicted to experimental stress provided the ability to judge the applicability of using J_{Ic} and Eq. (3) to predict the stress for crack initiation. The predictions were good (conservative) for 255 K and reasonable ($\sigma_{pred}/\sigma_{exp}$ ranged from 0.8 to 1.2) at 297 K.

A problem identified later[12] was that the value J_{Ic} = 39.6 kJ/m^2 used in the prediction at 297 K was for cleavage fracture whereas it was noted earlier that all specimens had crack initiation and penetration by dimple rupture. This suggested that the use of Eq. (3) would provide a nonconservative estimate of the stress for crack initiation if a larger value of J_{Ic} was used. Eq. (3) was evaluated using test data in Table 3 for 255 and 297 K to calculate J_{exp} corresponding to crack initiation. Figure 2 shows a plot of J_{exp} versus cF^2 and Eq. (3) provides good predictions for six specimens and is very conservative for the other four (one not plotted since J_{exp} > 800 kJ/m^2). The conservative values are associated with those specimens that experienced substantial crack tip blunting except for the specimen where J = 200 kJ/m^2. This specimen exhibited little or no crack tip blunting.

Figure 3 shows a plot of J_{exp} versus cF^2 with three nonconservative predictions based on using J_{Ic} = 39.6 kJ/m^2. If a higher value of J_{Ic}, such as that corresponding to the upper limit (250 kJ/m^2), is used, which is probably correct since no evidence of cleavage has been detected for crack initiation and penetration of the wall thickness for 31 specimens, then a substantial number of predictions are nonconservative. This suggests that Eq. (3) needs to be modified, but the apparent agreement in Figure 2 suggests the applicability of the equation. Therefore, the explanation could simply be that since these specimens did not satisfy the conditions for a J-controlled field, then Eq. (3) is not applicable. If this is correct, then criteria other than J are required.

Eq. (3) was used in Reference 13 to calculate J for surface-flawed specimens fabricated from annealed 304 SS. It was observed that J_{exp} increased with increasing cF^2 although J_{exp} remained less than the estimated value of J_{Ic} in Table 1. This suggests that Eq. (3) would provide nonconservative estimates of the stress for crack initiation which is the same as the results in Figure 3 for ASTM A710.

The results in Figure 3 and in Reference 13 are based on conditions at maximum crack depth. For both 304 SS and A710, tested at 297 K, it has been observed[14,15] that the maximum crack tip opening displacement (CTOD) as well as crack initiation occur at maximum crack depth regardless of a/2c. This differs from LEFM where the maximum K_I occurs at the free surface and at maximum crack depth for a/2c = 0.5 and 0.1 respectively. Therefore, it is possible that the geometry correction term needs to be modified to reflect this behavior as well as to predict the gradient in CTOD (J) around the circumference of the surface crack.

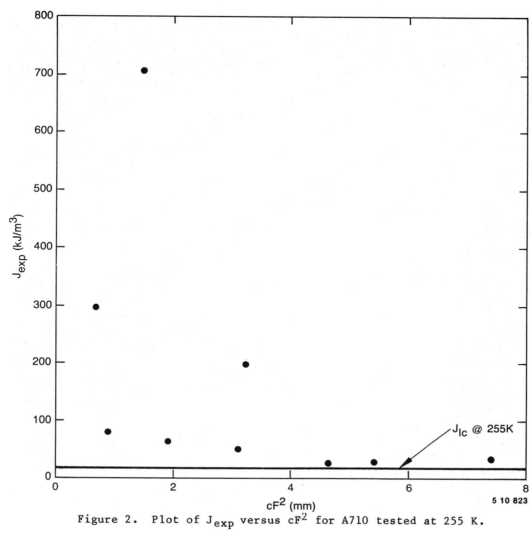

Figure 2. Plot of J_{exp} versus cF^2 for A710 tested at 255 K.

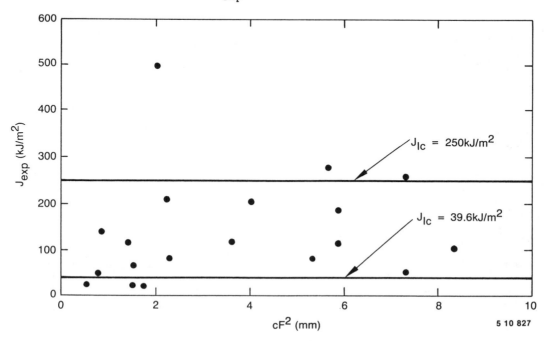

Figure 3. Plot of J_{exp} versus cF^2 for A710 tested at 297 K.

A further complication is the need to consider constraint effects.

The empirical approach presented here is based on the two-parameter fracture criterion (TPFC) of Newman.[4,5,16] The empirical approach is provided because the conditions for a J-controlled field were not met, which means that the use of J is questionable. The TPFC approach is intended to predict the peak load during a rising-load-to-failure experiment--e.g., instability, for materials less ductile than either ASTM A710 (at 297 K) or 304 SS. The TPFC approach has been used in Reference 17 to predict crack initiation and penetration for 304 SS. The approach used in Reference 17 was to plot K_{Ie} (the elastic stress intensity factor, equal to K_I) versus cF^2. The results for A710 (at 297 K) and 304 SS are shown in Figure 4. This figure shows a smooth trend of K_{Ie} versus cF^2 for A710, which could be generated by testing three to four specimens representing different cF^2 values. A similar observation may be made for the 304 SS. One of the limiting conditions identified for the TPFC approach is its sensitivity to thickness. Two different specimen thicknesses are plotted in Figure 4 for the A710 and there is no discernable difference. A similar observation may be made regarding the insensitivity of K_{Ie} versus cF^2 as a function of specimen width. Reference 17 showed a substantial difference in

K_{Ie} for a single specimen approximately 60% the thickness of the other specimens.

At the present time the limitations associated with extrapolating predictions from Figure 4 to structural applications are unknown.

SUMMARY AND CONCLUSIONS

A comparison of predicted and experimental results provided the basis for evaluating techniques for predicting the behavior of surface flaws exposed to LEFM and elastic-plastic or plastic conditions. These results are as follows:

LEFM – Figure 1 shows that only a few nonconservative predictions for crack initiation would occur if K_{Ic} was used in conjunction with Reference 1 solutions. The ratio K_{crit}/K_{Ic} ranged from 0.86 to 1.52 for tensile loads and from 1.12 to 2.14 for bending loads. For tensile loading, the use of Reference 1 provides predictions that are acceptable since the nonconservatism is minimal and the conservatism is reasonable. But, for bending loads the substantial conservatism associated with larger cF^2 values could mean the replacement of a useful component. Therefore, tests need to be conducted to verify the accuracy of the calculated K_{crit} value and modify, if required, the solutions in

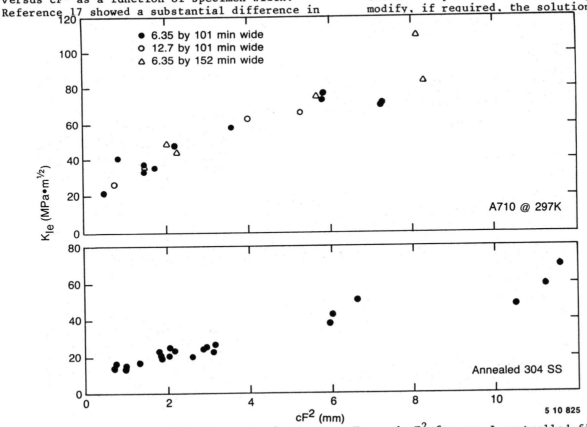

Figure 4. Empirical correlation between K_{Ie} and cF^2 for non-J-controlled field.

Reference 1 to reduce predictions for loads due to bending to acceptable values.

ELASTIC-PLASTIC AND PLASTIC - In all instances the cracks grew through the thickness before growing in the 2c direction for A710 tested at 297 K and for 304 SS.

J-INTEGRAL - Eq. (3), based on modifications suggested by Paris,[10] was used to calculate J_{exp} for comparison with J_{Ic}. Figure 2 shows that Eq. (3) provided good predictions for six specimens and very conservative predictions for the remaining four. These latter predictions are generally associated with substantial crack-tip blunting. The fracture mechanism was consistent (cleavage for the standard fracture toughness specimen (J_{Ic} = 18.6 kJ/m^2) and for the surface cracked specimens).

Figure 3 shows a plot of J_{exp} versus cF^2; there are three nonconservative predictions based on J_{Ic} = 39.6 k J/m^2 which is the lower limit given in Table 1. Some of the surface-cracked specimens satisfied the conditions for a J-controlled field. The fracture mechanism for the standard specimens was cleavage for J_{Ic} = 39.6 kJ/m^2 and dimple rupture for J_{Ic} = 250 kJ/m^2, but was entirely dimple rupture for the surface-cracked specimens. If a higher value of J_{Ic} was used with Eq. (3), then a larger number of specimens would be nonconservative, and none of the specimens would satisfy the conditions for a J-controlled field. Eq. (3) was used in Reference 13 to calculate J_{exp} for surface-flawed specimens fabricated from annealed Type 304 SS. None of these specimens satisfied the conditions for a J-controlled field. In all instances $J_{exp} < J_{Ic}$ which suggested that Eq. (3) would provide nonconservative predictions.

The geometry correction term needs to be modified to reflect the experimental observation that the maximum CTOD as well as crack initiation occur at maximum crack depth, regardless of a/2c. These modifications need to reflect the gradient in CTOD around the circumference of the surface crack.

EMPIRICAL APPROACH - The empirical approach is based on the TPFC[4,5,16] which was intended to predict the peak load during a rising-load-to-failure experiment. The TPFC approach was used for A710 (at 297 K) and for 304 SS, see Figure 4. Both plots of K_{Ie} versus cF^2 in Figure 4 show that each could be generated by testing three or four specimens representing different cF^2 values.

At the present time the limitations associated with extrapolating predictions from Figure 4 to structural applications are unknown.

ACKNOWLEDGMENT

This work was supported by the U.S. Department of Energy Office of Energy Research, Office of Basic Energy Sciences, under DOE Contract No. DE-AC07-76ID01570.

REFERENCES

1. Newman, J. C., Jr. and I. S. Raju, "Analysis of Surface Cracks in Finite Plate Under Tension or Bending Loads," NASA Technical Paper 1578, 1979.

2. "Standard Test Method for J_{Ic}, a Measure of Fracture Toughness, E813-81," 1985 Annual Book of ASTM Standards, Volume 03.01.

3. "Standard Test Method for Plane-Strain Fracture Toughness of Metallic Materials, E399-83," 1985 Annual Book of ASTM Standards, Volume 03.01.

4. Newman, J. C., Jr., "Fracture Analysis of Various Cracked Configurations in Sheet and Plate Materials," Properties Related to Fracture Toughness, ASTM STP 605, pp. 104-123, American Society for Testing and Materials (1976).

5. Newman, J. C., Jr., "Fracture Analysis of Surface and Through-Cracked Sheet and Plate," Engng Fracture Mech 5, pp. 667-689, (1973).

6. Reuter, W. G., S. D. Matthews, and F. W. Smith, "Critical Parameters for Ductile Fracture of Surface Flaws," Engng Fracture Mech 19, pp. 159-179 (1984).

7. Reuter, W. G. and A. K. Richardson, "Comparison of Fracture Toughness Measurement from Ferritic Steel Compact and Surface-Flawed Specimens," Nuclear Engineering and Design 79, pp. 255-266 (1984).

8. Shih, C. F., "Methodology for Plastic Fracture," EPRI NP-1735, Electric Power Research Institute, Palo Alto, CA, pp. 5-11, March 1981.

9. Hutchinson, J. W., A. Needleman, and C. F. Shih, "Fully Plastic Crack Problems in Bending and Tension," Fracture Mechanics, N. Perrine et al. (eds), University of Virginia Press, 1978.

10. Merkle, J. G., "Analysis, Explanation of Analytical Basis for Low Upper Shelf Vessel Toughness Evaluations," Resolution of the Task A-11 Reactor Vessel Materials Toughness Safety Issue, Appendix C, NUREG-0744, Vol. 2, Rev. 1, October 1982, p. C-37.

11. Reuter, W. G., "Comparison of Predicted Versus Experimental Stress for Initiation of Crack Growth in Specimens Containing Surface Cracks," to be published in ASTM STP 905, American Society for Testing and Materials, 1985.

12. Reuter, W. G., "Limits of Applicability of Models Predicting Initiation of Crack Growth for Surface Cracks," to be published in the Third Annual Proceedings of the DOE Engineering Research Review held at University of Pennsylvania, University Park, PA, October 1985.

13. Reuter, W. G., D. T. Chung and C. R. Eiholzer, "Evaluation of Plate Specimens Containing Surface Flaws Using J-Integral Methods," Elastic-Plastic Fracture: Second Symposium, Volume I--Inelastic Crack Analysis, ASTM STP 803, C. F. Shih and J. P. Gudes (eds), American Society for Testing and Materials, 1983, pp. I-480-I-502.

14. Reuter, W. G., H. L. Brown and T. A. Place, "Mechanism of Initiation, Subcritical Crack Growth and Instability for Annealed Type 304 Stainless Steel," Engng Fracture Mech, 15, Nos. 1-2, pp. 169-183, 1981.

15. Reuter, W. G. and J. S. Epstein, "Comparison of Experimental Results for A-710 Surface-Flawed Specimens to Predictions Based on the COD Design Curve," presented at the American Welding Society sponsored Fitness for Purpose in Welded Construction Symposium, Philadelphia, PA, May 13-16, 1985.

16. Newman, J. C. Jr., Fracture Analysis of Surface and Through-Cracks in Cylindrical Pressure Vessels, NASA TNX-73923, September 1976.

17. Reuter, W. G. and T. A. Place, "Estimating Structural Integrity, Using the TPFC Approach of Annealed Type 304 Stainless Steel Plate and Pipes Containing Surface Defects," Int. J. Press. Ves. and Piping 10, (1982) pp. 55-67.